Fascinating Life Sciences

This interdisciplinary series brings together the most essential and captivating topics in the life sciences. They range from the plant sciences to zoology, from the microbiome to macrobiome, and from basic biology to biotechnology. The series not only highlights fascinating research; it also discusses major challenges associated with the life sciences and related disciplines and outlines future research directions. Individual volumes provide in-depth information, are richly illustrated with photographs, illustrations, and maps, and feature suggestions for further reading or glossaries where appropriate.

Interested researchers in all areas of the life sciences, as well as biology enthusiasts, will find the series' interdisciplinary focus and highly readable volumes especially appealing.

More information about this series at http://www.springer.com/series/15408

Kristen J. Navara

Choosing Sexes

Mechanisms and Adaptive Patterns
of Sex Allocation in Vertebrates

Kristen J. Navara
Department of Poultry Science
The University of Georgia
Athens, Georgia
USA

ISSN 2509-6745 ISSN 2509-6753 (electronic)
Fascinating Life Sciences
ISBN 978-3-319-89057-9 ISBN 978-3-319-71271-0 (eBook)
https://doi.org/10.1007/978-3-319-71271-0

© Springer International Publishing AG 2018
Softcover re-print of the Hardcover 1st edition 2018
This work is subject to copyright. All rights are reserved by the Publisher, whether the whole or part of the material is concerned, specifically the rights of translation, reprinting, reuse of illustrations, recitation, broadcasting, reproduction on microfilms or in any other physical way, and transmission or information storage and retrieval, electronic adaptation, computer software, or by similar or dissimilar methodology now known or hereafter developed.
The use of general descriptive names, registered names, trademarks, service marks, etc. in this publication does not imply, even in the absence of a specific statement, that such names are exempt from the relevant protective laws and regulations and therefore free for general use.
The publisher, the authors and the editors are safe to assume that the advice and information in this book are believed to be true and accurate at the date of publication. Neither the publisher nor the authors or the editors give a warranty, express or implied, with respect to the material contained herein or for any errors or omissions that may have been made. The publisher remains neutral with regard to jurisdictional claims in published maps and institutional affiliations.

Printed on acid-free paper

This Springer imprint is published by Springer Nature
The registered company is Springer International Publishing AG
The registered company address is: Gewerbestrasse 11, 6330 Cham, Switzerland

This book is for my Dad, Dave Navara, who taught me that even when faced with the most daunting of tasks, you just get on them. And now that this daunting task is done, I promise to get going on fixing that garage door Dad! (Mom, I promise the next book is for you. I guess I should get on that!).

Acknowledgements

A heartfelt thanks to Mary Mendonça, who patiently listened to me whine and complain and get excited and just plainly obsess about writing this book for 2 full years. You are an absolute saint, and I can't tell you how much I appreciate your support in this and everything else. Thanks to my mother (Sandy Navara) and father (Dave Navara) who every once in a while would excitedly ask me, "How's that book coming?," after which I would guiltily run to school to write another chapter. Thanks to my students Sara Beth Pinson and Ashley Gam who helped generate the data that spurred many of the ideas behind these chapters, and Elizabeth Wrobel and Jay Curry who provided input on the crazy confusing concept of avian meiosis. Finally, thanks to my girls, Emmalyn and Adelise, for putting up with a distracted mommy who more than once responded with a "huh?" while her mind was sewn up in the organization of book chapters—and Emmy, yes you absolutely CAN write a book like mommy someday. Keep working on those sight words!

Contents

1	**Introduction to Vertebrate Sex Ratio Adjustment**		1
	1.1	How Should We Define the Many Types of Sex Allocation Found Among Vertebrates?	2
	1.2	Do Vertebrates Truly Allocate Sex Adaptively?	3
	1.3	Why and When Might Sex Allocation Be Advantageous?	6
	1.4	Do Mechanisms Exist That Allow for Adaptive Sex Allocation?	8
	1.5	The Purpose of This Book	9
	References		10
2	**It's a Boy! Evidence for Sex Ratio Adjustment in Humans**		13
	2.1	Where Are the Missing Females?	15
	2.2	Global Variation in Human Sex Ratios	16
	2.3	Influences of Ambient Temperature and Photoperiod on Natal Sex Ratios	17
	2.4	Racial Variation in Human Sex Ratios	20
	2.5	Influences of Socioeconomic Variables	22
	2.6	Influences of Diet and Malnutrition	23
	2.7	Influences of Stress	25
	2.8	Do All of These Cues Really Affect Human Sex Ratios?	27
	References		28
3	**Facultative Sex Ratio Adjustment in Nonhuman Mammals**		33
	3.1	Environmental and Social Factors Linked to Offspring Sex Ratios	34
		3.1.1 Sex Ratio Adjustment in Response to "Female Condition"	34
		3.1.2 Sex Ratio Adjustment in Response to Food Availability	37
		3.1.3 Female Dominance as a Driver of Sex Allocation Decisions	41
		3.1.4 Biasing Sex Ratios Based on Costs and Benefits to the Mother	43
		3.1.5 Sex Ratio Adjustment in Relation to Male Phenotype	46

	3.2	Can Males Be the Controllers of Sex Allocation?	47
	3.3	How Might All of These Factors Interact to Control Sex Ratios?	48
	References		51
4	**Potential Mechanisms of Sex Ratio Adjustment in Humans and Nonhuman Mammals**		**55**
	4.1	Differential Production of X- Versus Y-Bearing Sperm	55
	4.2	Differential Survival of X- Versus Y-Bearing Sperm	57
	4.3	Differential Survival and Motility of X- Versus Y-Bearing Sperm	58
	4.4	Sex-Specific Selection of Sperm in the Female Reproductive Tract	59
	4.5	Sex-Specific Implantation	61
	4.6	Sex-Specific Survival to Birth	62
	4.7	When During Development Are Sex Ratio Adjustments Really Happening?	63
	4.8	Can We Harness These Mechanisms to Artificially Control Offspring Sex?	64
	References		68
5	**The Bees Do It, but What About the Birds? Evidence for Sex Ratio Adjustment in Birds**		**71**
	5.1	Case Study: Extreme Sex Ratio Biases in Eclectus Parrots	72
	5.2	Can Birds Facultatively Adjust Offspring Sex Ratios?	73
	5.3	Case Study: Seasonal Variation in Offspring Sex Ratios in American Kestrels	74
	5.4	Seasonal Variation in Sex Ratios of Other Avian Species	76
	5.5	Variation in Sex Ratios Across the Laying Order	78
	5.6	Case Study: Male Attractiveness and Sex Allocation in Blue Tits	80
	5.7	Other Studies Relating Sex Ratios with Male Quality	82
	5.8	Case Study: Sex Ratio Adjustment and Food Supplementation in Kakapos	85
	5.9	Sex Ratio Adjustment Based on Female Quality and Food Availability	86
	5.10	Case Study: Seychelles Warblers Adjust Sex in Response to Helpers at the Nest	89
	5.11	Influences of Helpers on Sex Ratio Adjustment in Other Species	90
	5.12	How Do We Determine the Adaptive Significance of Avian Sex Ratio Adjustment?	91
	References		93

6	**Potential Mechanisms of Sex Ratio Adjustment in Birds**........		99
	6.1	How Is Sex Determined in Birds?......................	99
	6.2	Do Oocytes Have a Predetermined Sex?.................	103
	6.3	Do Factors During Rapid Yolk Deposition Influence Which Sex Chromosome Is Retained?...................	105
	6.4	Can Sex Ratios Be Altered via Direct Manipulation of Meiotic Segregation?.............................	106
	6.5	Can Sex Ratio Adjustment Occur via Disruption of the Normal Meiotic Process?.........................	112
	6.6	Can Sex Ratios Be Skewed After Meiotic Segregation?......	115
	6.7	Can Females Control Sex After Eggs Are Laid?............	117
	6.8	Which Would Be the Optimal Mechanism of Sex Ratio Adjustment?....................................	118
	References..		119
7	**Hormones Rule the Roost: Hormonal Influences on Sex Ratio Adjustment in Birds and Mammals**........................		123
	7.1	Links Between Hormones and Factors that Alter Sex Ratios...	124
	7.2	Evidence that Steroid Hormones Influence Avian Sex Ratios...	127
		7.2.1 Case Study: Peafowl.........................	127
		7.2.2 Case Study: Japanese Quail...................	127
		7.2.3 Influences of Corticosterone in Other Systems.......	128
		7.2.4 Influences of Testosterone....................	133
		7.2.5 Influences of Progesterone....................	136
		7.2.6 Influences of Estrogen.......................	136
	7.3	Evidence that Hormones Influence Sex Ratios in Mammals...	137
		7.3.1 Influences of Glucocorticoids..................	137
		7.3.2 Influences of Estrogen.......................	140
		7.3.3 Influences of Testosterone....................	142
		7.3.4 Evidence that Multiple Reproductive Hormones Act Together................................	145
	7.4	What About Nonsteroid Hormones?.....................	146
		7.4.1 Factors that Regulate Blood Glucose..............	146
		7.4.2 Leptin and Ghrelin..........................	147
	7.5	Conclusions.....................................	148
	References..		149
8	**What Went Wrong at Jurassic Park? Modes of Sex Determination and Adaptive Sex Allocation in Reptiles**.........		155
	8.1	The Range of Sex-Determining Systems in Reptiles.........	156
		8.1.1 Reptilian Species with Genetic Sex Determination....	157
		8.1.2 Genetic Control in the Absence of Heteromorphic Sex Chromosomes...........................	159
		8.1.3 True Temperature-Dependent Sex Determination.....	160

	8.2	Evidence for Adaptive Manipulation of Offspring Sex Ratios in Reptiles that Exhibit TSD	162
		8.2.1 Adaptive Models for TSD	162
		8.2.2 Case Study: Adaptive TSD in Painted Turtles?	166
		8.2.3 Evidence for Sex Ratio Adjustment in Other Turtles	170
		8.2.4 Case Study: Adaptive Sex Allocation via TSD in Jacky Dragons	171
		8.2.5 Do Viviparous Lizards Adaptively Allocate Sex via TSD?	173
		8.2.6 What Is Going on in Tuatara?	174
		8.2.7 Adaptive Sex Allocation in GSD Reptiles	176
		8.2.8 Conclusions	177
	References		178
9	**The Truth About Nemo's Dad: Sex-Changing Behaviors in Fishes**		183
	9.1	Patterns of Gonadal Differentiation in Fishes	184
		9.1.1 Gonochoristic Fishes	185
		9.1.2 Hermaphroditic Fishes	187
	9.2	Diversity in Sex-Determining Systems Among Fish Species	189
	9.3	Evidence for Adaptive Sex Ratio Adjustment in Gonochoristic Fishes	193
		9.3.1 Influences of Temperature on Sex Ratios in Fish	194
		9.3.2 Influences of pH on Sex Ratios in Fish	198
	9.4	Evidence for Adaptive Sex Ratio Adjustment in Hermaphroditic Fishes	201
		9.4.1 Adaptive Sex Change in Sequential Hermaphrodites	201
		9.4.2 Adaptive Sex Change in Simultaneous Hermaphrodites	207
	9.5	Concluding Remarks	209
	References		209
10	**Mechanisms of Environmental Sex Determination in Fish, Amphibians, and Reptiles**		213
	10.1	Potential Gene Targets in the Process of TSD	214
	10.2	Does TSD Occur via Epigenetic Modifications?	218
	10.3	The Role of Hormones in the Control of TSD	223
		10.3.1 Influences of Estrogens	223
		10.3.2 Influences of Androgens	226
		10.3.3 Influence of Glucocorticoids	227
	10.4	What About the Role of Thermal Fluctuations?	230
	10.5	Mechanisms of TSD in Frogs	231
	10.6	Mechanisms of Sex change in Hermaphroditic Fish	232
	10.7	What Do We Know and Where Do We Go from Here?	234
	References		236

Introduction to Vertebrate Sex Ratio Adjustment

> *Many see research into sex allocation as the jewel in the crown of evolutionary ecology.*
> Stuart West and Edward Herre (2002), in "Using Sex Ratios, Why Bother?"

What if we had the ability to design our children before they are born, to give them traits that would maximize their chances of thriving in the environment where they will live, or even to change our own traits as we pass through different life stages to maximize our own chances of surviving or reproducing? As it turns out, humans and other animals may be able to do just that, and without the use of fancy technology. The term "sex allocation" describes the distribution of resources between male and female reproductive function, and there is evidence in every vertebrate class that animals, including humans, have the ability to allocate sex according to the environmental and social conditions that surround them. For some species, this may mean producing more male or female offspring in response to social or environmental triggers. For others, it may mean choosing whether to produce sperm or eggs in a current reproductive attempt. In all cases, the ability to control whether individuals function as males or females can provide potent survival and reproductive benefits that may significantly influence fitness potentials at both the individual and population levels. Despite over a century of research focusing on theories and mechanisms of sex allocation in a variety of species, there are still many questions that have yet to be answered. In this introduction, I will highlight those questions and why it is important that we use comparative approaches, by examining mounting evidence that is emerging across all vertebrate classes, in our quest for answers.

1.1 How Should We Define the Many Types of Sex Allocation Found Among Vertebrates?

Sex allocation is a broad concept that involves parsing of energy that goes into male versus female reproductive output at a variety of developmental levels. How and when during development this parsing occurs differs dramatically among vertebrate species. In gonochoristic species, or species that differentiate into one sex and remain that sex for the rest of their lives, processes that influence sex allocation must occur relatively early in development. In birds and mammals, for example, sex is strictly determined by genetic input from the parents (i.e., genetic sex determination or GSD). Sex allocation, in this case, would occur either through a mechanism within one or both parents that influences the genetic makeup of the offspring prior to fertilization or through mechanisms that trigger sex-biased mortality either during gestation or beyond (see Chaps. 2–6). For these species, the term primary sex ratio adjustment is used to describe sex allocation processes that occur prior to fertilization and result in biases in the initial *production* of males and females, while secondary sex ratio adjustment is used to describe sex allocation that occurs after fertilization and results in biased *survival* of males and females. This could occur either during gestation or, in cases where parents care for offspring, during the period of parental care. Yet, sex allocation is not as easily classified into these categories for systems in which sex is determined either entirely or in part by the environment (i.e., environmental sex determination, or ESD). In these cases, the sex of the offspring either has the capacity to reverse during early embryonic or even post-hatch development or is not even determined until well after fertilization. In some gonochoristic fishes, for example, exposure of fry to high temperatures has the capacity to override the genetic influences on sex determination (see Chaps. 9 and 10), causing sex reversal. For many reptilian species, there is no initial genetic control of sex determination; sex is instead determined by the temperature at which the fertilized eggs are incubated (see Chaps. 8 and 10). How, then, do we define primary sex determination in these species? For reptiles, while sex is determined after fertilization, there is no sex change after that, so we could reasonably expand the definition of primary sex ratio adjustments to include processes that include adjustments up to the final differentiation into a male or a female, even if those processes occur after fertilization. However, how would this term be defined for animals like the fish described above, which are genetically one sex, but have the capacity to reverse during early development before they've ever reached a functional sexual phenotype? Would the definition of primary sex ratio adjustment include only the genetic contributions of the parents that defined the genetic sex of the offspring, or would it also include spawning in, for example, a warm environment that would override any genetic component and result in offspring that are more of one sex?

To make matters even more confusing, in some species, called hermaphrodites, sex is controlled and can change during adulthood by exposure to environmental and social pressure (see Chap. 9). In some species, individuals initially differentiate into one sex and then change sex during adulthood, while in others, individuals carry both ovarian and testicular tissue throughout their lives and actively decide whether to produce sperm or eggs in each reproductive attempt. Clearly in these cases, sex ratio adjustments are occurring well after fertilization or even early development, which would eliminate the use of the term primary sex ratio adjustment, but these adjustments are not happening as a result of sex-biased mortality, which would eliminate the term secondary sex ratio adjustments. As a result, in hermaphroditic species, the term sex reversal is used. Further, when examining how hermaphroditic individuals distribute resources to male versus reproductive function, the term to describe this is sex allocation. Thus, the term sex allocation is used both in the broad sense to encompass both parental and environmental manipulation of sex ratios and also in the very specific sense of parsing energy between the production of sperm and eggs.

The reason for the confusion in the implementation of these terms is because there is rarely cross talk among researchers examining sex allocation in different vertebrate classes. As a result, conversing about mechanisms and adaptive drivers of sex allocation in different vertebrate classes becomes difficult. Perhaps in the future, it would be helpful to divide the term sex allocation into parental sex allocation and individual sex allocation to further indicate the level at which the sex ratio manipulation is orchestrated. Further, rather than separating the modes of sex ratio adjustment according to developmental age, I suggest subdividing primary sex ratio adjustment into two groups: those that exhibit genetic sex ratio adjustment and those that exhibit environmental sex ratio adjustment. Using these terms, it is possible to have conversations about sex allocation that would be applicable to animals of all vertebrate groups.

1.2 Do Vertebrates Truly Allocate Sex Adaptively?

Questions about whether sex ratio patterns, either at the individual or population level, could be shaped by natural selection date back to Darwin's days when he wrote the following quote in the first edition of his book, *Descent of Man*:

> Let us now take the case of a species producing from the unknown causes just alluded to, an excess of one sex—we will say of males—these being superfluous and useless, or nearly useless. Could the sexes be equalised through natural selection?

In the same edition, Darwin theorized that in a scenario where there was an excess of one sex in the population, the more productive pairs would be those that produced offspring of the other sex and thus "*a tendency towards the equalization of the sexes would be brought about.*" While he, in his second edition, proclaimed to leave the intricate problem of explaining sex ratio theory to the future, his idea was actually very close to a now famous theory put forth by Fisher (1930); the theory posited that frequency-dependent selection should lead to the equal production of males and females. Later, Hamilton (1967) expanded upon this general idea with an example that was nicely summarized by Osborne (1996) (Fig. 1.1): If female births

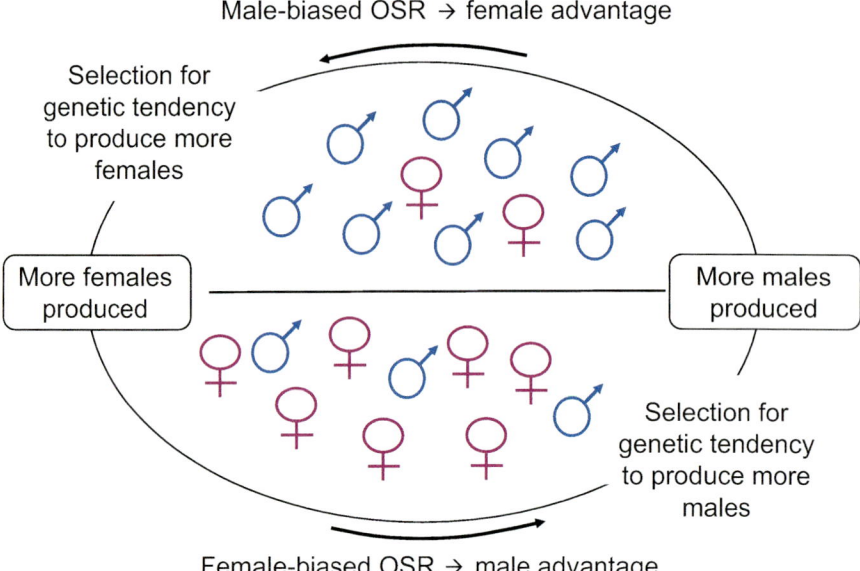

Fig. 1.1 Visual depiction of the ideas provided by Hamilton (1967) supporting the notion that population sex ratios should ultimately hover around 50:50 males:females. When operational sex ratios (OSRs) are female biased, selection should favor individuals with a genetic propensity to produce more male offspring. As this happens, however, the population OSR would become male biased, and the reverse selection process would occur. Thus, overall, the population sex ratio hovers around 50:50 males:females

are less common than males, this then shifts the mating advantage to individuals who are female, and those females will then go on to have more offspring. Thus, parents that are genetically predisposed to produce more females pass on their genes, and female births become even more common. This then, however, brings the sex ratio closer to 1:1, after which the advantage for producing males increases. The continuation of this process leads to stabilization of a 1:1 sex ratio overall. Indeed, Jull (1923) pointed out that overall, the sex ratios observed in farm stock are nearly always close to equality, despite the diverse conditions in which they are raised, a phenomenon that supports the ideas discussed by Fisher and Hamilton.

These theories do not exclude the possibility that adaptive patterns of sex ratio adjustment exist, particularly if patterns of adjustment are needed to bring the population sex ratio towards 1:1. However, Sven Krackow, one of the first to write a truly comprehensive review of the evidence for sex ratio adjustment in birds and mammals (Krackow 1995), also wrote a strong critical piece arguing *against* the idea that facultative adjustment of offspring sex ratios should be prevalent (Krackow 2002). He states that, for birds and mammals, the lack of a mechanism allowing for parental manipulation of offspring sex ratios is actually far more plausible than the presence of one. Indeed, mechanisms for sex ratio adjustment prior to conception still have not been definitively identified, and at least for mammals and birds, mechanisms that occur beyond conception would likely be too costly (Maynard Smith 1980). Others have argued that even selection for mechanisms prior to conception would not occur. For example, Williams (1979) points out that if mothers had the ability to differentiate X versus Y-bearing sperm and "chose" one over the other, a mechanism by which the sperm carrying the "unchosen" chromosome could overcome this process would be selected for, because the fitness benefit to the mother associated with sex ratio adjustment would be nowhere near the fitness benefit to the genes on the unselected chromosome (i.e., survival of the chromosome). Reiss (1987) addressed the possibility that an individual may be able to alter the process of meiotic segregation to produce more of the gamete containing a particular sex chromosome; he points out that such a mechanism would require that the "unchosen" sex chromosome be linked to an autosome that would, in essence, need to commit suicide in response to a parental signal. Simple mathematics shows that selection for this autosomal allele would require that the fitness benefit of producing the "chosen" sex chromosome would have to be at least three times that of producing the "unchosen" chromosome. Krackow (2002) further points out that selection of a particular sex chromosome would require "tagging" of that sex chromosome in some way, and "*as sex chromosomal genes have no fitness interest of being excluded from further replication, selection would favor genetic activity or conformational changes countering autosomal ability of sex discrimination.*" As a result, sex chromosomes that are not "tagged" in this way would be selected for.

If these suppositions are true, however, then why are there so many accounts of sex ratio skews in birds, mammals, and other vertebrate systems? Some have argued that the findings actually resulted from statistical Type I error resulting from very small or very large sample sizes, as well as our lack of knowledge about what a particular effect size means. For example, a majority of studies of sex ratios in human populations document variation in sex ratios as low as 0.1%, yet these differences are *statistically* significant because millions of individuals are included in the analyses. But is a sex ratio difference of 50.1 versus 50.2% *biologically* significant? Likewise, many sex ratio studies in birds include fewer than 20 individuals or focus on small numbers of potentially genetically related individuals in a single population. This could result in sex ratio skews that are actually spurious or biologically insignificant. Further, there is the question of whether demand for significant results has generated a publication bias within the field. Palmer (2000) indicates that selective reporting of significant results, particularly for studies of sex ratios, is widespread.

West and Sheldon (2002) conducted a stronger analysis that countered the idea that there is a bias in the publication of sex ratio studies towards results that are significant, and the fact remains that there are hundreds of studies showing sex ratio adjustments in vertebrates in response to environmental and social conditions. Further, the results of meta-analyses support the idea that adaptive adjustment of offspring sex ratios is, indeed, occurring in at least some of these systems (e.g., see Chaps. 3 and 5). In this book, I will review the evidence for adaptive sex allocation in birds, reptiles, mammals, and fishes and discuss studies that could help test for the presence of adaptive sex allocation in these systems.

1.3 Why and When Might Sex Allocation Be Advantageous?

Offspring sex is arguably the phenotypic trait with the greatest influence on the chances that the offspring will survive and successfully reproduce, because males and females differ in a suite of physiological and behavioral responses that influence how they may function in their environments. To understand the many ways in which this occurs, we can explore the many physiological and behavioral characteristics that differ between males and females. First, developmental rates often differ between males and females, sometimes starting well before phenotypic sex is determined. In some mammals, for example, there are sex differences in metabolic rates even as early as the blastocyst stage (Gutiérrez-Adán et al. 2000, 2006). Work in birds, reptiles, and fish have shown sexual dimorphism in rates of growth after hatch; sometimes males grow faster (Martins 2004; Parker 1992), and other times females do (Anderson et al. 1997; Haenel and John-Alder 2002). One can imagine how these differences in rates of development could make it more or less advantageous to produce a specific sex at a particular time in the breeding season, or when social dynamics make body size particularly important for breeding success, and this is an idea that has been explored in both birds (Daan et al. 1996 and see Chap. 5) and reptiles (Harlow and Taylor 2000; see Chap. 8).

Much of the sex ratio literature, especially in birds and mammals, has focused on the different energetic costs associated with producing high-quality males versus females. Not only do males have higher metabolic rates through development, they also can require higher amounts of energy if they must develop large armaments or ornaments as well as large body sizes. Trivers and Willard (1973), for example, highlighted the idea that, in a polygynous system in which males attract females or defend territories using costly ornaments or armaments, it may cost more energy to produce a male that is of high enough quality to win a large number of matings with females, but when resources are abundant and mothers pass down their good condition to offspring, the benefit of producing males likely outweighs the cost. This is the single most highly cited paper in the field of sex ratio adjustment and also the most widely tested. However, it is also important to consider the costs associated with the production of eggs versus sperm; eggs are generally considered to be much more costly (Williams 1975). Thus, perhaps in this same polygynous system, when food is limited and a female's condition is low, instead of producing more females as Trivers and Willard (1973) predicted, she would instead produce the sex that produces the less expensive gametes and donates less effort to parental care.

We must also consider that during adulthood, males and females differ substantially in both behavior and physiology. The presence of sex hormones, in particular, directs specific suites of behaviors in males and females and also exerts physiological effects that can greatly impact the survival of an individual. Perhaps the most obvious

1.3 Why and When Might Sex Allocation Be Advantageous?

influence is the effect of high circulating levels of testosterone on aggression in male members of every vertebrate class (e.g., Dixson 1980; Fernald 1976; Moore and Marler 1987; Wingfield et al. 1987). In general, males are much more likely to respond aggressively in response to a stressor or intrusion. A study in human females, on the other hand, indicated that the primary female response to a challenge is to "tend and befriend" rather than fight or flight (Taylor et al. 2000). The costs and benefits of these divergent reactions can differ based on the context in which an animal lives. An aggressive phenotype might, for example, benefit a polygynous male who has to fight to obtain as many mates as possible, but might instead serve as a detriment to a male that is born or hatches into an environment with high predator densities, because he might be more likely to react aggressively towards a predator and die rather than flee and survive. Testosterone is also a well-known immunosuppressor in vertebrates (Belliure et al. 2004; Grossman 1985; Hillgarth and Wingfield 1997; Marsh and Scanes 1994). As a result, the sexes differ in the way they fight pathogens and their abilities to do so, and those sex differences can be either minimized or exacerbated based on life history characteristics and environmental variables (Klein 2000). Thus, in addition to defining the costs of "maleness" or "femaleness" in the context of an energetic expenditure, it is also important to consider the costs of male and female traits in the context of mortality risks.

It is also unclear how we should parse out the potential costs and benefits in terms of offspring survival and reproductive success, and the potential costs and benefits in terms of parental survival to reproduce in the future. Work in primates and other mammals indicates that females may allocate sex according to how the production of a particular sex will affect their own competition for resources and mates in the future (i.e., The Local Resource Competition and Local Mate Competition hypotheses; see Cockburn et al. 2002; West 2009 for excellent reviews on these). Alternatively, if females help at the nest, the Local Resource Enhancement hypothesis predicts that sex may be allocated according to how much the presence of those helpers would benefit the mother's future reproductive output (see Chap. 3 for a discussion of all three hypotheses).

What we're missing from current analyses of the potential adaptive benefits associated with sex allocation is a broader scale approach (Fig. 1.2). To truly understand the ultimate costs and benefits of producing a particular sex in a fitness sense, we need to understand how the many different life history traits of animals, such as the mating system for example, influence sex allocation decisions. We also need to consider environmental influences and how sex allocation patterns may change from year to year based on environmental and social variation. Finally, we must also consider whether the mechanism of sex allocation, itself, is costly to parents. In his critique of adaptive sex allocation, Krackow (2002) pointed out that we cannot begin to understand whether adaptive sex allocation truly exists in vertebrates until we uncover physiological mechanisms by which it may occur.

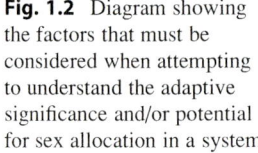
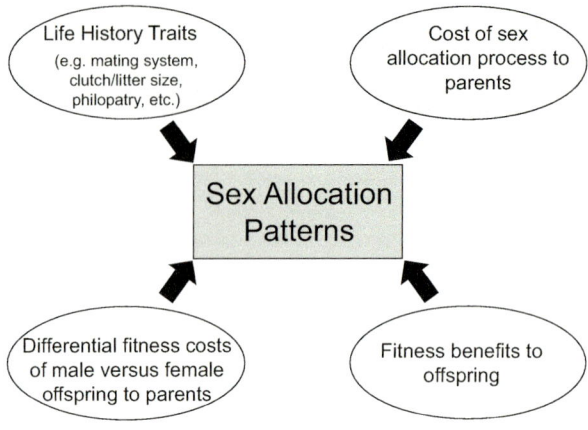

Fig. 1.2 Diagram showing the factors that must be considered when attempting to understand the adaptive significance and/or potential for sex allocation in a system

1.4 Do Mechanisms Exist That Allow for Adaptive Sex Allocation?

This leads us to the question of whether there is evidence that mechanisms for facultative adjustment of sex ratios do, in fact, exist. I have dedicated four chapters of this book to discussions of the multiple potential mechanisms by which different systems may accomplish facultative sex ratio skews. Vertebrates fall along a continuum of sex-determining mechanisms (Fig. 1.3), one end describing systems like mammals and birds where we see a gonochoristic or permanent form of sex determination that is governed by a genetic male/female "switch." Further down the line, there are systems where a genetic mechanism of sex determination exists, but can be overridden by environmental cues. Members of this group include reptiles and some fishes and amphibians. As we move along the continuum, we find the reptiles and fishes that do not determine sex genetically, but instead do so entirely based on environmental or social cues. Finally, at the far end, we have piscine systems that

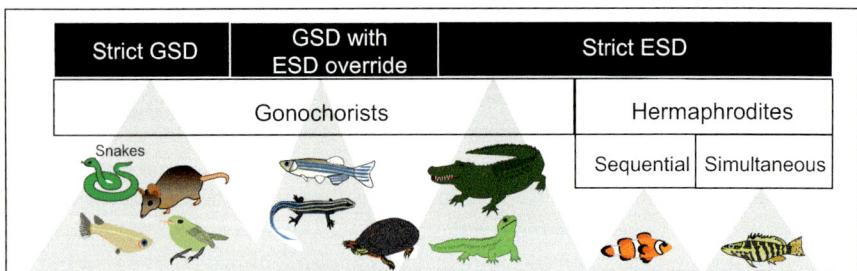

Fig. 1.3 Illustration of the continuum of sex-determining systems in vertebrates including examples of individuals within each vertebrate class that exhibit each category of sex determination. Gonadal sex determination is found in members of all vertebrate classes, while environmental sex determination is only found in reptiles, amphibians, and fish. Fish are the only vertebrate class to have representative species for all modes of sex determination

never truly become one sex and instead contain both testicular and ovarian tissue and thus have the capability to produce either sperm or eggs depending on the reproductive attempt and the cues surrounding them during that attempt. For the systems that exhibit ESD, it is clear that there are mechanisms available for biasing offspring or even adult sex ratios, and researchers are getting closer and closer to discovering the details of those mechanisms. For example, there is now overwhelming evidence that estrogen mediates the process of TSD in turtles, and genetic targets involved in the process have been identified (Matsumoto et al. 2013). In fishes, not only is it clear that some species have the ability to change sex from male to female or vice versa during adulthood, but it is now known that the mechanism involves estrogen as well (Guiguen et al. 2010). Aquaculturists are already using hormone treatments to purposefully skew the sex ratios of farmed fish (Devlin and Nagahama 2002). As researchers continue to unveil the details of the mechanisms that allow for sex allocation to occur in ESD systems, further understanding of the costs and benefits involved in the process will become possible (see Chap. 10).

As discussed above, it has been generally accepted that genetic systems of sex determination are much stricter and are less likely to contain mechanisms by which offspring sex can be determined in a way that is not too costly to parents. Yet, there is now evidence in pigs that the female reproductive tract can, in fact, identify and differentiate responses to X- and Y-bearing sperm in the reproductive tract (Almiñana et al. 2014, discussed in Chap. 4). In birds, treatment with hormones immediately prior to the completion of meiosis appears to stimulate which sex chromosome the egg contains at ovulation (Navara 2013; see Chap. 6), indicating that female birds may, in fact, be able to control the process of meiotic segregation. Work in mammals indicates that the hormone content of the fluid within the ovarian follicle can drive preferential selection of X- versus Y-bearing sperm (Grant and Chamley 2010; see Chap. 4). Finally, there are clear patterns in humans suggesting that male fetuses are more likely to suffer from prenatal mortality when conditions are challenging for the mother (Navara 2010; see Chap. 2). Thus, despite the skepticism that mechanisms of sex allocation would exist for systems that appears to be restricted within the framework of GSD, we continue to see evidence that mechanisms *do* in fact exist. I will highlight these potential mechanisms and the evidence that supports them in Chaps. 4, 6, 7, and 10.

1.5 The Purpose of This Book

When writing this book, I aimed to compile evidence for sex ratio adjustment and what we know about the mechanisms involved for a variety of vertebrate systems, because it is extremely important that we address the questions outlined above from a comparative standpoint. For example, in both hermaphroditic fishes and TSD reptiles, estrogen appears to play a large part in the control of sex determination and differentiation. While there is already cross talk among individuals working on species in these two vertebrate classes, more comparative analyses are warranted. Similarly, there is evidence in birds, humans, and nonhuman mammals that stress hormones may play a role in the process of sex ratio adjustment. Perhaps there is a common mechanism through which stress hormones act in these systems. In many cases, adaptive theories that have been developed for individuals in one group are not considered in another. And for some groups, there is a strong focus on mechanism but little focus on adaptive explanations, and vice versa. Through these chapters, I attempt to identify the "holes" that exist in the study of sex ratio biology in vertebrates in the hopes that this will help to further the advancement of the field.

References

Almiñana C, Caballero I, Heath PR, Maleki-Dizaji S, Parrilla I, Cuello C, Gil MA, Vazquez JL, Vazquez JM, Roca J (2014) The battle of the sexes starts in the oviduct: modulation of oviductal transcriptome by X and Y-bearing spermatozoa. BMC Genomics 15(1):1

Anderson D, Reeve J, Bird D (1997) Sexually dimorphic eggs, nestling growth and sibling competition in American kestrels Falco sparverius. Funct Ecol 11(3):331–335

Belliure J, Smith L, Sorci G (2004) Effect of testosterone on T cell-mediated immunity in two species of Mediterranean lacertid lizards. J Exp Zool A Ecol Genet Physiol 301(5):411–418

Cockburn A, Legge S, Double MC (2002) Sex ratios in birds and mammals: can the hypotheses be disentangled. In: Hardy ICW (ed) Sex ratios: concepts and research methods. Cambridge University Press, Cambridge, pp 266–286

Daan S, Dijkstra C, Weissing FJ (1996) An evolutionary explanation for seasonal trends in avian sex ratios. Behav Ecol 7(4):426–430

Devlin RH, Nagahama Y (2002) Sex determination and sex differentiation in fish: an overview of genetic, physiological, and environmental influences. Aquaculture 208(3):191–364

Dixson A (1980) Androgens and aggressive behavior in primates: a review. Aggress Behav 6(1):37–67

Fernald RD (1976) The effect of testosterone on the behavior and coloration of adult male cichlid fish (Haplochromis burtoni, Günther). Hormones 7(3):172–178

Fisher R (1930) Genetical theory of sex allocation. Clarendon Press, London

Grant VJ, Chamley LW (2010) Can mammalian mothers influence the sex of their offspring periconceptually? Reproduction 140(3):425–433

Grossman CJ (1985) Interactions between the gonadal steroids and the immune system. Science 227:257–262

Guiguen Y, Fostier A, Piferrer F, Chang C-F (2010) Ovarian aromatase and estrogens: a pivotal role for gonadal sex differentiation and sex change in fish. Gen Comp Endocrinol 165(3):352–366

Gutiérrez-Adán A, Oter M, Martínez-Madrid B, Pintado B, De La Fuente J (2000) Differential expression of two genes located on the X chromosome between male and female in vitro–produced bovine embryos at the blastocyst stage. Mol Reprod Dev 55(2):146–151

Gutiérrez-Adán A, Perez-Crespo M, Fernandez-Gonzalez R, Ramirez M, Moreira P, Pintado B, Lonergan P, Rizos D (2006) Developmental consequences of sexual dimorphism during pre-implantation embryonic development. Reprod Domest Anim 41(s2):54–62

Haenel GJ, John-Alder HB (2002) Experimental and demographic analyses of growth rate and sexual size dimorphism in a lizard, Sceloporus undulatus. Oikos 96(1):70–81

Hamilton WD (1967) Extraordinary sex ratios. Science 156(3774):477–488

Harlow PS, Taylor JE (2000) Reproductive ecology of the jacky dragon (Amphibolurus muricatus): an agamid lizard with temperature-dependent sex determination. Aust Ecol 25(6):640–652

References

Hillgarth N, Wingfield JC (1997) Testosterone and immunosuppression in vertebrates: implications for parasite-mediated sexual selection. In: Beckage NE (ed) Parasites and pathogens. Springer, Boston, pp 143–155

Jull MA (1923) Can sex in farm animals be controlled? Proc Am Soc Anim Nutr 1923(1):92–98

Klein SL (2000) Hormones and mating system affect sex and species differences in immune function among vertebrates. Behav Process 51(1):149–166

Krackow S (1995) Potential mechanisms for sex ratio adjustment in mammals and birds. Biol Rev 70(2):225–241

Krackow S (2002) Why parental sex ratio manipulation is rare in higher vertebrates (invited article). Ethology 108(12):1041–1056

Marsh JA, Scanes CG (1994) Neuroendocrine-immune interactions. Poult Sci 73(7):1049–1061

Martins TLF (2004) Sex-specific growth rates in zebra finch nestlings: a possible mechanism for sex ratio adjustment. Behav Ecol 15(1):174–180

Matsumoto Y, Buemio A, Chu R, Vafaee M, Crews D (2013) Epigenetic control of gonadal aromatase (cyp19a1) in temperature-dependent sex determination of red-eared slider turtles. PLoS One 8(6):e63599

Maynard Smith J (1980) A new theory of sexual investment. Behav Ecol Sociobiol 7(3):247–251

Moore MC, Marler CA (1987) Effects of testosterone manipulations on nonbreeding season territorial aggression in free-living male lizards, Sceloporus jarrovi. Gen Comp Endocrinol 65(2):225–232

Navara KJ (2010) Programming of offspring sex ratios by maternal stress in humans: assessment of physiological mechanisms using a comparative approach. J Comp Physiol B 180(6):785–796

Navara KJ (2013) Hormone-mediated adjustment of sex ratio in vertebrates. Integr Comp Biol 53 (6):877–887

Osborne MJ (1996) Darwin, Fisher, and a theory of the evolution of the sex ratio. http://www.economics.utoronto.ca/osborne/research/sexratio.pdf

Palmer AR (2000) Quasireplication and the contract of error: lessons from sex ratios, heritabilities and fluctuating asymmetry. Annu Rev Ecol Syst 31:441–480

Parker G (1992) The evolution of sexual size dimorphism in fish. J Fish Biol 41(sB):1–20

Reiss MJ (1987) Evolutionary conflict over the control of offspring sex ratio. J Theor Biol 125 (1):25–39

Taylor SE, Klein LC, Lewis BP, Gruenewald TL, Gurung RA, Updegraff JA (2000) Biobehavioral responses to stress in females: tend-and-befriend, not fight-or-flight. Psychol Rev 107(3):411

Trivers RL, Willard DE (1973) Natural selection of parental ability to vary the sex ratio of offspring. Science 179(4068):90–92

West S (2009) Sex allocation. Princeton University Press, Princeton

West SA, Herre EA (2002) Using sex ratios: why bother. In: Hardy ICW (ed) Sex ratios: concepts and research methods. Cambridge University Press, Cambridge, pp 399–413

West SA, Sheldon BC (2002) Constraints in the evolution of sex ratio adjustment. Science 295 (5560):1685–1688

Williams GC (1975) Sex and evolution, vol 8. Princeton University Press, Princeton

Williams G (1979) The question of adaptive sex ratio in outcrossed vertebrates. Proc R Soc Lond B Biol Sci 205(1161):567–580

Wingfield JC, Ball GF, Dufty AM, Hegner RE, Ramenofsky M (1987) Testosterone and aggression in birds. Am Sci 75(6):602–608

It's a Boy! Evidence for Sex Ratio Adjustment in Humans

> *Now by thy looks*
> *I guess thy message. Is the queen deliver'd?*
> *Say, ay: and of a boy.*
> From William Shakespeare's Henry VIII

Henry VIII went through six wives and defied a pope for want of a son among the many offspring that he fathered (Box 2.1). Humans have long been fascinated by the ratio of boys to girls in the next generation and the factors influencing it. Since Fisher (1930) first provided a theoretical framework explaining sex ratio patterns, scientists have struggled to understand variation in natal sex ratios among human populations. Birth sex ratios of humans change in relation to a staggering number of economic, social, and physiological factors, including marital status, social class, natural disasters, war, socioeconomic status, and psychological distress. Some studies suggest that humans could purposefully control the sexes of babies through specific sexual positions, by timing intercourse during specific points in the ovulatory cycle, or by eating specific foods. The exact time point during development when humans manipulate sex ratios is unclear, as some of the cues that stimulate biases occur prior to gestation while others are experienced during gestation. In this chapter, I will review evidence for sex ratio adjustment in humans in response to social and environmental variables and discuss the potential timing for this adjustment during offspring development.

Box 2.1 Henry VIII's Quest for a Son

Henry VIII was the king of England from 1509 until his death in 1547 and was the second Tudor king. He was well-known for the drastic measures he took to sire a son in order to continue the Tudor dynasty. Henry first married Katherine of Aragon, but after she failed to produce a son for him, he divorced her, provoking his excommunication by the Pope. He then married Anne Boleyn, who produced two daughters and a stillborn son. Frustrated with her inability to produce a viable son, Henry VIII executed her. He finally produced his son and heir, Edward, with his third wife, Jane Seymour.

A Gallup Poll conducted in 2011 indicated that gender preferences are far from gone. The statistics showed that, if Americans could produce only one child, 68% would have a particular gender preference for that child, and a majority of those would prefer a boy (Newport 2011). These preferences are even more extreme in other parts of the world, driving the practice of sex-selective abortion in many cases. More than 85% of abortions worldwide are performed in developing countries, many targeting baby girls simply because boys are considered to be optimal economically (Casey et al. 2013). As of 2007, it was estimated that approximately 100,000 abortions specifically driven by the female sex of the fetus are performed every year in India (MacPherson 2007). Similar practices have also been considered common in more developed countries such as China and, while less common, also exist in Europe and North America (Casey et al. 2013; Dubuc and Coleman 2007). This alone exhibits the need for a good understanding of what factors can affect the human sex ratio, perhaps to find methods of sex selection that can be used prior to conception, or at least very early in embryonic development.

2.1 Where Are the Missing Females?

It is well known that average birth sex ratios for humans are generally male biased, hovering at a stable 1.06 boys for every girl (or 51.5% boys). However, until recently, it was unclear whether this male bias represented the sex ratio at conception or whether losses later in gestation accounted for the "missing" females at birth. In 2015, Orzack and colleagues used the most comprehensive dataset ever assembled to determine sex ratio patterns at different developmental time points from conception through birth (Orzack et al. 2015), and their data may provide the answer to the question of why more males are produced than females (Fig. 2.1). To determine sex ratios around conception, they obtained 3–6-day-old embryos that were produced using assisted reproductive technologies (ART) and used fluorescence in situ hybridization (FISH) or array comparative genomic hybridization (aCGH) to karyotype the embryos. To compile sex ratio patterns at additional points through the remainder of gestation, they collected information from 41 studies that detailed the sexes of fetuses that had been aborted, analyzed samples that had been collected from women undergoing chorionic villi sampling and amniocentesis, and created a dataset containing sex for all US fetal deaths and live births for 1995–2004. What they found was that the sex ratio at conception was 50:50. When they included whether the karyotypes were normal or abnormal, the model had even more support; for abnormal karyotypes, the sex ratio at conception was 50.8%, while for normal karyotypes, the sex ratio was 49.3%.

This suggests that very early development may be slightly more hazardous for male than female embryos. But the sex differences in vulnerability then reverse. Analysis of induced abortion data, CVS data, and amniocentesis results showed that between 2 and 20 weeks, sex ratios rise continuously, and sex ratios collected for this gestational time frame via ultrasound suggest that the sex ratio rises dramatically between 12 and 20 weeks, reaching as high as the 65% males seen between 12 and 20 weeks of embryonic age? (Meagher and Davison 1996; Reece et al. 1987; Whitlow et al. 1999). This suggests that females are dying at a much higher rate

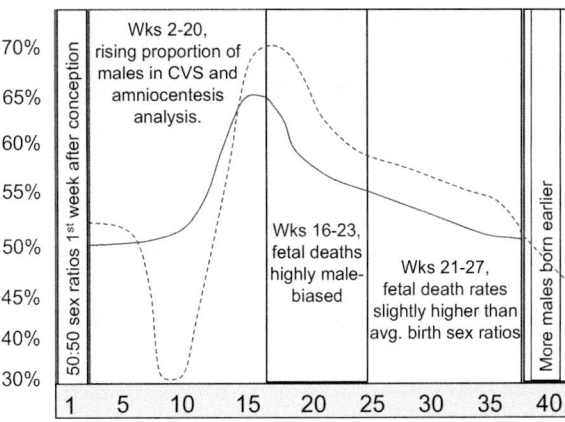

Fig. 2.1 A depiction of what we know about the progression of sex ratios from conception to birth as determined from compiled ART, CVS, amniocentesis, ultrasound, induced abortion, and fetal death data

than males at the end of the first trimester, and this female-biased vulnerability is likely what accounts for the slight male bias at birth.

However, why isn't the sex ratio at birth as high as 65%? Fetal death data may help shed light on this question. At the start of the second trimester, the mortality rates appear to flip again; a study incorporating over 1.6 million births to Norwegian women showed that sex ratios of fetuses born between 16 and 19 weeks of gestation reach 71.3%. Sex ratios of births after that point then remain higher than those of births at term (55–56% males), suggesting that males are more likely to suffer complications that lead to very early preterm birth (Challis et al. 2013). Further, a study of over 450,000 gestations in Scotland showed that males had a significantly higher risk of stillbirth between 28 and 43 weeks (Smith 2000). Thus, it appears that the high sex ratio at the end of the second trimester is compensated for by the slow loss of males through the second and third trimesters, resulting in an average sex ratio of approximately 51.5% males.

2.2 Global Variation in Human Sex Ratios

While the average birth sex ratio is consistently male biased, the ratio of boy to girl children produced varies widely across the globe when split among individual countries (Fig. 2.2) (Navara 2009). Countries such as China and India produce sex ratios skewed drastically towards boy babies, exceeding, in some cases, 1.11 boys for every girl produced (or 52.6% boys). The opposite is seen in countries located in southern and mid-Africa, where substantially fewer boy babies are produced compared to the global average. The Central African Republic, for example, produced a *female*-biased sex ratio (48.8% boys). Countries located in Europe and North and South America tend to produce ratios near the global average (51.5% boys). Interestingly, the natal sex ratio follows a latitudinal gradient, with countries closer to the equator producing fewer boys (Navara 2009).

Why this variation in natal sex ratios among individual countries? Perhaps the most parsimonious explanation is that the visible variation in natal sex ratios stems from the practices of sex-specific abortion mentioned above. Indeed, the countries with the highest ratios of boys to girls are China, India, and South Korea, some of the most well-known countries for practicing sex-specific abortion (Hesketh and Xing 2006). One study conducted in rural China in 2001, a year within the span of data collection, showed that sex-selective abortion was prevalent after prenatal sex screening by ultrasound and that 25% of all female pregnancies were aborted compared to 2% of male pregnancies (Junhong 2001). However, when those countries known for practicing sex-specific abortion are removed from the analyses, the latitudinal gradient in natal sex ratios persists at a global level. This latitudinal trend is not as clear when examining natal sex ratios at the continental level, however. When data from only within Asia or Europe are analyzed, women living at more southern latitudes instead produced more males (Grech 2013; Grech et al. 2002), while women living in North America produce more males at higher latitudes (Grech et al. 2002). The reasons behind these disparities remain unknown.

2.3 Influences of Ambient Temperature and Photoperiod on Natal Sex Ratios

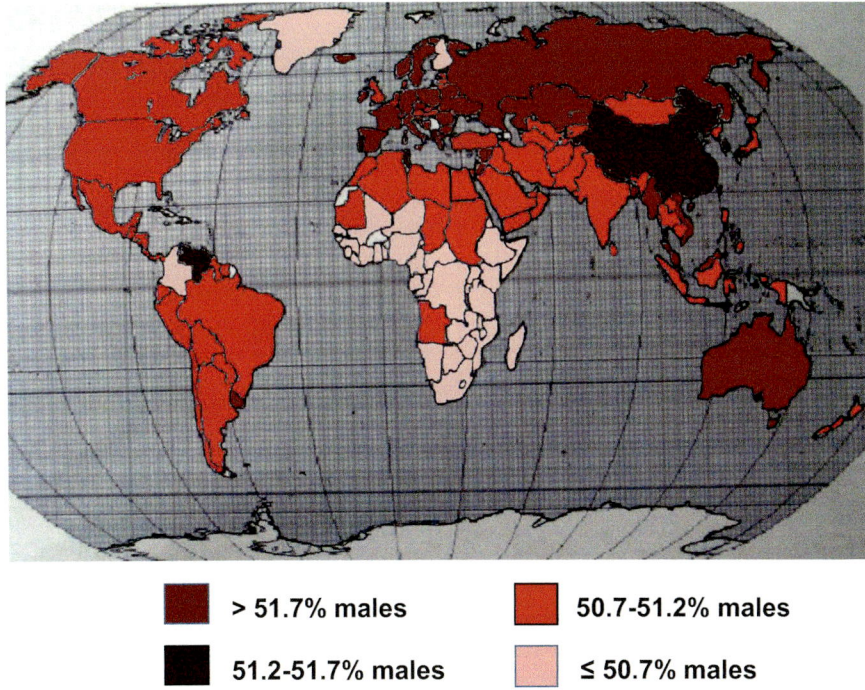

Fig. 2.2 World map showing the average percentage of male offspring born for each country from 1997 to 2006. Figure reproduced from Navara (2009)

Regardless, when examined globally, natal sex ratios still show striking patterns, particularly in sub-Saharan Africa where significantly fewer males are produced than would be expected according to the global average (Garenne 2002; Navara 2009).

2.3 Influences of Ambient Temperature and Photoperiod on Natal Sex Ratios

Some have attempted to explain the global variation we see in natal sex ratios by looking for influences of environmental variables that are related to latitude, such as ambient temperature and photoperiod. Not surprisingly, the global sex ratios shown in Fig. 2.2 were also positively correlated with annual variation in day length recorded for each country from 2000 to 2008; however, because this and other variables related to latitude are highly correlated, it remains impossible to distinguish which variable may have contributed most to the observed global pattern in natal sex ratios. To my knowledge, no other study has addressed the idea that photoperiod and its related physiological influences may trigger skews in natal sex ratios produced by humans, but there is one study in a hamster showing that, despite constant temperatures, winter-like day lengths caused Siberian hamsters to produce

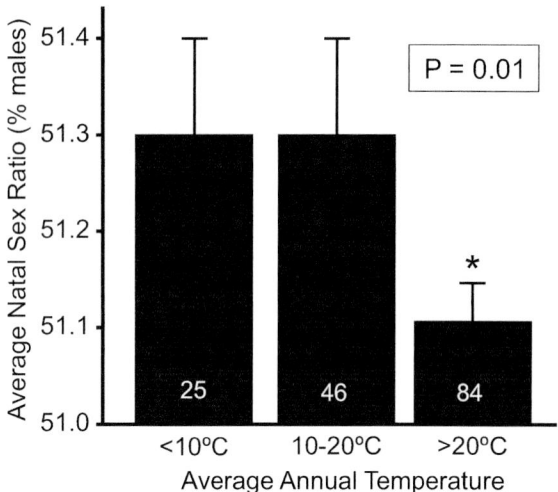

Fig. 2.3 Global relationship between average natal sex ratios from 1997 to 2006 and average temperature for each country. The number of countries included in each temperature range is noted in the individual bars. An asterisk denotes the group that is significantly different

more male offspring (Navara et al. 2010). Humans do appear to vary conception rates in response to photoperiodic influences (Cagnacci and Volpe 1996). Perhaps mammals (including humans) possess the physiological machinery that would allow adjustment of natal sex ratios in response to changing photoperiods as well. This remains to be determined.

In addition to photoperiod, the global proportion of males is also positively correlated to the average ambient temperature for each country (Fig. 2.3), and a study conducted in Northern Finland similarly showed that natal sex ratios were male biased in warmer years and female biased in cold years (Helle et al. 2008). If temperature is affecting sex ratios, at what point during development is this happening? In the USA, the temperature in the month before conception predicted the likelihood of having a male, with more males being produced when temperatures were warmer (Meyers 2012). Lerchl (1998) documented a similar effect in Germany, where warmer temperatures 10–11 months prior to birth predicted an increased likelihood of producing male babies. On the other hand, Catalano et al. (2008) contributed evidence that temperature can also exert influences during gestation; Scandinavian women produced significantly fewer male babies when temperatures during gestation were cold. This effect would result from sex-specific vulnerability to adverse conditions within the uterus. Perhaps both mechanisms are at work to influence human natal sex ratios in response to fluctuations in temperature. However, the relationships between ambient temperature are not likely that simple. While Finnish women produced more males when temperatures were hotter in the concurrent year, sex ratio was *inversely* related to the temperatures from the *previous* year (Helle et al. 2008). Also, studies conducted in Australia and New Zealand showed no relationship between ambient temperature and the sex ratio at birth (Dixson et al. 2013a, Dixson et al. 2013b).

If temperature does in fact influence human sex ratios, we would expect to see substantial seasonal variation in sex ratios as a result. Human conceptions follow a clear seasonal rhythm, with more babies being conceived near the vernal equinox (Roenneberg and Aschoff 1990). It has been suggested that this pattern represents the vestiges of an adaptive mechanism by which human babies are born during spring and summer when resources are abundant. Conception rates are also specifically related to temperature; the highest rates of conception occur when temperatures are between 5 and 20 °C. If the influences of temperature on conception rates are more potent in the process of producing a boy versus a girl, this could ultimately influence natal sex ratios. However, the results of studies examining the potential for seasonal influences on natal sex ratios do not provide overwhelming support. Some of the most convincing evidence comes from studies on US women; the highest proportions of male offspring were born in early summer, and the lowest in winter (James 1984, 1987; Slatis 1953; Lyster 1971). Similarly, Cagnacci and colleagues (2003) showed that natal sex ratios in Modena, Italy, were not related to the season of birth, but were instead related to the season of conception, with the proportion of males peaking in October. This would put the highest rate of male births at June, or early summer. These Italian and US results follow with annual temperature patterns where male births appear to be more likely when temperatures are warmer. Yet, the variation in natal sex ratios among seasons is extremely small (about 0.5%), and several other studies failed to find a seasonal pattern in the proportion of males that were born.

Perhaps the relationships between temperature and the proportion of male babies produced result not from mild temperature fluctuations but are instead driven by extreme fluctuations in temperatures. In Germany, the USA, and Finland, where significant relationships between sex ratios and temperatures were seen, temperatures change by 19 °C, 23 °C, and 24 °C, respectively, over the course of the year. In comparison, the two countries where no effect was found only see fluctuations of 9° and 10° across the year. Alternatively, perhaps it is not the temperature change but is instead how hot *or* how cold the region gets at any one point that alters sex ratios. Finland, Germany, and the USA reach temperatures as low as -7 °C, -1 °C, and 1 °C, respectively, while Australia and New Zealand only get as low as 11 °C. It may be that extreme cold exerts a temperature-related stress on developing babies during gestation (Catalano et al. 2008), causing more vulnerable males to die (McLachlan and Storey 2003). A study in Japan, however, supports influences of extremes in *both* directions. Fukuda et al. (2014) showed that the degree of temperature changes from year to year is negatively correlated with the proportion of boys produced, and also positively correlated with the proportion of male fetal deaths. In fact, both a very hot summer and a very hot winter in the region induced significant declines in natal sex ratios 9 months later as well as significant increases of fetal death rates immediately. So, extreme variation in temperature from a given optimal mean could exert influences on sex ratios prior to conception, or could prove detrimental to fetuses in a sex-specific manner. It seems there is promising evidence that temperature can be a player in the determination of natal sex ratios in humans. More work needs to be done to determine how prominent of a role it plays and how it may act.

2.4 Racial Variation in Human Sex Ratios

Another well-demonstrated factor related to natal sex ratios that may also contribute to global and latitudinal patterns is racial background, because, on average, the lowest natal sex ratios are found near the equator where human skin colors are significantly darker (Jablonski and Chaplin 2010). When examining populations with heterogeneous cultural groups, including the USA, Wales, South Africa, and the West Indies, black populations produce lower natal sex ratios compared to white populations (Ciocco 1938; Erickson 1976; Strandskov and Roth 1949; Teitelbaum and Mantel 1971; Visaria 1967). Garenne (2002) showed that, within Africa, there are segments where populations produce low, average, and high natal sex ratios, just as we see in other parts of the world, and the lowest natal sex ratios were produced in Southern Africa, primarily in Bantu populations. It remains unclear what practices within this culture may potentially influence natal sex ratios, and this finding does not explain why we see low sex ratios in black populations across the world where lifestyles are very different than those experienced by Bantu populations.

The high prevalence of interracial unions in heterogeneous populations provides an opportunity to examine whether these low sex ratios in black populations are primarily of maternal or paternal origin, or both. Khoury et al. (1984) examined births with interracial parentage and found that the influence likely lies with the father; black fathers produced significantly lower sex ratios than white fathers, while white and black mothers produced similar sex ratios (Fig. 2.4). This same effect was not seen, however, in a smaller study conducted in Hawaii (Morton et al. 1967), and another study corroborated that paternal influences drive the natal sex

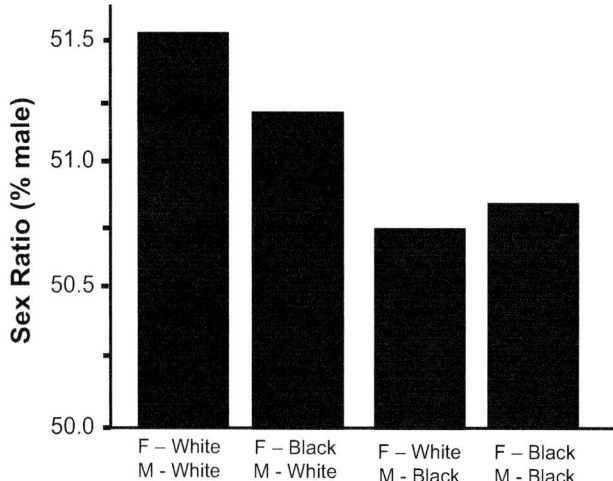

Fig. 2.4 Sex ratios produced by couples of mixed race. Sex ratios of offspring from black males were significantly lower than those from white males. Figure reproduced from Khoury et al. (1984)

ratios, but black fathers did not produce lower sex ratios than white fathers in this study (52.0 vs. 51.6%). More work is needed to determine whether the racial differences in natal sex ratios are indeed driven by a paternal factor. Further, while black populations produce low natal sex ratios, people of Asian descent appear to produce high natal sex ratios (James 1985; Morton et al. 1967; Visaria 1967). While it may be tempting to attribute this trend to the practice of sex-specific abortion that has been historically present in some Asian populations, these studies were conducted in the USA and other areas where racial distributions are heterogeneous and practices of sex-specific abortion are known to be relatively rare.

So what, then, may underlie such consistent racial variation in human sex ratios? Gellatly (2009) conducted a population genetic model to determine whether sex ratios are heritable. The model suggests that the inheritance of an autosomal gene with polymorphic alleles affects the offspring sex ratio via the male reproductive system. This idea would thus explain the Khoury et al.'s (1984) results showing that racial variation in sex ratios appears to lie with the male. However, it is also possible that intrinsic racial differences in physiology underlie these small but significant differences in sex ratios produced by couples of different races. For example, there are clear differences in stress-induced cardiovascular reactivity (Anderson 1989), psychological responses to large-scale disturbances (Elliott and Pais 2006), and even nutrient absorption (Beydoun et al. 2008) among humans of different races. If stress and/or nutrient intake are, indeed, active regulators of offspring sex ratios, this could explain the racial variation we see. Perhaps sex ratios are influenced by a physiological mechanism that also determines skin color. In their article describing the evolution of human skin color, Jablonski and Chaplin (2000) provided global measures of skin reflectance for different countries around the world. When average sex ratios (calculated using data from 1997 to 2006) are compared for countries with low, medium, and high skin reflectance, there is significant relationship between skin reflectance and the average natal sex ratio (Fig. 2.5). It is now well known that the mechanisms responsible for producing

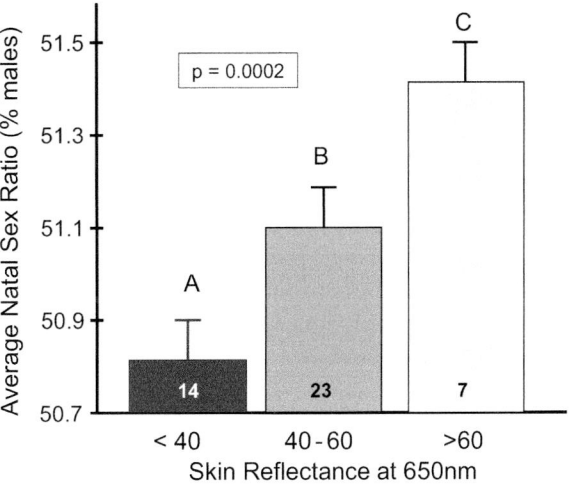

Fig. 2.5 Global relationship between average natal sex ratios from 1997 to 2006 and skin color as measured by reflectance at 650 nm in Jablonski and Chaplin (2000)

melanin-based skin color are part of a complex pleiotropic web that influences many body systems, including the stress axis, the metabolic axis, and the immune system (Ducrest et al. 2008). Thus, it is possible that racial differences in physiology underlie sex ratio variation. More work needs to be done to examine this possibility.

2.5 Influences of Socioeconomic Variables

Adaptive theory suggests that resource availability should be a significant predictor of offspring sex ratios (Trivers and Willard 1973), and the economic stress hypothesis, posited by Catalano (2003), predicts that natal sex ratios should decrease when the economic environment declines. More than 25 studies have shown biases towards female offspring in response to low socioeconomic status (reviewed in Lazarus 2002), though a very large study (<1 million births) in Scotland showed no relationship (Rostron and James 1977). Due to low sample sizes and inconsistent measures of socioeconomic status in some of these studies, results must be interpreted with caution (Lazarus 2002). However, the influences of socioeconomic status on sex ratios could help explain some of the racial variation we see; Teitelbaum and Mantel (1971) showed that natal sex ratios were lowest for black individuals in the lowest socioeconomic class. However on a global scale, natal sex ratios were unrelated to a measure of social stability, calculated using a principal component including the gross domestic product (GDP), unemployment rates, and an instability index (Navara 2009), suggesting, at least at the global level, that socioeconomic status is not a direct driver of the sexes of babies produced by humans.

Still, we see changes in human sex ratios in response to large-scale economic upheavals. During the economic collapse of East Germany, for example, natal sex ratios were significantly lower than those documented for West Germany at the same time (Catalano 2003); however, a recent reanalysis suggests that this may be due to higher levels of random variation that resulted from a substantially decreased birth rate at this time (Schnettler and Klüsener 2014). However, a similar effect of socioeconomic variables on sex ratio has also been documented in response to economic changes on a smaller scale; Catalano and Bruckner (2005) examined how sex ratios of births related to private consumption of goods and services and found that when the Swedish economy contracted, the proportion of males produced decreased, while more males were produced when economic times were good. At the opposite end of the spectrum, Cameron and Dalerum (2009) showed that US billionaires produced an abnormally high proportion of sons, though this relationship could have been driven by one billionaire who produced 61 offspring. More work needs to be done to determine whether small-scale variation in socioeconomic status at the individual level influences the sexes of offspring produced by humans, and how this might contribute to global and racial variation in natal sex ratios.

2.6 Influences of Diet and Malnutrition

It was suggested long ago that food and nutrient availability should influence the ratio of male to female offspring produced by both humans and other animals. Relationships between the natal sex ratio and food availability, nutritional content of food, and disorders that alter food intake have been documented in humans. Interestingly, the predicted direction of the skew in response to malnutrition has changed over time. In 1958, Ploss (1858) posited that malnutrition would selectively favor survival of male over female fetuses. Indeed, he found that, in Saxony, more male than female babies were born during years of famine. However, the US depression of 1929 was not associated with a change in the natal sex ratio (Ciocco 1938).

In 1973, Trivers and Willard (1973) suggested that, for species in which males produce costly ornaments to attract mates, mothers in better condition should favor the production of sons over daughters. This idea gained widespread support, and as a result, the predicted direction of sex ratio skews in response to malnutrition reversed, and it is now believed that decreased food availability should instead lead to the production of daughters. There is more support for this idea. For example, both the Bangladesh famine of 1974 and the Great Leap Forward Famine in China in the late 1950s–early 1960s corresponded to significant decreases in the proportion of female babies born (Hernández-Julián et al. 2014; Song 2012). Similarly, Ethiopian women that were in a better nutritional state according to their body mass indices produced a higher proportion of male babies, while thinner Italian women produced a lower proportion of boys (Cagnacci 2004). Given these patterns, we would also expect that conditions that significantly decrease food intake would also influence natal sex ratios, which is indeed the case; women who suffer from anorexia nervosa or bulimia produce significantly lower proportions of male offspring compared to women with no eating disorders, or those with disorders that involve binge eating (Bulik et al. 2008). Maternal celiac disease diagnosed prior to pregnancy also reduces the natal sex ratio, though paternal celiac disease diagnosed at the same point was instead linked to increased natal sex ratios (Khashan et al. 2010). Overall, however, there is convincing evidence showing a link between maternal food intake and natal sex ratios. Whether this link is a direct one is still not known.

We also don't know whether the patterns observed are due to the presence or absence of particular dietary components in the diet. In 1975, Lorrain suggested that alkaline diets cause an excess of girls while acidic diets lead to an excess of boys (Lorrain 1975). He followed 100 couples who had previously had at least three children, all boys or all girls. He found that women who had boys had eaten a diet rich in sodium and poor in alkaline ash, while the reverse was true for those that produced girls. He then provided experimental diets; 30 women received a diet rich in alkaline ash, and 24 of those (80%) produced boys. He assigned a diet poor in alkaline ash to 20 women, and 16 (80%) produced girls. The precise methods used in this study have not, to my knowledge, been tested further in humans.

Four years later, Stolkowski and Choukroun (1981) suggested that the ionic balance, produced by the ratio of sodium and potassium:calcium and magnesium,

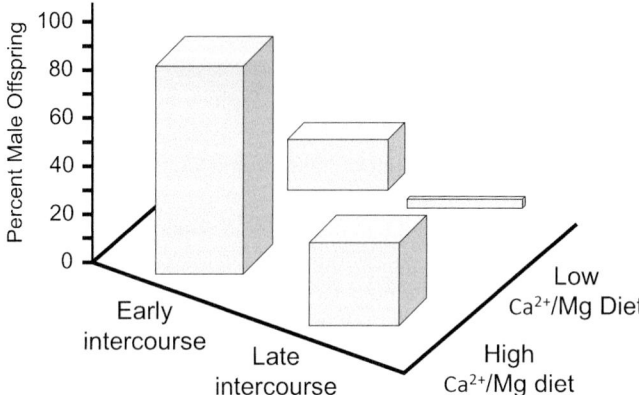

Fig. 2.6 Influences of high dietary calcium and magnesium in combination with intercourse timing substantially earlier than ovulation. Figure adapted from Noorlander et al. (2010)

may influence the sex ratio at conception. High ratios produce male offspring, while low ratios produce female offspring. Using this method in two published studies, they obtained the predicted sex in >80% of births (Stolkowski and Choukroun 1981; Stolkowski and Lorrain 1979). In 1958, a 2-year clinical trial was carried out to test this method, and 45 of 58 patients produced the expected sex, though 75% of patients abandoned the diet regimen during the course of the study. This dietary influence was further tested in conjunction with timing intercourse well before ovulation, which has been suspected as an alternative method by which couples may achieve daughters (Noorlander et al. 2010). Diets given were high in calcium and magnesium, and blood samples were collected to confirm high levels of these ions as a result of the diet. This diet increased the proportion of females produced by the study women, but only when intercourse was timed substantially early compared with ovulation. Similarly, the timing of intercourse only influenced sex ratios when blood levels of calcium and magnesium were also high (Fig. 2.6) (Noorlander et al. 2010). A recent study further supports the influences of ionic balance on offspring sex ratios. Edessy et al. (2016) provided women with a diet high in calcium and magnesium or a diet high in sodium and potassium. The women on the calcium and magnesium diet produced 28% males, while those on the sodium and potassium diet produced 76% males. More work must be done to determine how these dietary components might work to influence offspring sex.

Finally, in 2008, Mathews and colleagues asked women to put together food diaries from before and during their pregnancies and conducted analyses examining which dietary components were significantly related to the natal sex ratios produced by these women (Mathews et al. 2008). They found that high nutrient intake was linked to higher natal sex ratios, and a forward stepwise regression selected potassium as the only predictor of offspring sex. In addition, breakfast cereal, when consumed regularly, was the only food that predicted offspring sex; women who ate more cereal were more likely to produce boys. These relationships were only present in the pre-conception diet, and not in diets recorded at 16–28 weeks of pregnancy. It has been suggested, however, that multiple comparisons in this study may have resulted in false positives (Young et al. 2009), as hundreds of comparisons were conducted without correction for this high number of statistical comparisons. In addition, pre-pregnancy diets were not recorded during the study, but were recalled via memories of the study participants, while the diets recorded during pregnancy were recorded daily. Thus, there is potential that the memories of these diets were not accurate. More work needs to be done in this arena.

2.7 Influences of Stress

Many of the potential influences on human sex ratios discussed thus far, such as dramatic temperature changes, food limitations, and low socioeconomic status, are generally perceived as stressful challenges. Economic upheavals, for example, disrupt physical and mental health, induce psychological stress, and reduce access to critical resources, all of which can cause transgenerational effects on offspring physiology (Cohen et al. 2006; Lupien et al. 2000). It is possible that the stressful nature of these events in general may be the overarching cue responsible for biasing human sex ratios after exposure to these stimuli. Indeed, sex ratio biases in response to other types of stressful encounters have been extensively documented. For example, large-scale natural disturbances such as earthquakes, floods, or hurricanes bring with them significant lifestyle disruptions and high levels of stress (Ironson et al. 1997; Song et al. 2008). Not surprisingly, they are also often followed by decreases in natal sex ratios. For example, Fukuda et al. (1998) showed that sex ratios declined 9 months after the Kobe earthquake, suggesting an influence on initial production of male versus female fertilizations. Sadaat (2008) showed a similar effect; sex ratios declined 11 months after a severe earthquake hit Southern Iran, while sex ratios remained stable in areas of Iran that were unaffected by the earthquake. Following the March 2011 earthquake in eastern Japan, there was a similar decline in the natal sex ratio, but this time it was observed 7 months after the earthquake, and only lasted for one month (Hamamatsu et al. 2014). This decline was strongest in areas of Japan that were not disaster stricken, suggesting that the decline may have represented natural variation in the sex ratio at that time. However, Torche and Kleinhaus (2012) used a semi-experimental design to show that exposure to the Tarapaca earthquake that hit Chile in 2005 resulted in a significant decline in the natal sex ratio, particularly when exposure occurred during the third month of gestation. Thus, overall, exposure to earthquakes and the resulting lifestyle disruptions either before pregnancy or during early gestation results in a decrease in the proportion of male babies produced.

These effects are not restricted to earthquakes. There was also a drop in the proportion of male babies 320 days after the London smog of 1952 and the Brisbane flood of 1965 (Lyster 1974). There were changes in natal sex ratios in the USA following Hurricane Katrina, but in this case, sex ratio spiked instead of declining

(Grech and Scherb 2015). The sex ratio spike only occurred in the three states that received the highest rainfall (Alabama, Mississippi, and Louisiana), and the increase in the portion of males followed a dose response with the amount of rainfall each state received. This spiked occurred 8–10 months after the hurricane and did not occur in states that did not receive significant rainfall during this event. The authors suggest that the spike could have been triggered by either natural or man-made radiation contained within the rainfall, which may explain why this event did not trigger the predicted decline in sex ratios that results from other stressful encounters.

Perhaps some of the most convincing evidence that stress itself influences human sex ratios is the evidence that severe life events, in the absence of socioeconomic disruptions or reduced access to resources, influence natal sex ratios. For example, sex ratios in both California and New York declined 3–4 months following the September 11 attacks on New York City (Catalano et al. 2005a, 2006), an effect that likely resulted from increases in male fetal deaths following the attacks (Bruckner et al. 2010). The results shown in California are particularly relevant, because individuals in California experienced stress as a result of the event (Schuster et al. 2001), but did not experience the lifestyle disruptions that those located in New York City likely did. Smaller scale events exert a similar effect on natal sex ratios. A study conducted in Denmark showed that stressful life events, such as the death of the mother's partner or a serious illness in the family, were associated with decreases in the proportions of male births (Hansen et al. 1999). Highly stressful jobs can exert similar influences. In a UK study that included over 16,000 births to working mothers from 2000 to 2005, jobs held by the mothers were ranked (1–10) according to how stressful those job types have been perceived to be. Mothers with jobs ranked as "low stress" (1 or 2) produced male-biased sex ratios (53.68%), while those with jobs ranked as "high stress" (9 or 10) produced female-biased sex ratios (47.05%). The latter ratio was driven by the ninth ranked category that included jobs in trade, transport, and construction. Overall, as the ranking of job stress increased, the probability of producing a boy decreased, but this effect was mediated by the partner's income, as the relationship between job stress and natal sex ratio only existed when the partner's income level was low (Ruckstuhl et al. 2010). The authors suggest that the status of the father may have overcome the influences of stress on the natal sex ratio. This is not the only case where the father's status or job potentially influenced the proportion of boys produced. Lyster (1982) found a female bias (34.6% males) in the 130 children produced by abalone divers, and Snyder (1961) found a female-biased sex ratio (37.2% male) in 94 children of fighter pilots compared to children of 128 men that were pilots of other plane types (52.5% male) or were not pilots at all (60.2% male). The authors suggest that the stressful nature of the job contributed to a reduction in the proportions of male babies born. It is unclear, however, whether this effect resulted from the stress directly experienced by the father or indirectly experienced by the mother. Regardless, these studies contribute to the evidence that high levels of stress can induce reductions in the proportion of male babies produced.

Even levels of day-to-day anxiety in the absence of major life events show correlations to the proportion of male babies that are born; a questionnaire administered to over 6000 Danish women during early pregnancy showed that the level of psychological disturbance affected the sexes of offspring produced, with women reporting greater levels of psychological disturbance producing fewer boys (Obel et al. 2007). Similarly, a Swedish study using medical data from 1974 to 1996 showed that daily doses of antidepressants/anxiolytics dispensed during the first and second months of gestation were inversely related to the sex of the baby that was born (Catalano et al. 2005b). This suggests that reducing stress via chemical intervention may alter either the sex ratio at fertilization or by influencing the early survival of male blastocysts or embryos.

2.8 Do All of These Cues Really Affect Human Sex Ratios?

How do we make sense of the many different variables that appear to influence sex ratios? In the interest of length, we have not even addressed all of the variables reported to relate to offspring sex ratios, including operating sex ratios (Lummaa et al. 1998), exposure to alcohol and lead (Dickinson and Parker 1994), cigarette smoking (Fukuda et al. 2002), cancer (Fukuda et al. 2002), and many more. Is it possible that each variable influences sex ratios via independent mechanisms? In an evolutionary sense, it would be unlikely that a species would evolve several unrelated ways of altering the sexes of offspring produced. Instead, we would predict that a central mechanism of sex ratio manipulation influences offspring sex and that each of the variables found to significantly relate to human sex ratios would interact with this mechanism somehow. As mentioned above, the stress axis responds to many of the changes and events that have been shown to relate to offspring sex ratios. Perhaps the stress axis acts as a central regulator that directs downstream influences on the process of sex determination in mammals. We will discuss in Chap. 8 the evidence suggesting that hormones may regulate this process; however, future work needs to examine how each variable described above may elicit influences on *both* sex ratios and related physiological or hormonal processes.

There is also the question of whether the small differences in sex ratios (generally <1%) truly represent adaptive manipulation of offspring sex. Gellatly (2009) argues that genetic influences can actually explain many of the relationships described above, eliminating the possibility that sex ratios are being adjusted in relation to factors such as resource availability. However, in a compilation of behavioral, toxicological, virological, gastroenterological, and epidemiological data, James (2006) suggests that, as long as the complexity and associated constraints produced by the primate endocrine system are taken into account, adaptive explanations can be used to explain the sex ratio variation we see in humans. Some updated meta-analyses would greatly help to inform the field about whether or not we are seeing adaptive sex ratio manipulation in humans.

References

Anderson NB (1989) Racial differences in stress-induced cardiovascular reactivity and hypertension: current status and substantive issues. Psychol Bull 105(1):89

Beydoun MA, Gary TL, Caballero BH, Lawrence RS, Cheskin LJ, Wang Y (2008) Ethnic differences in dairy and related nutrient consumption among US adults and their association with obesity, central obesity, and the metabolic syndrome. Am J Clin Nutr 87(6):1914–1925

Bruckner TA, Catalano R, Ahern J (2010) Male fetal loss in the US following the terrorist attacks of September 11, 2001. BMC Public Health 10(1):273

Bulik CM, Von Holle A, Gendall K, Lie KK, Hoffman E, Mo X, Torgersen L, Reichborn-Kjennerud T (2008) Maternal eating disorders influence sex ratio at birth. Acta Obstet Gynecol Scand 87 (9):979–981

Cagnacci A (2003) The male disadvantage and the seasonal rhythm of sex ratio at the time of conception: Reply. Hum Reprod 18(11):2492–2494

Cagnacci A (2004) Reply to 'Influences of maternal weight on the secondary sex ratio of human offspring'. Hum Reprod 19(10):2425–2426

Cagnacci A, Volpe A (1996) Influence of melatonin and photoperiod on animal and human reproduction. J Endocrinol Investig 19(6):382–411

Cameron EZ, Dalerum F (2009) A Trivers-Willard effect in contemporary humans: male-biased sex ratios among billionaires. PLoS One 4(1):e4195

Casey SB, Waxman DB, Pedagno AT (2013) No girls allowed: sex-selective abortion and a guide to banning it in the United States. Regent JL Pub Pol'y 5:111

Catalano RA (2003) Sex ratios in the two Germanies: a test of the economic stress hypothesis. Hum Reprod 18(9):1972–1975

Catalano RA, Bruckner T (2005) Economic antecedents of the Swedish sex ratio. Soc Sci Med 60 (3):537–543

Catalano R, Bruckner T, Gould J, Eskenazi B, Anderson E (2005a) Sex ratios in California following the terrorist attacks of September 11, 2001. Hum Reprod 20(5):1221–1227

Catalano R, Bruckner T, Hartig T, Ong M (2005b) Population stress and the Swedish sex ratio. Paediatr Perinat Epidemiol 19(6):413–420

Catalano R, Bruckner T, Marks AR, Eskenazi B (2006) Exogenous shocks to the human sex ratio: the case of September 11, 2001 in New York City. Hum Reprod 21(12):3127–3131

Catalano R, Bruckner T, Smith KR (2008) Ambient temperature predicts sex ratios and male longevity. Proc Natl Acad Sci 105(6):2244–2247

Challis J, Newnham J, Petraglia F, Yeganegi M, Bocking A (2013) Fetal sex and preterm birth. Placenta 34(2):95–99

Ciocco A (1938) Variation in the sex ratio at birth in the United States. Hum Biol 10(1):36

Cohen S, Schwartz JE, Epel E, Kirschbaum C, Sidney S, Seeman T (2006) Socioeconomic status, race, and diurnal cortisol decline in the coronary artery risk development in young adults (CARDIA) study. Psychosom Med 68(1):41–50

Dickinson H, Parker L (1994) Do alcohol and lead change the sex ratio? J Theor Biol 169 (3):313–315

Dixson BJ, Haywood J, Lester PJ, Ormsby DK (2013a) Ambient temperature variation does not influence regional proportion of human male births in New Zealand. J Roy Soc N Z 43 (2):57–74

Dixson BJ, Haywood J, Lester PJ, Ormsby DK (2013b) Feeling the heat?: substantial variation in temperatures does not affect the proportion of males born in Australia. Hum Biol 85 (5):757–767

Dubuc S, Coleman D (2007) An increase in the sex ratio of births to india-born mothers in England and Wales: evidence for sex-selective abortion. Popul Dev Rev 33(2):383–400

Ducrest A-L, Keller L, Roulin A (2008) Pleiotropy in the melanocortin system, coloration and behavioural syndromes. Trends Ecol Evol 23(9):502–510

Edessy M, El Rashedy MI, El Batal K, Ahmed S, Badawy M, Oun AE, Bendary A (2016) Pre-conceptional maternal diet and fetal sex pre-selection. Int J Curr Res Med Sci 2(12):8–13

Elliott JR, Pais J (2006) Race, class, and Hurricane Katrina: social differences in human responses to disaster. Soc Sci Res 35(2):295–321

Erickson JD (1976) The secondary sex ratio in the United States 1969–71: association with race, parental ages, birth order, paternal education and legitimacy. Ann Hum Genet 40(2):205–212

Fisher RA (1930) The genetical theory of natural selection: a complete, variorum edition. Oxford University Press, Oxford

Fukuda M, Fukuda K, Shimizu T, Møller H (1998) Decline in sex ratio at birth after Kobe earthquake. Hum Reprod 13(8):2321–2322

Fukuda M, Fukuda K, Shimizu T, Andersen CY, Byskov AG (2002) Parental periconceptional smoking and male: female ratio of newborn infants. Lancet 359(9315):1407–1408

Fukuda M, Fukuda K, Shimizu T, Nobunaga M, Mamsen LS, Andersen CY (2014) Climate change is associated with male: female ratios of fetal deaths and newborn infants in Japan. Fertil Steril 102:1364–1370

Garenne M (2002) Sex ratios at birth in African populations: a review of survey data. Hum Biol 74:889–900

Gellatly C (2009) Trends in population sex ratios may be explained by changes in the frequencies of polymorphic alleles of a sex ratio gene. Evol Biol 36(2):190–200

Grech V (2013) Secular trends and latitude gradients in sex ratio at birth in Asia during the past 60 years. Pediatr Int 55(2):219–222

Grech V, Scherb H (2015) Hurricane Katrina: influence on the male-to-female birth ratio. Med Princ Pract 24(5):477–485

Grech V, Savona-Ventura C, Vassallo-Agius P (2002) Unexplained differences in sex ratios at birth in Europe and North America. BMJ 324(7344):1010–1011

Hamamatsu Y, Inoue Y, Watanabe C, Umezaki M (2014) Impact of the 2011 earthquake on marriages, births and the secondary sex ratio in Japan. J Biosoc Sci 46(06):830–841

Hansen D, Møller H, Olsen J (1999) Severe periconceptional life events and the sex ratio in offspring: follow up study based on five national registers. BMJ 319(7209):548–549

Helle S, Helama S, Jokela J (2008) Temperature-related birth sex ratio bias in historical Sami: warm years bring more sons. Biol Lett 4(1):60–62

Hernández-Julián R, Mansour H, Peters C (2014) The effects of intrauterine malnutrition on birth and fertility outcomes: evidence from the 1974 Bangladesh famine. Demography 51(5):1775–1796

Hesketh T, Xing ZW (2006) Abnormal sex ratios in human populations: causes and consequences. Proc Natl Acad Sci 103(36):13271–13275

Ironson G, Wynings C, Schneiderman N, Baum A, Rodriguez M, Greenwood D, Benight C, Antoni M, LaPerriere A, Huang H-S (1997) Posttraumatic stress symptoms, intrusive thoughts, loss, and immune function after Hurricane Andrew. Psychosom Med 59(2):128–141

Jablonski NG, Chaplin G (2000) The evolution of human skin coloration. J Hum Evol 39(1):57–106

Jablonski NG, Chaplin G (2010) Human skin pigmentation as an adaptation to UV radiation. Proc Natl Acad Sci 107(Suppl 2):8962–8968

James W (1984) Seasonality in the sex ratio of US Black births. Ann Hum Biol 11(1):67–69

James WH (1985) The sex ratio of oriental births. Ann Hum Biol 12(5):485–487

James WH (1987) The human sex ratio. Part 1: A review of the literature. Hum Biol 59:721–752

James WH (2006) Possible constraints on adaptive variation in sex ratio at birth in humans and other primates. J Theor Biol 238(2):383–394

Junhong C (2001) Prenatal sex determination and sex-selective abortion in rural central China. Popul Dev Rev 27(2):259–281

Khashan AS, Henriksen TB, McNamee R, Mortensen PB, McCarthy FP, Kenny LC (2010) Parental celiac disease and offspring sex ratio. Epidemiology 21(6):913–914

Khoury MJ, Erickson JD, James LM (1984) Paternal effects on the human sex ratio at birth: evidence from interracial crosses. Am J Hum Genet 36(5):1103

Lazarus J (2002) Human sex ratios: adaptations and mechanisms, problems, and prospects. In: Hardy I (ed) Sex ratios: concepts and research methods. Cambridge University Press, New York, pp 287–311

Lerchl A (1998) Seasonality of sex ratio in Germany. Hum Reprod 13(5):1401–1402

Lorrain J (1975) Pre-conceptional sex selection. Int J Gynaecol Obstet 13(3):127–130

Lummaa V, Merilä J, Kause A (1998) Adaptive sex ratio variation in pre–industrial human (Homo sapiens) populations? Proc R Soc Lond B Biol Sci 265(1396):563–568

Lupien SJ, King S, Meaney MJ, McEwen BS (2000) Child's stress hormone levels correlate with mother's socioeconomic status and depressive state. Biol Psychiatry 48(10):976–980

Lyster W (1971) Three patterns of seasonality in American births. Am J Obstet Gynecol 110 (7):1025–1028

Lyster W (1974) Altered sex ratio after the London smog of 1952 and the Brisbane flood of 1965. BJOG Int J Obstet Gynaecol 81(8):626–631

Lyster W (1982) Altered sex ratio in children of divers. Lancet 320(8290):152

MacPherson Y (2007) Images and icons: harnessing the power of the media to reduce sex-selective abortion in India. Gend Dev 15(3):413–423

Mathews F, Johnson PJ, Neil A (2008) You are what your mother eats: evidence for maternal preconception diet influencing foetal sex in humans. Proc R Soc Lond B Biol Sci 275 (1643):1661–1668

McLachlan JC, Storey H (2003) Hot male: can sex in humans be modified by temperature? J Theor Biol 222:71–72

Meagher S, Davison G (1996) Early second-trimester determination of fetal gender by ultrasound. Ultrasound Obstet Gynecol 8(5):322–324

Meyers MC (2012) Associations between climate, latitude, fertility and the decline of the US sex ratio at birth

Morton NE, Chung CS, Mi M-P (1967) Genetics of interracial crosses in Hawaii. Monogr Hum Genet 3:1

Navara KJ (2009) Humans at tropical latitudes produce more females. Biol Lett 5(4):524–527

Navara KJ, Workman JL, Oberdick J, Nelson RJ (2010) Short day lengths skew prenatal sex ratios toward males in Siberian hamsters. Physiol Biochem Zool 83(1):127–134

Newport F (2011) Americans prefer boys to girls, just as they did in 1941. http://www.gallup.com/poll/148187/americans-prefer-boys-girls-1941.aspx

Noorlander A, Geraedts J, Melissen J (2010) Female gender pre-selection by maternal diet in combination with timing of sexual intercourse–a prospective study. Reprod BioMed Online 21 (6):794–802

Obel C, Henriksen TB, Secher NJ, Eskenazi B, Hedegaard M (2007) Psychological distress during early gestation and offspring sex ratio. Hum Reprod 22(11):3009–3012

Orzack SH, Stubblefield JW, Akmaev VR, Colls P, Munné S, Scholl T, Steinsaltz D, Zuckerman JE (2015) The human sex ratio from conception to birth. Proc Natl Acad Sci 112(16):E2102–E2111

Ploss (1858) Ueber die das Geschlechtsverhältniss der Kinder bedingenden Ursach. Monatschr f Geburtsk и Frauen 12:321–360

Reece EA, Winn HN, Wan M, Burdine C, Green J, Hobbins JC (1987) Can ultrasonography replace amniocentesis in fetal gender determination during the early second trimester? Am J Obstet Gynecol 156(3):579–581

Roenneberg T, Aschoff J (1990) Annual rhythm of human reproduction: II. Environmental correlations. J Biol Rhythm 5(3):217–239

Rostron J, James WH (1977) Maternal age, parity, social class and sex ratio. Ann Hum Genet 41 (2):205–217

Ruckstuhl KE, Colijn GP, Amiot V, Vinish E (2010) Mother's occupation and sex ratio at birth. BMC Public Health 10(1):1

Saadat M (2008) Decline in sex ratio at birth after Bam (Kerman Province, Southern Iran) earthquake. J Biosoc Sci 40(06):935–937

Schnettler S, Klüsener S (2014) Economic stress or random variation? Revisiting German reunification as a natural experiment to investigate the effect of economic contraction on sex ratios at birth. Environ Health 13(1):117

Schuster MA, Stein BD, Jaycox LH, Collins RL, Marshall GN, Elliott MN, Zhou AJ, Kanouse DE, Morrison JL, Berry SH (2001) A national survey of stress reactions after the September 11, 2001, terrorist attacks. N Engl J Med 345(20):1507–1512

Slatis HM (1953) Seasonal variation in the American live birth sex ratio. Am J Hum Genet 5(1):21

Smith GC (2000) Sex, birth weight, and the risk of stillbirth in Scotland, 1980–1996. Am J Epidemiol 151(6):614–619

Snyder RG (1961) The sex ratio of offspring of pilots of high performance military aircraft. Hum Biol 33(1):1–10

Song S (2012) Does famine influence sex ratio at birth? Evidence from the 1959–1961 great leap forward famine in China. Proc R Soc Lond B: Biol Sci rspb20120320

Song Y, Zhou D, Wang X (2008) Increased serum cortisol and growth hormone levels in earthquake survivors with PTSD or subclinical PTSD. Psychoneuroendocrinology 33(8):1155–1159

Stolkowski J, Choukroun J (1981) Preconception selection of sex in man. Isr J Med Sci 17(11):1061–1067

Stolkowski J, Lorrain J (1979) Preconceptional selection of fetal sex. Int J Gynaecol Obstet 18(6):440–443

Strandskov H, Roth J (1949) A comparison of rural and urban birth sex ratios for the total, the "white" and the "colored" US populations. Am J Phys Anthropol 7(1):91–100

Teitelbaum MS, Mantel N (1971) Socio-economic factors and the sex ratio at birth. J Biosoc Sci 3(01):23–42

Torche F, Kleinhaus K (2012) Prenatal stress, gestational age and secondary sex ratio: the sex-specific effects of exposure to a natural disaster in early pregnancy. Hum Reprod 27(2):558–567

Trivers RL, Willard DE (1973) Natural selection of parental ability to vary the sex ratio of offspring. Science 179(4068):90–92

Visaria PM (1967) Sex ratio at birth in territories with a relatively complete registration. Eugen Q 14(2):132–142

Whitlow B, Lazanakis M, Economides D (1999) The sonographic identification of fetal gender from 11 to 14 weeks of gestation. Ultrasound Obstet Gynecol 13(5):301–304

Young SS, Bang H, Oktay K (2009) Cereal-induced gender selection? Most likely a multiple testing false positive. Proc R Soc Lond B Biol Sci 276(1660):1211–1212

Facultative Sex Ratio Adjustment in Nonhuman Mammals

3

> *Let us now take the case of a species producing from the unknown causes just alluded to, an excess of one sex—we will say of males—these being superfluous and useless, or nearly useless. Could the sexes be equalised through natural selection?*
>
> Charles Darwin, Decent of Man (first edition)

Just as the quest for a method to skew sex ratios in humans has continued for centuries, there have also been concurrent efforts to understand whether and how *nonhuman* mammals adjust offspring sex. Studies examining sex ratio patterns in agricultural and laboratory animals date back to the late 1800s, and surprisingly, many of the mechanisms we today believe may underlie the ways in which mammals adjust sex ratios were initially proposed at the turn of the nineteenth century, when our understanding of the XY system of sex determination in mammals had only just been conceived (reviewed in Parkes 1926). We now have thousands of studies examining whether mammals adjust sex ratios adaptively and how this might occur. Yet, we're not much closer to answering those questions than we were a hundred years ago. In this chapter, I address adaptive hypotheses that may explain patterns of sex ratio adjustment in mammals, discuss evidence that mammals do adjust offspring sex ratios in an adaptive manner, and review the factors that may influence both males and females to skew offspring sex in a particular direction. Figure 3.1 contains a diagram of the adaptive theories that may explain observed biases in mammalian sex ratios. Below, I will highlight some evidence that mammals adaptively adjust offspring sex ratios in response to a variety of conditions and discuss how the observed patterns fit into those adaptive theories.

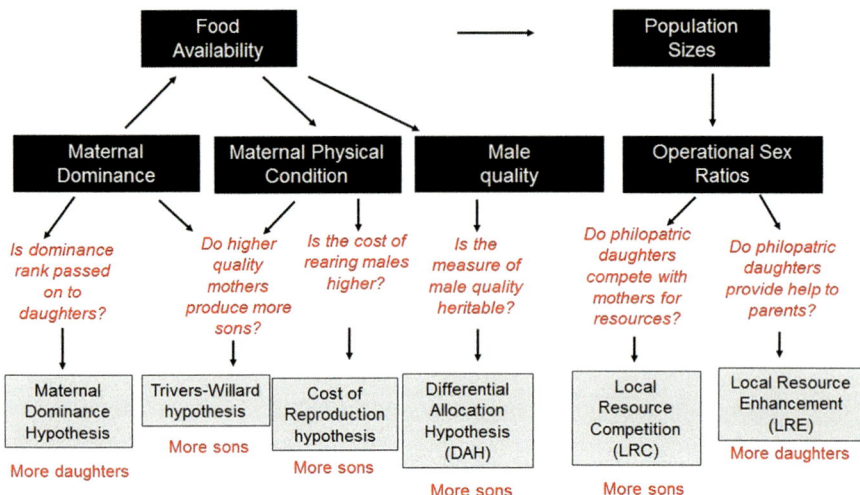

Fig. 3.1 A summary of the adaptive hypotheses most commonly used to explain sex allocation patterns observed in mammalian species

3.1 Environmental and Social Factors Linked to Offspring Sex Ratios

The body of work examining the many influences on offspring sex ratios in mammals is extremely large and continues to grow. Before we can understand whether mammals adjust offspring sex ratios in an adaptive manner and how that occurs, we must first determine which environmental, social, and physiological factors emerge time and time again as potential modulators of offspring sex ratios and then examine how those factors interact. It is unlikely that individual mechanisms for adjusting sex ratios evolved for each individual factor that has been shown to influence offspring sex ratios. Instead, we would expect that a few distinct mechanisms evolved, and that those mechanisms are influenced in some centrally regulated way by the individual factors that are shown to influence offspring sex ratios. First I will summarize these factors and then later talk about how they may interact.

3.1.1 Sex Ratio Adjustment in Response to "Female Condition"

In Chap. 2, I briefly introduced one of the most widely cited hypotheses in sex ratio biology, generated by Robert Trivers and Dan Willard, now called the Trivers–Willard hypothesis (Trivers and Willard 1973). They posited a link between maternal condition and offspring sex ratios, such that mothers in good condition produce more male offspring. The reasoning, as was presented using caribou as a theoretical example, is that when males are in better condition, they can produce better quality traits (such as large body size or large antlers) that would help them obtain more mates, while females, even those in good condition, are constrained to a fixed number of offspring due to limitations on number of eggs they can produce. To the contrary, when male offspring are in bad condition, and cannot produce the traits necessary to obtain a mate, their reproductive success falls to near zero, and in this case, it is more profitable to produce female offspring. Many have attempted to test for evidence of this idea in a variety of animal systems; however, in many cases, these systems violate the assumptions of the original model. Trivers and Willard (1973) state the following requirements for this model: (1) Mothers in good condition should produce offspring in good condition, and vice versa, (2) the influences on offspring condition must last into adulthood, and (3) slight advantages in condition must benefit reproductive success in males more than in females. The third assumption requires that the variance in reproductive success in general is greater for males than for females, and this generally occurs in polygynous species for which males show little to no parental investment. Thus, while tests of Trivers and Willard have been attempted in, for example, monogamous bird species, the high level of parental investment in these cases restricts the potential for males to obtain many mates during the breeding season and thus restricts the overall variance in reproductive success for males. On the other hand, over 90% of mammals exhibit polygynous mating systems that comply with the assumptions of the Trivers–Willard model (Clutton-Brock 1989), so mammalian systems are in general better for testing this hypothesis.

The next issue, however, is understanding what is meant by "good condition" because this descriptor is completely subjective. In many animal studies, the term "condition index" is used to describe the proportion of muscle and fat relative to skeletal size; a higher condition index is expected to be an indicator of good overall condition, because that individual was healthy enough to successfully obtain enough food required to produce large quantities of fat and muscle (Hayes and Shonkwiler 2001). However, studies attempting to test the Trivers–Willard hypothesis have used various measures of condition, including but not limited to age, weight, dominance, and litter size, all of which could have different implications in terms of the fitness of mother and offspring. For the Trivers–Willard hypothesis to hold, the critical criterion is that the measure of maternal condition ultimately relates to a critical quality in offspring that puts those offspring at the high end of a highly variable fitness spectrum, and the ability of this adaptive pattern of sex allocation to evolve depends on how the measure of maternal condition translates into a fitness effect in the offspring of a particular sex. For example, if the measure of quality is passed on genetically, then the ability to bias sex ratios in relation to this measure of quality would require a genetic link between offspring sex and the quality that ultimately increases fitness. In this case, we would not expect that environmental conditions (such as food availability, see below) would relate significantly to offspring sex ratios. If, however, the quality indicator is simply a predictor of what offspring will encounter in the postnatal environment, then the ability to adjust sex ratios would not have to be linked to that measure of quality, but

> **The story of sex ratio adjustment in red deer**
>
> Red deer are considered to be the classic poster children for the Trivers-Willard model of sex ratio adjustment in mammals, because more dominant mothers produce a higher proportion of male offspring (Clutton-Brock 1986), and this a flurry of studies examining the same relationship in other ungulate species, with a finding triggered majority showing relationships in the same direction. Initially, however, red deer were held up as an example supporting the Local Resource Competition hypothesis, because females are philopatric, sharing territories with mothers and siblings, those breeding in large groups have lower reproductive success, and the overall population sex ratio is male-biased (Clutton-Brock 1982). To make matters more complicated, in 1999, Loeske Kruuk found that the relationship between aggression and dominance did not hold up when population densities were high, suggesting that male susceptibility to suboptimal conditions may overcome Trivers-Willard patterns of sex allocation related to female condition. Thus red deer have now become examples for three models of sex allocation: the Trivers-Willard hypothesis, the Local Resource Competition hypothesis, and the disadvantaged male hypothesis.

Fig. 3.2

to a physiological response to environmental conditions, making the web of potential variables that contribute to the process of sex ratio adjustment much wider.

Work in red deer was some of the first that provided support for the Trivers–Willard hypothesis of sex allocation (Clutton-Brock et al. 1984) (Fig. 3.2) Since then, Sheldon and West (2004) conducted a meta-analysis to test whether the available evidence on sex allocation in ungulate species does, in fact, support the Trivers–Willard hypothesis of sex allocation. Using 37 studies of 18 ungulate species, they showed a weak but positive relationship between maternal condition and offspring sex ratios, suggesting that females in better condition do, indeed, produce more male offspring; however, when they split the studies into physical and behavioral measures of quality, only the behavioral measures (dominance) strongly predicted offspring sex ratios, while physical measures of quality showed little to no relationship with offspring sex ratios. Cameron (2004) went a step further and conducted a similar meta-analysis on studies in all mammals except for humans and also broke down the studies based on specific indicators of maternal quality. She showed that, when 422 studies containing all the different measures of maternal quality were considered, only 34% showed positive support for the Trivers–Willard hypothesis, a weak but significant indication that females in good condition generally produce more males. But why is the correlation not stronger? Cameron also showed that each individual measure differed substantially in its ability to predict offspring sex ratios (Fig. 3.3) and that when these measures were taken in relation to the timing of conception strongly influenced whether a significant correlation was found. When body weight, body condition, or food availability are

3.1 Environmental and Social Factors Linked to Offspring Sex Ratios

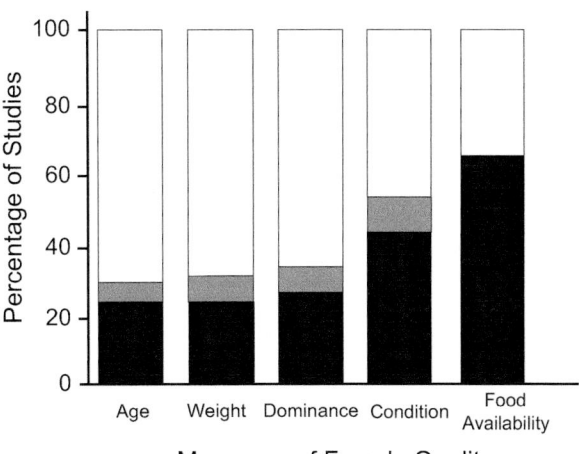

Fig. 3.3 Percentages of studies that tested for evidence of sex allocation according to the Trivers–Willard hypothesis and showed (black) significant sex ratio skews in the predicted direction, (gray) significant sex ratio skews in the opposite direction, or (white) no significant sex ratio skews. This figure was adapted from Cameron (2004)

measured specifically around the time of conception, 74% of the studies support the predictions of the Trivers–Willard hypothesis. However, can food availability truly be used as a direct proxy for maternal condition? When only body condition and weight are considered, fewer than 50% of studies show significant relationships with offspring sex ratios. So, are sex ratio adjustments really being made according to maternal condition, as Trivers and Willard predicted? Perhaps instead the links we see between maternal condition and offspring sex ratios are simply a secondary link that occurs due to the ultimate influence of food availability on offspring sex. In this case, we would expect that the effects of changing food availability on maternal condition, as well as the links between maternal condition and sex ratios, would be highly variable among species, and this is exactly what we see. Further, Cameron and Linklater (2007) found a dramatic relationship in feral horses between the *change* in maternal condition, measured as a visual assessment of body fat distribution, and offspring sex ratios; 80% of females whose condition improved through conception produced a son. This strongly indicates that sex ratios respond to changing conditions surrounding the female, rather than an inherent trait carried by the female, herself. For this reason, studies testing for sex ratio adjustments in response to factors that are known to influence maternal condition are important pieces of the sex ratio puzzle.

3.1.2 Sex Ratio Adjustment in Response to Food Availability

If food availability and/or maternal nutrition are the true drivers underlying adaptive manipulation of sex ratios in mammals, we would expect that experimental food restriction would potently alter the sex ratios of mammalian offspring. Despite a large amount of evidence showing that factors likely to influence access to critical food resources alter mammalian sex ratios, there are only a few studies that directly manipulate overall food availability, and these show mixed support for the idea that females may alter sex ratios in response to their access to food. Golden hamsters, for example, reduced the proportion of male offspring they produced by 10% when experiencing a food restriction (75% of an *ad libitum* diet) during pregnancy. Smaller litter sizes in this group indicate that sex ratios were altered via a preferential loss of male offspring during gestation (Labov et al. 1986). However, laboratory mice showed the opposite pattern; when food was readily available, they produced *fewer* male offspring compared to when food was restricted (though this was only a difference between 47% and 50% males) (Zamiri 1978). In a study of domestic mice, sex ratios were highest in a group that received moderate amounts of food (80% of *ad libitum*) compared with either high (*ad libitum*) or low (65% of *ad libitum*) amounts of food.

Perhaps when sex ratio adjustments are seen in response to food limitations or other related variables, the changes occur due the stressful nature of the limitation, rather than the limitation itself. Meikle and Drickamer (1986) intermittently limited food provided to both wild and laboratory strains of female house mice for either 1 or 2 weeks prior to mating. Control females of each strain produced relatively high proportions of male offspring (wild: 61%, captive: 56%) while those given 1 week of food stress produced significantly fewer male offspring (Wild: 40%, Captive: 46%). However, 2 weeks of the food limitation did not influence sex ratios in wild mice (59% males), and while sex ratios were lower in the laboratory strain given this treatment (52%), this sex ratio was not significantly different from those produced by control animals. The authors suggest that the food limitation acted as a stress upon females, and that after 1 week of the stress, females were able to acclimate to it.

Overall, these studies suggest that the *perception* of food availability does not appear to be a strong driver of sex ratio adjustment in one direction or the other, yet studies that test influences of more specific dietary components appear to support the idea that the composition of the maternal diet can influence offspring sex ratios (Fig. 3.4). Rivers and Crawford (1974) showed that mice fed a low calorie diet had smaller litters and produced only ~25% males compared to ~50% produced by control mice. Similarly, fallow deer given a high energy diet in the winter preceding puberty produced 75% males compared to 46% produced by those on a low energy diet (Enright et al. 2001). These studies suggest that the energy content of food may be more important than the presence or absence of food itself in the manipulation of offspring sex ratios.

These results provoke the question of whether it is the energy limitation or excess that drives changes in sex ratios or whether the source responsible for the energy and/or particular nutrient components of those high energy diets are the key to determining whether more male or female offspring are produced. There is now growing evidence for the latter. Nutritional factors that influence offspring sex ratios in different mammalian systems have now been summarized in two nice reviews (Rosenfeld 2011; Rosenfeld and Roberts 2004), but I will approach this information differently by focusing on the specific dietary components that have been shown to influence offspring sex ratios in mammals. Sources of dietary energy include carbohydrates and sugars, fats, and proteins. To my knowledge, there are

3.1 Environmental and Social Factors Linked to Offspring Sex Ratios

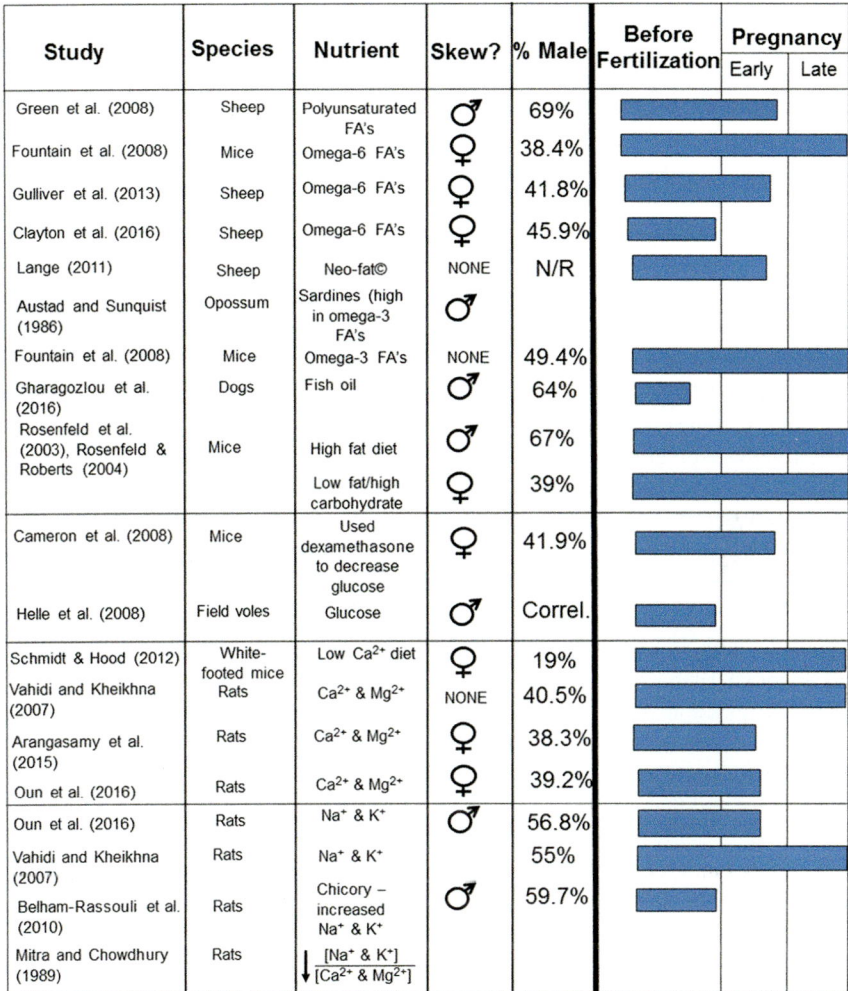

Fig. 3.4 Summary of studies examining effects of different dietary components on sex ratios in mammals

currently no studies that test the influences of dietary protein content on offspring sex ratios. However, a few studies suggest that glucose may play a role in the process of sex ratio adjustment. Field voles with high circulating levels of glucose prior to conception produced male-biased litters (Helle et al. 2008). Further, in an experimental study of mice, the addition of dexamethasone to drinking water for 3 days surrounding conception decreased plasma glucose levels and also caused females to produce fewer sons (41.9% males compared to 53.5% males produced by control mice). The findings of these studies mesh nicely with studies in dairy cows showing that male and female blastocysts have different glucose requirements very early in development (Tiffin et al. 1991), suggesting that glucose levels in the diet could determine survival at these stages in a sex-specific manner. However, in the field vole study, testosterone concentrations also correlated with sex ratios, making it impossible to determine whether testosterone or glucose was the primary driver of the skew. Additionally, in the mouse study, dexamethasone is a synthetic glucocorticoid, and its administration is often used to downregulate the stress axis. Given that both stress and testosterone have been implicated as triggers of sex ratio skews (see sections below and Chap. 8), it remains unclear whether glucose is exerting a direct effect, or whether the observed relationships between sex ratios and glucose are merely coincidental. More work is needed to test this relationship.

Dietary fat also appears to have links with offspring sex ratios, though the directions of the sex ratio skews vary. Dogs supplemented with fish oil, which is high in omega-3 fatty acids, produced significantly more male pups at birth (64%) compared to those that were not (46.3%) (Gharagozlou and Youssefi 2016), and similarly, ewes given a diet supplemented with polyunsaturated fatty acids starting 4 weeks prior to breeding and lasting through 13d post-estrus also produced significantly more male offspring (Green et al. 2008). These effects are not restricted to ruminants; in a study of mice, diets high in fat acted to prevent a decline in the proportions of male offspring that normally occurred when conditions were crowded (Dama et al. 2011), and opossums supplemented with sardines, which are high in omega-3 fatty acids, produced more male offspring (Austad and Sunquist 1986). However, there are just as many studies showing the opposite effect. For example, mice fed a diet supplemented with omega-6 (but not omega-3) fatty acids from 4 weeks old through the end of the study instead produced significantly more *female* offspring (38% males) compared to controls (Fountain et al. 2008), and three additional studies in ewes showed that feeding diets high in omega-6 fatty acids led to the production of more *female* offspring (Clayton et al. 2012, 2016; Gulliver et al. 2013). Interestingly, these authors showed that ewes fed the diets high in omega-6 fatty acids also had a shorter time to conception and suggested that the timing of conception relative to ovulation may be the actual mechanism responsible for the sex ratio skews. As for the reasons for the variation in the direction of the sex ratio skews among studies, Gulliver et al. (2013) speculated that the diet high in omega-6 could have decreased the absorption of other fatty acids, such as omega-3 fatty acids, that have been previously shown to induce a skew towards males in opossums and dogs (Austad and Sunquist 1986; Gharagozlou and Youssefi 2016). Because the diets supplied in all of these cases altered not only omega-6 fatty acids but also other components of the diet, including other fatty acids, it is important to conduct further studies testing the effects of these fatty acids individually. Also, high fat diets can increase levels of circulating glucose (Folmer et al. 2003), and determining whether fatty acids are acting directly or through another mediator is also important.

As discussed in Chap. 2, some early studies in humans indicated that the ionic balance of the diet at the time of conception may influence whether an X- or a Y-bearing sperm successfully reaches and fertilizes the egg. In particular, the ratio of [sodium and potassium] to [calcium and magnesium] in the diet at the time of conception appeared to be a particularly good indicator of which sex would be produced; a higher ratio resulted in more male offspring whereas a lower ratio resulted in more female offspring. This idea has support in nonhuman mammals as well. In five studies using rats as a model, diets high in sodium and potassium stimulated the production of significantly more male offspring, and in the four of these studies that also tested diets high in calcium and magnesium, higher levels of dietary calcium and magnesium led to the production of more female offspring (Arangasamy et al. 2015; Behnam-Rassouli et al. 2010; Mitra and Chowdhury 1989; Oun et al. 2016; Vahidi and Sheikhha 2007). Thus, the studies in rodents support the findings in humans. The authors of these studies suggest two potential mechanisms by which the ionic balance could influence sex ratios. First, changing the charge balance in the reproductive tract could alter vaginal PH, which could differentially affect survival of X- versus Y-bearing sperm. Alternatively, slight alterations in the electric charge of the egg itself could make it differentially permeable to X- versus Y-bearing sperm, which have slight differences in electrical charge between them. These ideas are discussed in more detail in Chap. 4. Arangasamy et al. (2015) showed that, in addition to altering sex ratios, increasing the dietary content of calcium and magnesium reduced testosterone concentrations (known to stimulate male biases in humans—see Chap. 2) in circulation and also altered the presence of proteins and the expression of genes (GDF-9 and BMP-15) in the ovarian follicles. Perhaps these changes underlie a mechanism by which oocytes become differentially susceptible to X- versus Y-bearing sperm. Additional studies testing the influences of the dietary ionic balance on the oocyte itself would be helpful.

Overall, there is a good amount of evidence in both humans and nonhuman mammals that the nutritional content of food can influence the sexes of offspring produced. Why this does not translate into consistent sex ratio skews when food is restricted overall remains unknown, and whether these changes occur during or after conception, or perhaps both, remains to be seen. Additionally, the field would benefit from an additional understanding of how these dietary components interact with one another as well as with other factors hypothesized to influence mammalian sex ratios, such as vaginal PH, ovulatory timing, and levels of reproductive hormones. From an adaptive standpoint, the story does not seem to be a simple link whereby the mere presence or absence of food influences sex ratios via alterations of maternal condition. Instead, it appears that the discrepancies we see among studies testing for relationships between sex ratios and maternal body condition occur because sex ratio adjustments may instead be directly impacted by specific dietary components. This makes the formation of an adaptive hypothesis even more complicated.

3.1.3 Female Dominance as a Driver of Sex Allocation Decisions

While there's an obvious potential link between dietary quality/content and maternal condition measured as body composition, we must also consider that a more distant link with maternal behavior, dominance in particular, may play a role in sex ratio adjustment. If dominance hierarchies influence access to critical resources and/or influence the competitive capabilities of sons, then according to the Trivers–Willard hypothesis, females of higher dominance rank should produce a higher proportion of male offspring. This idea has been tested in two major groups of mammals; ungulates and primates. In their meta-analysis of studies examining the Trivers–Willard hypothesis in ungulates, Sheldon and West (2004) found that the correlation between maternal dominance and offspring sex ratios was highly significant; across studies, more dominant mothers produced higher proportions of male offspring. Cameron's analysis of all mammals supported this finding (Cameron 2004); across studies, dominance was significantly related to sex ratios in the same manner. However, while the relationship is highly significant among ungulates, the relationship when all mammals are considered is much weaker, with only about 30% of studies testing for an effect of maternal dominance on offspring sex ratios showing significant relationships.

It is possible that the inclusion of primate studies weakened the relationship seen in Cameron's meta-analysis, because while primates were the first system in which the connection between maternal dominance and offspring sex ratios were observed, the patterns observed in primates are often contradictory (Fig. 3.5). For example, in the very first published link between maternal rank and offspring sex ratios produced by primates, Jeanne Altman (1980) showed that high-ranking baboon mothers produced more *females*, which is the opposite of what we would expect if mothers were biasing sex ratios according to the predictions of the Trivers–Willard hypothesis. Two additional studies in captive primates, one in bonnet macaques and the other in rhesus macaques, supported this finding; once again, higher ranking mothers produced more daughters (Silk et al. 1981; Simpson and Simpson 1982), and the authors suggested that this pattern still supported the

Sex ratios of offspring produced by dominant mothers in primate species			
Species	More ♂	More ♀	No bias
Macaques (24 studies, 9 species)	8	4	13
Baboons (6 studies*, 5 species)	0	1	5
Other Cercopithicines (2 studies, 2 species)	0	0	2
Non-cercopithicine primates (5 studies, 4 species)	1	0	5

One baboon studies was a meta-analysis of data from 11 populations of baboons

Fig. 3.5 Summary of primate studies that have addressed the question of whether maternal dominance correlates with offspring sex ratios. A majority do not support such a linkage

premise behind the Trivers–Willard hypothesis in that perhaps daughters can repay the maternal investment most effectively in these species (Simpson and Simpson 1982). Indeed, in some primate species, a female passes on her dominance rank to her daughters, and thus for a high-ranking mother, the fitness return of producing a daughter is greater than that of producing a son. This idea has been termed the Maternal Dominance Hypothesis (Grant 1996). However, when sex ratio patterns were tested in wild rhesus monkeys, Meikle et al. (1984) found that high-ranking mothers instead produced a greater proportion of *sons*, just as the original form of the Trivers–Willard hypothesis predicts. This pattern was supported by work in captive pigtail macaques, where mothers treated for bite wounds (an indicator of low dominance) produced fewer male infants (Sackett 1981), and in long-tailed macaques where high-ranking females also produced more sons (Van Schaik et al. 1989). Still other studies showed no relationship between female rank and offspring sex ratios (Berman 1988; Small and Hrdy 1986).

Because of these inconsistencies, the ability of primates to allocate sex in an adaptive manner has been called into question. In a meta-analysis of studies testing for a link between maternal dominance and offspring sex ratios in primates, Brown and Silk (2002) showed no difference between the sex ratios of high- versus low-ranking females, refuting the idea that primates adjust sex ratios adaptively according to the Trivers–Willard hypothesis. A few years later, Silk et al. (2005) showed in a meta-analysis including studies conducted in baboons that the available evidence does not support either a link between maternal rank or population sizes and sex ratios, thus refuting the predictions of the Trivers–Willard hypothesis in these baboon species. However, Schino (2004) suggested that if sex ratio variation were purely random, we would expect no consistency in patterns of sex ratio adjustment within a single population over time, and using a meta-analysis, they instead show that there is, in fact, a temporally consistent relationship between maternal dominance rank and birth sex ratios within primate populations. Additionally, authors also found that there was no overall relationship between dominance rank and sex ratios in their analyses; however, this relationship did emerge when there was low resource availability and high sexual dimorphism. These results suggest that we must consider multiple environmental and social factors when testing the predictions of adaptive hypotheses.

3.1.4 Biasing Sex Ratios Based on Costs and Benefits to the Mother

Rather than using sex ratio adjustment as a means to optimize offspring quality, it is also possible that sex ratios are adjusted in response to the predicted role that those offspring will play in the environment and the costs and benefits that the *mother* will experience as a result of producing more of a particular sex. These ideas are discussed nicely in two books that address sex ratio evolution (Cockburn et al. 2002; West 2009). In 1983, Silk (1983) expanded on an idea originally proposed for primates by (Clark 1978) called the Local Resource Competition (LRC) hypothesis. Under this idea, in systems where daughters remain close to their natal group and compete for resources as adults, females should limit the production of daughters. This would be particularly important for those species in which low-ranking daughters would experience frequent harassment, as is the case in many primates (reviewed in van Schaik and Hrdy 1991). Indeed, a meta-analysis of primate species shows strong support for the LRC hypothesis; females consistently biased sex ratios towards the dispersing sex (Silk and Brown 2008). van Schaik and Hrdy (1991) suggested that there may be potential for primate populations to follow *both* the Trivers–Willard hypothesis and the LRC hypothesis simultaneously, and which model is followed depends upon surrounding environmental conditions; when resources are plentiful, producing more males, according to the Trivers–Willard hypothesis, may be more beneficial in an adaptive sense, while producing more females, as predicted by the LRC hypothesis, may occur when resources are limited, and competition for those resources will likely be fierce in the next generation.

Support for the LRC hypothesis is certainly not limited to primates. Using his work in red deer as an example, Clutton-Brock et al. (1982) provided support for the LRC hypothesis in ungulates, and since that time, studies showed that white-tailed deer, mule deer, and roe deer also allocate sex in patterns consistent with the LRC hypothesis (Caley and Nudds 1987; Hewison et al. 1999). In each of these species, females are philopatric and adopt home ranges that may generate competition with their mothers and female siblings. When resources are limited, we would predict that sex ratios would be biased towards the dispersing sex, which is, in this case, males. This is, in fact, what is observed in these species. The LRC hypothesis has also been well addressed in marsupials (Fig. 3.6), and Robert and Schwanz (2011) provide an excellent review of the widespread support for this hypothesis in marsupial models and how evidence in marsupials supports hypotheses generated in primates that both the LRC and Trivers–Willard may operate simultaneously. Perhaps the most well-known support for this in marsupials comes from data on antichinus, in which a proportion of females survives to breed for a second season. More male offspring were produced in years when a higher proportion of females survived to breed again, perhaps because females are responding to the probability that they would compete with any philopatric daughters that they produced (Cockburn et al. 1985, and reviewed in Cockburn 1989). A similar idea was demonstrated in common brushtail possums, though rather than adjusting sex ratios according to food availability, females of this species adjusted sex ratios according to the availability of den sites. Female possums control a number of den sites in which they shelter during the day. When females had fewer den sites, they produced a higher proportion of male offspring. By doing this, they may be limiting future competition for den sites (Johnson et al. 2001). Evidence for sex ratio biases according to the LRC hypothesis has also been found in rodent species; in a study examining the influences of habitat fragmentation on sex ratios produced by root voles, females inhabiting a habitat fragment alone produced more female-biased litters compared to those sharing a fragment with other reproducing female voles (Aars et al. 1995).

3.1 Environmental and Social Factors Linked to Offspring Sex Ratios 45

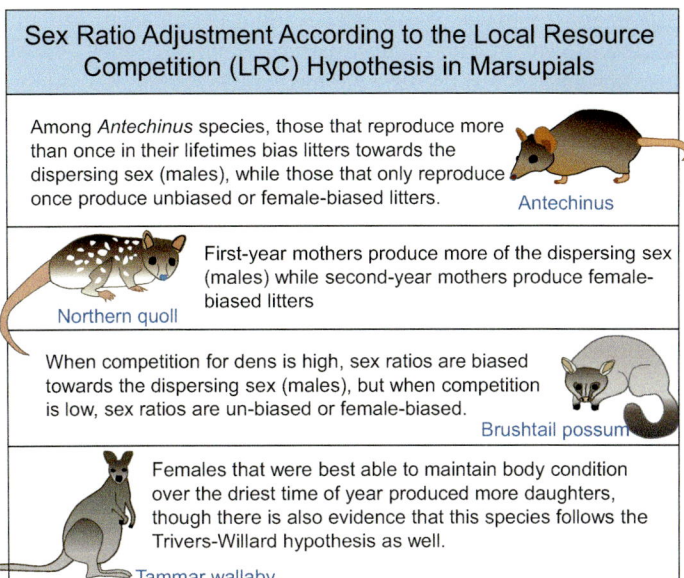

Fig. 3.6 Evidence supporting the Local Resource Competition (LRC) Hypothesis of sex allocation in marsupial species. The first three examples were summarized in Robert and Schwanz (2011) and the final example was reported in Schwanz and Robert (2014)

Along similar lines, if philopatric females provide a benefit to mothers, perhaps in the form of help raising offspring, we would instead predict that females would produce *female*-biased sex ratios. This idea is called the Local Resource Enhancement (LRE) hypothesis, an idea also credited to Trivers and Willard (1973). Support for this idea in mammalian systems is scant. African wild dogs live in communal family groups, and helpers to the breeding pair are predominantly male relatives. In a study of this system, birth sex ratios were male-biased, the pattern that would be predicted according to the LRE hypothesis. Female Townshend voles breed in family units that increase breeding success and survival. At low densities, when competition for resources is low, females produce 65–70% female offspring, which also complies with the predictions of the LRE hypothesis.

Myers (1978) suggested that when resources are limited, females should adjust the production of offspring to favor the cheaper sex. This would serve to minimize the risk of complete failure for the current reproductive attempt and potentially save energy for or reduce costs that would hinder future reproductive attempts. Studies in bighorn sheep indicate that they may follow this pattern. In this system, the birth of a son decreases the survival of the lamb born after it, while the birth of a daughter does not (Berube et al. 1996). Additionally, in this system, when females were older, the sex they produced depended on environmental conditions. They reproduced every year but minimized fitness costs associated with this high reproductive rate by producing more daughters. When conditions were poor, they reduced their reproductive rates, but produced a higher proportion of sons. There is also evidence that females of this species favor maintenance of their own body mass over the growth of their lambs, earning them the label "selfish mothers" (Festa-Bianchet and Jorgenson 1998). Thus, while lambs may adjust sex ratios according to similar cues to those predicted by the Trivers–Willard hypothesis, the result is a direct fitness effect on the *mother*, rather than an indirect fitness effect acting through offspring. Similar patterns have been documented in other mammalian species including rhesus macaques, which provide richer milk to male offspring (Hinde 2009) and suffer greater influences on later fertility when rearing those sons (Bercovitch et al. 2000). As a result, the ultimate driver of sex allocation in these cases may be the costs that producing a son or a daughter exerts on the mother.

3.1.5 Sex Ratio Adjustment in Relation to Male Phenotype

Thus far, we have been focusing on the potential influences of maternal quality on offspring sex ratios. Yet, the potential influences of male phenotype have been largely ignored in discussions of sex ratio adjustment in mammalian species, despite the fact that male quality is a major factor relating to offspring sex ratios in birds and fishes (Booksmythe et al. 2015, and see Chaps. 5 and 7). The attractiveness or quality of the male may directly influence the fitness of offspring if the traits that control that attractiveness or measure of quality are heritable. To explain how male quality may relate to offspring sex ratios, Nancy Burley created the Differential Allocation Hypothesis (DAH), which states that if male attractiveness is heritable, females should bias offspring sexes towards males when mated to an attractive male, because resulting sons would also carry those genes for attractiveness and would thus have a high likelihood of attracting mates, ultimately producing more offspring (Burley 1981, 1988). This idea was generated with birds in mind and, until fairly recently, was completely ignored in discussion of mammalian sex ratio adjustment. In 2006, Gomendio and colleagues were the first to test whether offspring sex ratios in a mammal related to a measure of male quality; in red deer, male fertility, as well as the percentage of normal spermatozoa, was positively correlated with the proportion of males produced, such that the most fertile male had the highest proportion of male offspring. These findings support the DAH, particularly given that high fertility rates are likely heritable and would thus result in offspring with higher reproductive success as well (Gomendio et al. 2007).

Since then, only two additional studies have been conducted to test this idea. Using reindeer, Røed et al. (2007) showed that male mass was significantly related to offspring sex ratios; the authors manipulated the group composition of male reindeer during 10 consecutive ruts (or mating seasons) such that for each rut, the average body mass of the males increased. By the 10th rut, the average body mass was double that of the original group. Doubling the average mass in this way stimulated the production of significantly more male offspring produced by the group, and at the individual level, the proportion of males produced was significantly correlated with male body mass; heavier males produced a higher proportion of male offspring. This finding supports the DAH given that male body size is one of the most favorable traits in mammals due to its importance in male–male competition (Røed et al. 2007) and it is also heritable as well (Kruuk et al. 2000).

The third study to examine this idea was conducted in wild bighorn sheep. Males are polygynous, and in each estrus, females are defended by a dominant male that ultimately receives 60% of the matings. Subordinate males attempt to break up those pairings and receive 40% of the matings as a result (reviewed in Douhard et al. 2016). Douhard et al. (2016) used reproductive success as a measure of male quality, because males with high reproductive success have a suite of preferable traits, including higher dominance ranks, larger horn volumes, and larger body masses, and the latter two traits are heritable. The proportion of male offspring produced was positively correlated with annual reproductive success of the sire; the range was between about 30% to about 60% male offspring.

Why has the DAH been largely ignored in mammals? It is possible that this is because in most mammalian systems in which sex ratios have been studied, male–male competition prevails over female mate choice, and the DAH was specifically designed for species in which females are choosy. However, it is now clear that, even in the absence of choosiness, females can still adjust sex ratios according to qualities that they observe in their mates. Thus, while we now have evidence for this idea in three ungulate species, we need additional work in other species to see if male quality may be a potent driver of sex allocation as it appears to be in other systems.

3.2 Can Males Be the Controllers of Sex Allocation?

To date, nearly all discussion of adaptive sex allocation places the female in the driver's seat of these processes. This makes sense in systems like birds in which the female is the heterogametic sex and has ultimate control over determining the sex of offspring. Also, even in mammals, females are the focus of adaptive theories explaining sex allocation because they are, in theory, the sex that contributes the most to the reproductive attempt, both in the form of large gametes and parental care. However, given that mammalian males are heterogametic, it makes sense to at least consider the possibility that they, too, may play a role in controlling the sex of offspring in an adaptive manner.

Edwards and Cameron (2014) provide an excellent review examining the available evidence that the ratio of X:Y-bearing sperm can be manipulated in response to changing environmental or social conditions. Given that sperm and their functional characteristics tend to be extremely variable between and within populations, there is much potential for that variation to be harnessed in an adaptive mechanism to manipulate offspring sex, and a recent study in mice showed that the ratio of X:Y-bearing sperm, itself, varies significantly among samples (Edwards et al. 2016). It has now been demonstrated that the ratio of X:Y-bearing sperm is altered in humans and mice by exposure to environmental contaminants, heat stress, sexual rest, age, and diet (reviewed in Edwards and Cameron 2014), though the variation in these proportions is generally small, ranging from 48 to 51% Y-bearing sperm.

Can male mammals adjust these proportions in an adaptive manner? Work in mice, conducted by Edwards and colleagues, showed that when coital rate, which can indicate male attractiveness, was experimentally manipulated, the seminal plasma accompanying sperm had higher concentrations of glucose. The authors suggest that, since high peri-conceptional glucose concentrations in the uterine environment may be important for successful male development, this may be a way to stimulate production of male offspring, though whether the contents of the seminal plasma reach the location of fertilization is unknown. In addition, males that had mated the most had a higher proportion of X-bearing sperm, and the ratio of X:Y-bearing sperm was inversely related to the glucose concentrations in seminal plasma. This is the opposite of what would be predicted if more attractive (i.e., more sexually active) males adaptively produced more male offspring, but the authors suggest that since the semen samples were collected after the mice had mated, the findings may actually indicate depletion of Y-bearing sperm during the matings that took place before semen collection. Unfortunately, there has yet been no test of whether male coital rates actually stimulate adjustments of the offspring sex ratio in mice, and thus this idea remains mostly speculation at this point.

Still, this work as well as studies conducted in humans that show potential influences of male job stress on offspring sex ratios (see Chap. 2 for details) call for additional work in this area. If males can, indeed, trigger biases in offspring sex ratios, then adaptive hypotheses underlying sex ratio variation produced by pairs will likely become even more complicated.

3.3 How Might All of These Factors Interact to Control Sex Ratios?

These summarized studies together indicate that a single adaptive hypothesis cannot possibly explain the sex ratio variation we see in mammals. If this is true, does this mean that different mechanisms evolved to manipulate offspring sex ratios in response to different environmental and social variables, or that there is a complex physiological mechanism that integrates each of these factors and allows mammals to adjust sex ratios according to the adaptive hypothesis that fits the circumstances? Ungulates, primates, and marsupials are by far the most well-studied systems in which adaptive hypotheses of sex allocation are considered. In all three systems, we see examples where individuals appear to follow a Trivers–Willard pattern of sex allocation under normal conditions, but when competition becomes more extreme or when resource availability becomes limited, they appear to instead follow the predictions of the LRC hypothesis: a recent demonstration of this was shown in bridled nailtail wallabies, where females in better condition produce more male offspring in the wild, supporting the Trivers–Willard hypothesis, but when brought into more dense populations in captivity, the reverse pattern is seen where females in worse condition produce a son, and the population as a whole produces more male offspring than is seen in the wild. This alternate pattern instead supports the LRC hypothesis (Moore et al. 2015). In other species, we see changes in the patterns of sex

3.3 How Might All of These Factors Interact to Control Sex Ratios?

allocation based on age. Common brushtail wallabies produce significantly more male offspring during their first reproductive attempt (Isaac et al. 2005), and the authors suggest that this pattern complies with a common prediction of the LRC hypothesis, that females should produce higher and higher proportions of the philopatric sex as they age, and the likelihood of encountering them as future competitors decreases. However, after that first reproductive attempt, sex ratios in this species instead relate to maternal condition in a direction predicted by the Trivers–Willard hypothesis, with females in better condition producing more male offspring.

While it appears easy to link an observed pattern of sex allocation with an adaptive hypothesis that may explain it, it is more difficult to conceive of a single mechanism that would allow patterns of sex allocation to follow one adaptive hypothesis in a given set of environmental conditions and another when conditions change. Figure 3.7 illustrates three potential theoretical mechanisms by which this may occur. Consider a situation in which a limited dietary component (factor 1) was the trigger for producing more males, and mothers with more of that dietary

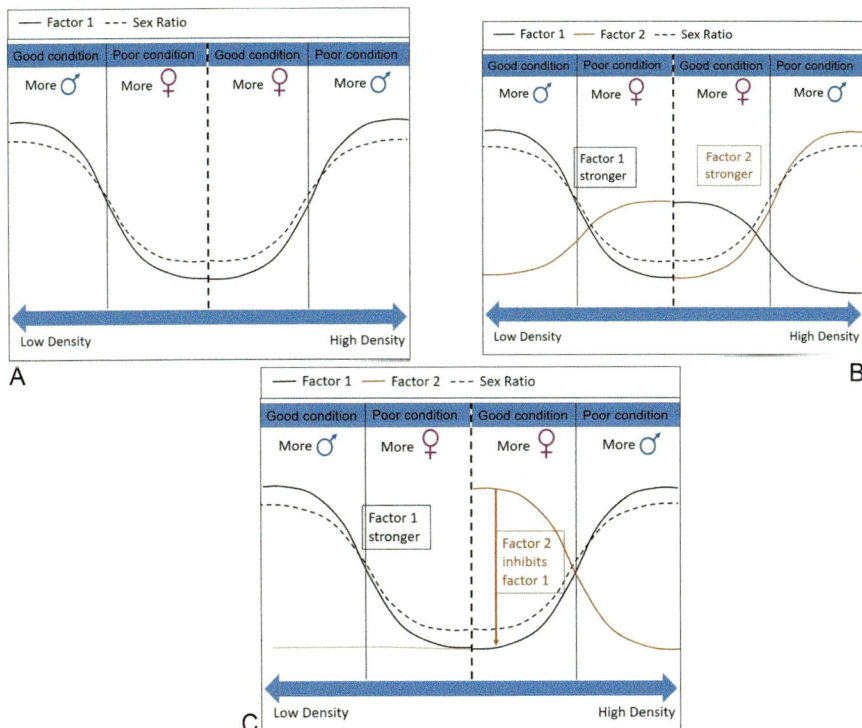

Fig. 3.7 Models illustrating three potential ways that one (**a**) or two (**b** and **c**) factors could control whether sex ratios are adjusted according to the Trivers–Willard hypothesis or the Local Resource Competition (LRC) hypothesis in species that appear to substitute one or the other depending on surrounding conditions. See the text for in-depth explanations of these models

component are also in better condition and produce higher quality offspring. We would then postulate that these females are allocating sex according to the predictions of the Trivers–Willard hypothesis. However, when conditions get more crowded, resources, including the dietary component that triggers the sex ratio skews, get more limited, and in this case, we would expect females in the lowest condition to produce even more dramatic female biases in the offspring they produce; however instead the opposite happens. Females in bad condition produce more *males* despite having the lowest access to the dietary component that triggered the production of more males previously. This would require a mechanism where a single factor that triggeres the production of more males then switches its influence to produce more females instead. (Fig. 3.7a). It seems very unlikely, however, that a single factor would have the capability to exert opposite influences on the reproductive system and/or developing offspring if it were working in isolation.

Perhaps, instead, sex ratios are controlled by a combination of factors, and different factors have a more or less potent effect on sex ratios in different environmental conditions. Consider an example where the hypothetical dietary factor discussed above normally triggers the production of more males when females are in good condition, but when population densities increase, and females are exhibiting higher levels of stress and aggression, perhaps a second factor related to those behaviors exerts an even stronger effect on sex ratios, triggering sex ratio skews in the opposite direction. Under this scenario, sex ratios would closely align with the dietary component (factor 1) when population densities are low to normal, but above a certain threshold population density, sex ratios would align with the second trigger (factor 2) (Fig. 3.7b).

In a third scenario, consider that factor 1 and factor 2 may not be independent of one another. If factor 2 is only present at high levels when population densities are high, and factor 2 inhibits factor 1, then we may expect to see the opposite patterns of sex allocation at high densities compared to those we see at low densities. To put this in a real context and make it easier to understand, consider a scenario in which, at low population densities, females in good condition enjoy good territories without many challenges, since there are enough good quality breeding sites available for everyone in the population. In contrast, at high densities, good quality females still win the best territories, but also have to defend them frequently, leading to the highest levels of aggression in females that are in good condition and also reside in populations with high densities. If the aggressive behaviors or a physiological mediator associated with those aggressive behaviors acted to inhibit factor 1, then we would see the opposite pattern of sex allocation compared to what would be predicted according to the Trivers–Willard hypothesis.

All of these ideas are highly speculative, but they point to the need to understand how multiple physiological factors may interact to influence offspring sex ratios from a mechanistic standpoint. It is unlikely that a single factor serves as the "holy grail" trigger of sex ratio adjustment. A single trigger would be likely to trigger more extreme variation in sex ratios than have been documented in mammalian studies. It is more likely that a physiological network provides a system of checks and balances to allow for fine-scale manipulations of offspring sex ratios in response to a variety of environmental and social variables and in patterns that match an optimal strategy for a particular context. Therefore, an understanding of these mechanisms is critical for understanding how sex ratio adjustment functions in an adaptive context.

While there is abundant evidence that many mammalian species produce sex ratios that deviate from the 50:50 ratio of males:females that would be expected based on Mendelian inheritance, there are still many questions regarding the reasons driving the directions of those sex ratio skews, whether there really is an adaptive basis to sex allocation or if the observed patterns simply result from sampling bias and chance, and what the underlying mechanisms might be. The data suggest that there are multiple mechanistic targets for sex allocation, that many physiological factors may interact to influence these targets, and that the optimization of sex ratios to surrounding environmental and social conditions would need to harness these factors to adjust sex ratios according to the adaptive strategy that provides the largest fitness benefit. More mechanistic work in this area is badly needed to unravel the complicated story behind sex allocation in mammals.

References

Aars J, Andreassen HP, Ims RA (1995) Root voles: litter sex ratio variation in fragmented habitat. J Anim Ecol 64:459–472

Altman J (1980) Baboon mothers and infants. Harvard University Press, Cambridge

Arangasamy A, Selvaraju S, Parthipan S, Somashekar L, Rajendran D, Ravindra J (2015) Role of calcium and magnesium administration on sex ratio skewing, follicular fluid protein profiles and steroid hormone level and oocyte transcripts expression pattern in Wistar rat. Indian J Anim Sci 85(11):1190–1194

Austad SN, Sunquist ME (1986) Sex-ratio manipulation in the common opossum. Nature 324:58–60

Behnam-Rassouli M, Aliakbarpour A, Hosseinzadeh H, Behnam-Rassouli F, Chamsaz M (2010) Investigating the effect of aqueous extract of Chicorium intybus L. leaves on offspring sex ratio in rat. Phytother Res 24(9):1417–1421

Bercovitch FB, Widdig A, Nürnberg P (2000) Maternal investment in rhesus macaques (Macaca mulatta): reproductive costs and consequences of raising sons. Behav Ecol Sociobiol 48 (1):1–11

Berman C (1988) Maternal condition and offspring sex ratio in a group of free-ranging rhesus monkeys: an eleven-year study. Am Nat 131(3):307–328

Berube CH, Festa-Bianchet M, Jorgenson JT (1996) Reproductive costs of sons and daughters in Rocky Mountain bighorn sheep. Behav Ecol 7(1):60–68

Booksmythe I, Mautz B, Davis J, Nakagawa S, Jennions MD (2015) Facultative adjustment of the offspring sex ratio and male attractiveness: a systematic review and meta-analysis. Biol Rev 92:108–134

Brown GR, Silk JB (2002) Reconsidering the null hypothesis: is maternal rank associated with birth sex ratios in primate groups? Proc Natl Acad Sci 99(17):11252–11255

Burley N (1981) Sex ratio manipulation and selection for attractiveness. Science 211 (4483):721–722

Burley N (1988) The differential-allocation hypothesis: an experimental test. Am Nat 132 (5):611–628

Caley MJ, Nudds TD (1987) Sex-ratio adjustment in Odocoileus: does local resource competition play a role? Am Nat 129(3):452–457

Cameron EZ (2004) Facultative adjustment of mammalian sex ratios in support of the Trivers-Willard hypothesis: evidence for a mechanism. Proc R Soc London, Ser B 271(1549):1723–1728

Cameron EZ, Linklater WL (2007) Extreme sex ratio variation in relation to change in condition around conception. Biol Lett 3(4):395–397

Cameron EZ, Lemons PR, Bateman PW, Bennett NC (2008) Experimental alteration of litter sex ratios in a mammal. Proc R Soc Lond B Biol Sci 275(1632):323–327

Clark AB (1978) Sex ratio and local resource competition in a prosimian primate. Science 201 (4351):163–165

Clayton E, Gulliver C, Wilkins J, King B, Meyer R, Friend M (2012) Increasing the proportion of female lambs by supplementary feeding oats high in omega-6 fatty acids at joining. In: Proceedings of the 27th annual conference of the Grassland Society of NSW Inc., Wagga Wagga, NSW, pp 107–113

Clayton E, Wilkins J, Friend M (2016) Increased proportion of female lambs by feeding Border Leicester× Merino ewes a diet high in omega-6 fatty acids around mating. Anim Prod Sci 56 (5):824–833

Clutton-Brock TH (1989) Review lecture: mammalian mating systems. Proc R Soc Lond B Biol Sci 236(1285):339–372

Clutton-Brock T, Albon S, Guinness F (1982) Competition between female relatives in a matrilocal mammal. Nature 300(5888):178–180

Clutton-Brock T, Albon S, Guinness F (1984) Maternal dominance, breeding success and birth sex ratios in red deer. Nature 308(5957):358–360

Cockburn A (1989) Sex-ratio variation in marsupials. Aust J Zool 37(3):467–479

Cockburn A, Scott MP, Dickman CR (1985) Sex ratio and intrasexual kin competition in mammals. Oecologia 66(3):427–429

Cockburn A, Legge S, Double MC (2002) Sex ratios in birds and mammals: can the hypotheses be disentangled. Sex ratios: concepts and research methods. Cambridge University Press, Cambridge, pp 266–286

Dama MS, Singh NMP, Rajender S (2011) High fat diet prevents over-crowding induced decrease of sex ratio in mice. PLoS One 6(1):e16296

Douhard M, Festa-Bianchet M, Coltman DW, Pelletier F (2016) Paternal reproductive success drives sex allocation in a wild mammal. Evolution 70(2):358–368

Edwards AM, Cameron EZ (2014) Forgotten fathers: paternal influences on mammalian sex allocation. Trends Ecol Evol 29(3):158–164

Edwards A, Cameron E, Pereira J, Ferguson-Smith M (2016) Paternal sex allocation: how variable is the sperm sex ratio? J Zool 299:37–41

Enright W, Spicer L, Kelly M, Culleton N, Prendiville D (2001) Energy level in winter diets of fallow deer: effect on plasma levels of insulin-like growth factor-I and sex ratio of their offspring. Small Rumin Res 39(3):253–259

Festa-Bianchet M, Jorgenson JT (1998) Selfish mothers: reproductive expenditure and resource availability in bighorn ewes. Behav Ecol 9(2):144–150

Folmer V, Soares JC, Gabriel D, Rocha JB (2003) A high fat diet inhibits δ-aminolevulinate dehydratase and increases lipid peroxidation in mice (Mus musculus). J Nutr 133(7):2165–2170

Fountain ED, Mao J, Whyte JJ, Mueller KE, Ellersieck MR, Will MJ, Roberts RM, MacDonald R, Rosenfeld CS (2008) Effects of diets enriched in omega-3 and omega-6 polyunsaturated fatty acids on offspring sex-ratio and maternal behavior in mice 1. Biol Reprod 78(2):211–217

Gharagozlou F, Youssefi R (2016) Akbarinejad V Effects of diets supplemented by fish oil on sex ratio of pups in bitch. In: Veterinary research forum, vol 2. Faculty of Veterinary Medicine, Urmia University, Urmia, Iran, p 105

Gomendio M, Malo AF, Garde J, Roldan ER (2007) Sperm traits and male fertility in natural populations. Reproduction 134(1):19–29

Grant VJ (1996) Sex determination and the maternal dominance hypothesis. Hum Reprod 11 (11):2371–2375

Green MP, Spate LD, Parks TE, Kimura K, Murphy CN, Williams JE, Kerley MS, Green JA, Keisler DH, Roberts RM (2008) Nutritional skewing of conceptus sex in sheep: effects of a maternal diet enriched in rumen-protected polyunsaturated fatty acids (PUFA). Reprod Biol Endocrinol 6(1):21

Gulliver C, Friend M, King B, Wilkins J, Clayton E (2013) A higher proportion of female lambs when ewes were fed oats and cottonseed meal prior to and following conception. Anim Prod Sci 53(5):464–471

Hayes JP, Shonkwiler JS (2001) Morphometric indicators of body condition: worthwhile or wishful thinking. In: Speakman JR (ed) Body composition analysis of animals. A handbook of non-destructive methods. Cambridge University Press, Cambridge, pp 8–38

Helle S, Laaksonen T, Adamsson A, Paranko J, Huitu O (2008) Female field voles with high testosterone and glucose levels produce malebiased litters. Anim Behav 75(3):1031–1039

Hewison AM, Andersen R, Gaillard J-M, Linnell JD, Delorme D (1999) Contradictory findings in studies of sex ratio variation in roe deer (Capreolus capreolus). Behav Ecol Sociobiol 45(5):339–348

Hinde K (2009) Richer milk for sons but more milk for daughters: sex-biased investment during lactation varies with maternal life history in rhesus macaques. Am J Hum Biol 21(4):512–519

Isaac JL, Krockenberger AK, Johnson CN (2005) Adaptive sex allocation in relation to life-history in the common brushtail possum, Trichosurus vulpecula. J Anim Ecol 74(3):552–558

Johnson CN, Clinchy M, Taylor AC, Krebs CJ, Jarman PJ, Payne A, Ritchie EG (2001) Adjustment of offspring sex ratios in relation to the availability of resources for philopatric offspring in the common brushtail possum. Proc R Soc Lond B Biol Sci 268(1480):2001–2005

Kruuk LE, Clutton-Brock TH, Slate J, Pemberton JM, Brotherstone S, Guinness FE (2000) Heritability of fitness in a wild mammal population. Proc Natl Acad Sci 97(2):698–703

Labov JB, William Huck U, Vaswani P, Lisk RD (1986) Sex ratio manipulation and decreased growth of male offspring of undernourished golden hamsters (Mesocricetus auratus). Behav Ecol Sociobiol 18(4):241–249

Lange SE (2011) Effect of Neofat on offspring sex ratios in Rambouillet and Suffolk sheep. Angelo State University

Meikle D, Drickamer LC (1986) Food availability and secondary sex ratio variation in wild and laboratory house mice (Mus musculus). J Reprod Fertil 78(2):587–591

Meikle D, Tilford B, Vessey S (1984) Dominance rank, secondary sex ratio, and reproduction of offspring in polygynous primates. Am Nat 124(2):173–188

Mitra J, Chowdhury M (1989) Glycerylphosphorylcholine diesterase activity of uterine fluid in conditions inducing secondary sex ratio change in the rat. Gamete Res 23(4):415–420

Moore EP, Hayward M, Robert KA (2015) High density, maternal condition, and stress are associated with male-biased sex allocation in a marsupial. J Mammal 96(6):1203–1213

Myers JH (1978) Sex ratio adjustment under food stress: maximization of quality or numbers of offspring? Am Nat 112(984):381–388

Oun AE, Bakry S, Soltan S, Taha A, Kadry E (2016) Preconceptional minerals administration skewed sex ratio in rat offspring. Res Obstet Gynecol 4(1):11–15

Parkes A (1926) The mammalian sex ratio. Biol Rev 2(1):1–51

Rivers J, Crawford M (1974) Maternal nutrition and the sex ratio at birth. Nature 252:297–298

Robert KA, Schwanz LE (2011) Emerging sex allocation research in mammals: marsupials and the pouch advantage. Mammal Rev 41(1):1–22

Røed KH, Holand Ø, Mysterud A, Tverdal A, Kumpula J, Nieminen M (2007) Male phenotypic quality influences offspring sex ratio in a polygynous ungulate. Proc R Soc Lond B Biol Sci 274(1610):727–733

Rosenfeld CS (2011) Periconceptional influences on offspring sex ratio and placental responses. Reprod Fertil Dev 24(1):45–58

Rosenfeld CS, Roberts RM (2004) Maternal diet and other factors affecting offspring sex ratio: a review. Biol Reprod 71(4):1063–1070

Rosenfeld CS, Grimm KM, Livingston KA, Brokman AM, Lamberson WE, Roberts RM (2003) Striking variation in the sex ratio of pups born to mice according to whether maternal diet is high in fat or carbohydrate. Proc Natl Acad Sci 100(8):4628–4632

Sackett GP (1981) Receiving severe aggression correlates with fetal gender in pregnant pigtailed monkeys. Dev Psychobiol 14(3):267–272

Schino G (2004) Birth sex ratio and social rank: consistency and variability within and between primate groups. Behav Ecol 15(5):850–856

Schmidt CM, Hood WR (2012) Calcium availability influences litter size and sex ratio in white-footed mice (Peromyscus leucopus). PLoS One 7(8):e41402

Schwanz LE, Robert KA (2014) Proximate and ultimate explanations of mammalian sex allocation in a marsupial model. Behav Ecol Sociobiol 68:1085–1096

Schwanz LE, Robert KA (2016) Costs of rearing the wrong sex: cross-fostering to manipulate offspring sex in Tammar wallabies. PLoS One 11(2):e0146011

Sheldon BC, West SA (2004) Maternal dominance, maternal condition, and offspring sex ratio in ungulate mammals. Am Nat 163:40–54

Silk JB (1983) Local resource competition and facultative adjustment of sex ratios in relation to competitive abilities. Am Nat 121(1):56–66

Silk JB, Brown GR (2008) Local resource competition and local resource enhancement shape primate birth sex ratios. Proc R Soc Lond B Biol Sci 275(1644):1761–1765

Silk JB, Clark-Wheatley CB, Rodman PS, Samuels A (1981) Differential reproductive success and facultative adjustment of sex ratios among captive female bonnet macaques (Macaca radiata). Anim Behav 29(4):1106–1120

Silk JB, Willoughby E, Brown GR (2005) Maternal rank and local resource competition do not predict birth sex ratios in wild baboons. Proc R Soc Lond B Biol Sci 272(1565):859–864

Simpson M, Simpson A (1982) Birth sex ratios and social rank in rhesus monkey mothers. Nature 300:440–441

Small MF, Hrdy SB (1986) Secondary sex ratios by maternal rank, parity, and age in captive rhesus macaques (Macaca mulatta). Am J Primatol 11(4):359–365

Tiffin G, Rieger D, Betteridge K, Yadav B, King W (1991) Glucose and glutamine metabolism in pre-attachment cattle embryos in relation to sex and stage of development. J Reprod Fertil 93(1):125–132

Trivers RL, Willard DE (1973) Natural selection of parental ability to vary the sex ratio of offspring. Science 179(4068):90–92

Vahidi A, Sheikhha M (2007) Comparing the effects of sodium and potassium diet with calcium and magnesium diet on sex ratio of rats' offspring. Pak J Nutr 6(1):44–48

van Schaik CP, Hrdy SB (1991) Intensity of local resource competition shapes the relationship between maternal rank and sex ratios at birth in Cercopithecine primates. Am Nat 138:1555–1562

Van Schaik C, Netto W, Van Amerongen A, Westland H (1989) Social rank and sex ratio of captive long-tailed macaque females (Macaca fascicularis). Am J Primatol 19(3):147–161

West S (2009) Sex allocation. Princeton University Press, Princeton

Zamiri MJ (1978) Effects of reduced food intake on reproduction in mice. Aust J Biol Sci 31(6):629–640

Potential Mechanisms of Sex Ratio Adjustment in Humans and Nonhuman Mammals

> *And yet, in each human coupling, a thousand million sperm vie for a single egg. Multiply those odds by countless generations, against the odds of your ancestors being alive; meeting; siring this precise son; that exact daughter...*
> — Alan Moore

The steps to producing viable mammalian offspring, from the point of gamete production through birth, are numerous and complex. At each developmental stage, challenges exist for the gametes and the conceptuses, and at each stage, there is evidence for differential susceptibility to these challenges based on the sex chromosomes present. If humans and nonhuman mammals could use these sex-based differences in susceptibility to control which sex prevails, this would allow for evolution of facultative sex ratio manipulation in mammalian systems. Many of the studies discussed in the two previous chapters provide evidence about when during development sex ratio adjustment may be occurring, and it appears that not one but several developmental time periods may be targets for adaptive manipulation of sex ratios in mammals. In addition, while we tend to assume that, despite the XY system of sex determination in mammals, the control of sex ratios lies within the female, there are also possibilities for control by the male. By working through the complex processes involved in fertilization through the successful parturition of a baby, we can identify all of the potential time points at which sex ratios may potentially be adjusted (Fig. 4.1).

4.1 Differential Production of X- Versus Y-Bearing Sperm

The initial determination of whether X or Y sperm are produced clearly lies in the male. Spermatogenesis and its ingrained processes of meiosis occur continuously in men from puberty through death, and every meiotic segregation event results in the

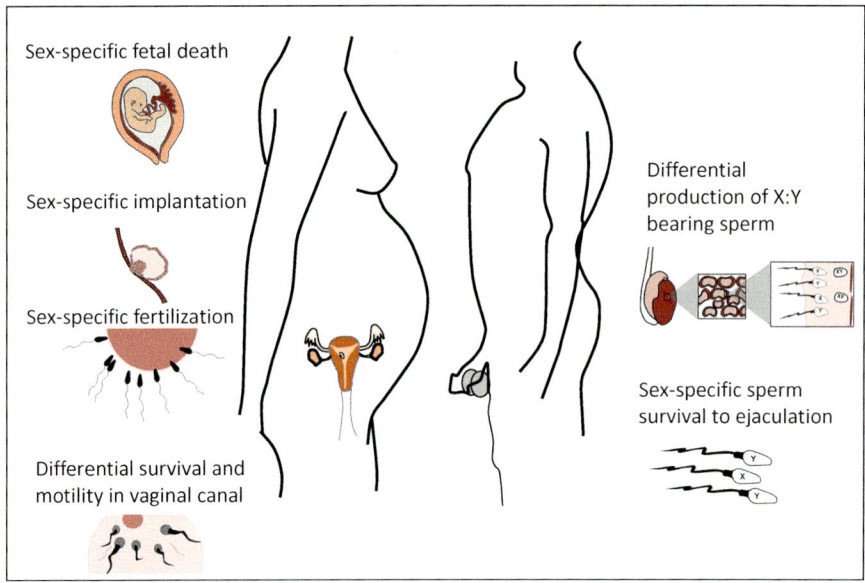

Fig. 4.1 Potential targets of sex ratio manipulation in mammals. While the figure shows human targets, these targets would be expected to exist in most mammals. Figure adapted from Navara (2010)

production of equal numbers of X and Y sperm. Any aberration from this proportion would likely occur through an aneuploidy event, which ultimately would result in high rates of miscarriage, fetal death, and infant abnormality; none of these have been indicated in studies in which sex ratio skews were documented. Thus, it is more likely that sex ratio skews occurring at the stage of gamete production in the male result from changes in growth and differentiation rates of X- versus Y-bearing sperm, or differential survival of sperm carrying a particular sex chromosome in the male reproductive tract.

There is evidence for the latter, as the Y chromosome is particularly vulnerable to DNA deletions, especially when induced as a result of oxidative damage (Aitken and Krausz 2001). Indeed, exposure to toxins alters the proportion of X:Y-bearing sperm in semen; exposure of men to boron in the workplace reduced the proportion of Y-bearing sperm (Robbins et al. 2008), while exposure to persistent organochlorine pollutants (POPs) increased this proportion (Tiido et al. 2005). Chaudhary and colleagues (2014) examined the proportion of X- to Y-bearing sperm in human ejaculates in relation to diet, season, and paternal occupation. They showed that there were significantly more X-bearing sperm in ejaculates overall; however, this proportion did not change with a vegetarian diet, between winter and summer seasons, or between professional and labor-heavy occupations. These findings cast doubt on the potential for observed patterns of sex ratio adjustment to result from alterations in the ratio of X- to Y-bearing sperm; however, more studies should be done because there have not been previously documented sex ratio skews in response to two of the three specific criteria tested. For example, there is no evidence that a vegetarian diet skews human natal sex ratios. Instead, natal sex ratios have been linked to alkalinity and ionic balance of the diet, as well as the total intake of nutrients (see Chap. 2). It is unclear whether a vegetarian diet would influence these factors. Similarly, there has been no documentation of skewed sex ratios resulting from labor-heavy as opposed to professional jobs. Previously documented paternal influences on sex ratios appear to be related either to the presence of extreme stress in the occupation or chemical exposure. Thus, it would be interesting to look at the proportions of X- to Y-bearing sperm in these groups that have been specifically related to biases in natal sex ratios. This test has been done in pygmy hippos. Saragusty et al. (2012) used fluorescence in situ hybridization (FISH) to test the ratio of X:Y-bearing sperm in males from a population where birth sex ratios were female biased at 42% males. The ratio closely matched the birth sex ratios, suggesting that males may possess a mechanism to bias the production and/or survival of X- versus Y-bearing sperm. At this point, however, evidence for the role of chromosomal ratios in the ejaculate to drive skews in natal sex ratios is still scarce in other mammalian species.

4.2 Differential Survival of X- Versus Y-Bearing Sperm

In 1960, Landrum Shettles noted, using phase-contrast microscopy, that there were two distinct morphological populations of sperm: one population with round heads and the other with oval heads (Shettles 1960b). Because there were more of the round-headed sperm, and there are more boys born in the general population, he concluded that those sperm must contain the Y chromosome while the larger oval-headed sperm must contain the X chromosome (Shettles 1960a). Later studies in which the X and the Y chromosomes were molecularly identified revealed that X- and Y-bearing sperm are not in fact obviously identifiable through a microscope, though this did spur a field of research examining morphological and functional differences between X- and Y-bearing sperm. The sex chromosome carried by a sperm cell influences several physical characteristics of the sperm. The X chromosome is larger and, because of that, contains more genes than the Y chromosome (1100 on the X compared to a few dozen on the Y) (Graves 1995; Ross et al. 2005). In addition, the head, perimeter, and area of the head of sperm are, in fact, slightly larger when the sperm is carrying an X chromosome. The lengths of the neck and tail are also greater for X-bearing sperm. Together, these characteristics make X-bearing sperm heavier than Y-bearing sperm (Cui 1997). X-bearing sperm also have a larger net negative charge on the cell surface than Y-bearing sperm (Ishijima et al. 1991), and some studies also show that Y-bearing sperm migrate faster through a medium (Rohde et al. 1973), though this point is controversial and methods to select sperm of a particular sex based on this idea do not appear to produce reliable and repeatable results. However, there are enough differences between sperm that carry X versus Y chromosomes to generate potential for natural sex selection to occur at the level of the sperm.

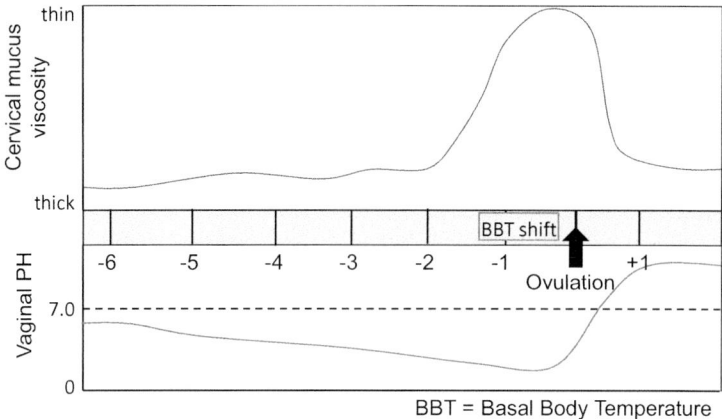

Fig. 4.2 Timing of changes in cervical mucus viscosity, vaginal pH, and the basal body temperature shift relative to ovulation. Numbers represent days prior to or after ovulation

4.3 Differential Survival and Motility of X- Versus Y-Bearing Sperm

Perhaps instead of differential *production* of X- and Y-bearing sperm, it is the differential *survival* of X- versus Y-bearing sperm in the female reproductive tract that influences sex ratios, or the ability of those sperm to successfully traverse the female reproductive tract and reach the egg. The vaginal environment undergoes changes during the ovulatory process that could significantly influence sperm survival, perhaps in a sex chromosome-specific way (Fig. 4.2). For example, the pH of the vaginal environment is acidic and becomes more and more so up to the point of ovulation, after which the pH shifts to become more alkaline (Guerrero 1975). Both in vitro and in vivo studies have shown that human sperm are vulnerable to extremely acidic or basic environments (Masters 1960; Muschat 1926); motility ceases at a pH of less than 6 and higher than 10 (Muschat 1926). Unterberger (1930) suggested that a more alkaline environment favored Y-bearing sperm, because he noticed that women using alkaline douches produced an excess of males. However, in an in vitro experiment, Diasio and Glass (1971) tested whether the acidity of the media in which sperm were held would influence their ability to migrate, but found no significant influence. Additionally, Masters (1960) showed that male ejaculate acts to decrease the vaginal acidity, and Fox and colleagues confirmed this finding using continuous monitoring of the vaginal environment using radiotelemetry (Fox et al. 1973). They found that within 8s of ejaculation, the vaginal environment went from a pH of 4.3 to a PH of 7.2. This would mean that X- and Y-bearing sperm must be differentially susceptible to very small differences in pH for sex-biased selection to occur via this mechanism.

Yet, there are still other changes that occur within the female reproductive tract that could differentially affect the survival and motility of X- and Y-bearing sperm. For example, the cervical mucus not only changes in acidity, but also in viscosity (Moghissi 1966), with the highest viscosity occurring at ovulation and the lowest viscosity just prior and just after ovulation. Spermatozoa containing a Y chromosome migrate more readily through a variety of media, including cervical mucus, so Y selection should increase with the viscosity of the cervical mucus (Ericsson 1994; Pyrzak 1994). This would result in a greater proportion of males prior to and after ovulation, when cervical mucus is thick, producing a U-shaped pattern of sex ratios across the ovulatory cycle (Jongbloet 2003). Indeed, this is what has been found. Guerrero (1974) showed data that agreed with a high proportion of male conceptions early in the cycle, but he found a U-shaped pattern of sex ratios in relation to the timing of the ovulatory cycle; 68% of babies conceived before the thermal shift that indicates ovulation were male compared to only 44% conceived on the day of the shift. The ratio then returned to a male bias after this point. James (2000) conducted a statistical analysis using 11 studies that tested sex ratios on the most fertile or "other" days and found that there is evidence that sex ratios are higher on days other than the most fertile day. Martin (1997) also provided a nice review incorporating some of these studies and arrived at the same conclusion. Thus, the idea the sex selection may occur through the differential ability of X- versus Y-bearing sperm to traverse the cervical mucus appears plausible.

4.4 Sex-Specific Selection of Sperm in the Female Reproductive Tract

While it was previously thought that the oviduct and the target egg were only passively involved in the process of determining which sperm fertilized the egg, it is now becoming clear that both may be actively involved in the process of sperm selection. Work in mice, pigs, and cows indicates that sperm announce their arrival into the reproductive tract by altering the expression of genes within the oviduct (reviewed in Holt and Fazeli 2016). These changes appear to be controlled by the sperm themselves, and not the seminal plasma. The oviduct responds not only by preparing for arrival of the sperm and the future implantation of the embryo but also appears to protect the sperm until it is time for fertilization. For example, heat shock protein 70 A8 is a soluble component of the epithelial membrane in the oviduct that can bind to the sperm cell surface and promote survival by increasing membrane fluidity and potential repairing membrane damage (Moein-Vaziri et al. 2014). Given the responsiveness of the oviduct to sperm entry, there is potential for the evolution of mechanisms that would allow females to select specific sperm for fertilization, based on genes carried by the sperm or, more specifically, by the sex chromosome that is present. Almiñana et al. (2014) used microarrays in pigs to demonstrate that >200 transcripts were either up- or downregulated in the oviduct when a Y-bearing sperm was present. The most representative of these were genes involved in signal transduction and immunological responses. Because Y-bearing sperm carry male-specific antigens that can potentially be targeted by the female immune system, the authors suggested that pigs may have the potential to target sperm carrying the Y chromosome to induce natural sex selection. However, there are still many questions. For example, if sperm carrying the Y chromosome upregulate immunological reactions against them, why do we not see very low proportions of male offspring on a fairly regular basis? More work needs to be done to examine the potential for the oviduct to select sperm carrying a specific sex chromosome.

Along similar lines, any differences in susceptibility of the oocyte to fertilization by sperm that carry X versus Y chromosomes could also exert significant effects on the sex ratios at conception. In the Orzack et al. (2015) study discussed earlier, embryos that were produced by assisted reproductive technologies were equally likely to be male and female, suggesting that, at least in normal culture conditions, the likelihood that eggs are penetrated by X- and Y-bearing sperm is similar. However, changes in the conditions surrounding the woman could potentially alter how the egg reacts to sperm of a given sex, particularly given the results in pigs showing an ability to differentiate between X- and Y-bearing sperm (Alminana et al. 2014). Work in heifers supports this idea. Ideta et al. (2009) conducted a study using 25 superovulated Holstein heifers, 13 of which underwent a stress protocol involving frequent transrectal examination and blood collection during the follicular phase. Oocytes were then collected from the animals and underwent fertilization in vitro. Fertilized oocytes that were collected from stressed heifers produced 67% female embryos, compared to 50% from control animals. This suggests that the susceptibility of the oocyte to X- versus Y-bearing sperm can change based on the experiences and physiology of the female. Given this result in heifers, it is surprising that human ART embryos do not show a sex ratio skew given the stressful nature of fertility treatments and the challenges of infertility overall.

However, stress is not the only factor that appears to influence the susceptibility of the egg to sperm carrying a particular sex chromosome. In a review addressing whether mammalian sex ratios can be modified around the time of conception, Valerie Grant and Lawrence Chamley (2010) discuss two studies that provide evidence that testosterone in the follicular fluid may influence which sperm penetrate the egg. In two studies (García-Herreros et al. 2010; Grant and Irwin 2005), ova were collected from heifers and the follicular fluid in which the ova had developed was collected and assayed for testosterone concentrations. In the Grant and Irwin (2005) study, follicles producing eggs that were subsequently fertilized by Y-bearing sperm had significantly higher testosterone levels in the follicular fluid. Garcia-Herreros and colleagues (García-Herreros et al. 2010) showed a nonsignificant trend in the same direction ($p = 0.06$) using the same study design. When the results of the two studies are combined, there is a significant relationship between concentrations of testosterone in the follicular fluid and the sex of the resulting embryo after fertilization (Grant and Chamley 2010). Macaulay and colleagues (2013) experimentally tested this idea by adding physiologically relevant concentrations of testosterone or androstenedione to oocyte maturation media and observing the influences on oocyte cleavage rate and sex. They did not find an effect of testosterone on offspring sex, and androstenedione influenced sex only in conjunction with the physical material used during culture. Instead, it appears that the sexes of embryos depended more on the quality of the oocyte and its cleavage rate. Thus, it is possible that, in the correlational studies, the concentration of testosterone in the follicular fluid was related to something else that influenced oocyte quality and, ultimately, whether it was penetrated by an X- or a Y-bearing sperm. This idea could potentially explain the links seen between decreased maternal quality and sex ratios. Perhaps as maternal quality declines, oocyte quality declines, and low-quality oocytes are more susceptible to being fertilized by an X-bearing sperm, producing more female offspring. These concepts need to be examined further, and work needs to be done to see whether similar relationships between follicular hormone content, oocyte quality, and sexes produced exist in humans.

4.5   Sex-Specific Implantation

Another mechanism that lies under control of the female reproductive tract is the potential for the uterus to be amenable to implantation in a sex-specific manner. That is, if the female uterus can detect whether the blastocyst contains the XX or XY combination, there may be a sex-specific response that affects the success of the implantation process. After fertilization, the fertilized ovum or zygote transitions through the morula stage (12–16 cells) and travels through the fallopian tube into the uterus. At this time, the cell mass is encapsulated in the zona pellucida. Over the next 3 days, the morula continues to divide and forms an inner fluid-filled cavity, at which point it is then called the blastocyst. At 6–7 days after ovulation, the blastocyst hatches from the zona pellucida and begins initial adhesion to the uterine wall. The first steps of implantation rely on the interaction of uterine projections, called pinopodes, with microvilli on the outer surface of the blastocyst. This process is regulated by growth factors and cytokines produced by both the blastocyst and the uterine wall. Failure to synchronize the molecular processes involved results in the failure of implantation. Due to the complex communication between the blastocyst and uterus, there may be potential for the uterus to respond to the blastocyst differently based on the sex chromosomes it carries. In fact, work in red deer showed sex-specific levels of trophoblast interferon and suggested that this dimorphism triggered sex-specific blastocyst loss when mothers were in poor condition (Flint et al. 1997). To my knowledge, this idea has never been tested further, but future work should be done to determine whether XX and XY blastocysts produce dimorphic concentrations of substances that would differentially influence blastocyst survival, and whether the presence of an XX or XY blastocyst induces differential gene expression in the uterine wall, particularly focusing on genes controlling the production of growth factors, cytokines, other immunoregulatory molecules, and those responsible for responses to steroid hormones, which are also involved in the process. But how could this mechanism control how external cues alter the implantation of one sex over another? The answer to this question brings up even more questions. For example, are XX and XY blastocysts equally successful at implanting in a normal situation when we would not expect a sex ratio bias? This is currently unknown. If thresholds for the production of or sensitivity to the required molecular signals are different based on whether an XX versus an XY blastocyst is implanting, then small modulations of hormones that regulate the production of these molecules could generate the slight adjustments necessary to impact whether the blastocyst of a particular sex can implant. Of course, these ideas are completely untested. However, one can see where the potential exists; if the oviduct can recognize sperm carrying a particular sex chromosome, why not the uterus?

4.6 Sex-Specific Survival to Birth

Male and female offspring differ developmentally from very early on (Gutiérrez-Adán et al. 2006). Microarray analysis revealed that over 600 genes were differentially expressed in male and female mouse blastocysts prior to implantation (Kobayashi et al. 2006). Differences in gene expression between male and female blastocyst were also found in bovine embryos as well (Gutiérrez-Adán et al. 2000). Gutiérrez-Adán et al. (2000) hypothesized that sexually dimorphic gene expression may provide a mechanism by which mammals can control sex ratio; this would occur through sex-specific survival under different levels of oxidative stress. Pérez-Crespo et al. (2005) showed that heat stress induces the production of less hydrogen peroxide in female embryos than it does in males, and this is likely regulated by glucose-6-phosphate dehydrogenase, a gene that is expressed at higher levels in female embryos. When activity of this gene was inhibited, the sex differences in the production of hydrogen peroxide under heat stress disappeared, and male and female levels were similar at the blastocyst stage. Given that stress appears to be a potent regulator of sex ratios in humans (Navara 2010), and stressors of many kinds can induce oxidative stress (Pérez-Crespo et al. 2005), this may be a plausible mechanism by which sex ratio skews may occur in humans.

Susceptibility to oxidative stress is not the only way in which male and female embryos differ. For example, human male embryos have higher cleavage rates as well as higher pyruvate and glucose uptakes and lactic acid production (Ray et al. 1995). These indicate higher metabolic activity in male embryos compared to females. One could see where food limitation could be more detrimental to a male embryo that has higher energy demands, resulting in sex ratio skews. I have shown that the proportion of males at birth is very tightly linked to the amount of weight gain during gestation; at higher gains, a higher proportion of males are born (Fig. 4.3). I further showed that this is likely due to higher rates of fetal deaths for males at 20 weeks of gestation and perhaps before, when weight gains are low, because when I looked at the sexes of fetuses that had died at this time, a higher proportion of them were male when weight gain was low. Perhaps males are more susceptible to an energy limitation at this time, and likely even earlier, in gestation.

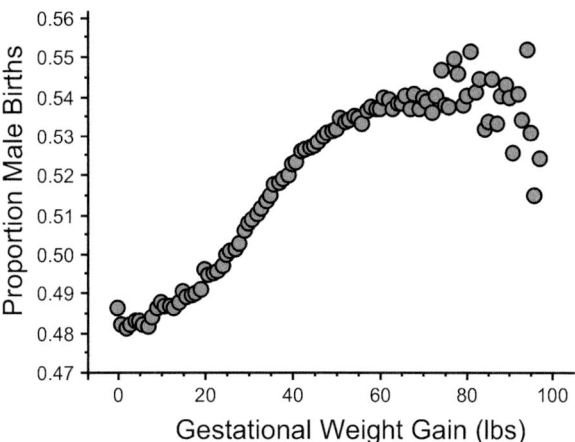

Fig. 4.3 The proportion of male births for women of all races combined in relation to gestational weight gain. Each data point represents the average proportion of male births for a given weight gain calculated using data collected by the Center for Disease Control from 1990 to 2012. Figure reproduced from Navara (2009)

This idea could explain why lower female condition and food availability has been linked with female-biased sex ratios in mammals (Trivers and Willard 1973).

Even more specifically, the amount of glucose available can affect male and female embryos differently. For bovine embryos, high glucose levels can even block female development at the time the blastocoel forms (Larson et al. 2001). This would result in a male-biased sex ratio if glucose levels were high. Work in field voles supports this; females with high concentrations of glucose in their blood produce smaller litters that were male biased (Helle et al. 2008).

Because of the sexual dimorphism in the developmental mechanisms and requirements throughout development, males and females are differentially susceptible to adverse conditions. Many of the influences on human sex ratios, such as that extreme cold temperatures experienced during gestation influence sex ratios produced by Scandinavian women (Catalano et al. 2008), show that this sexual dimorphism can result in sex ratio skews. Even late in gestation, there are reports indicating that high levels of stress can lead to a higher proportion of male stillbirths (Catalano et al. 2006). Thus, there is a good amount of support for the idea that, in humans, sex ratio skews can result from sex-specific embryonic or fetal death.

4.7 When During Development Are Sex Ratio Adjustments Really Happening?

There is evidence that sex ratio adjustments occur in humans and other mammals after exposure to changing environmental and social conditions at all developmental time points. For example, Fig. 4.4 shows studies in which stress influenced sex ratios in humans, and when during development these stressful stimuli were experienced and likely influenced offspring sex. If all human studies documenting sex ratio skews were compiled, it may be possible to pinpoint when during development most of the sex selection occurs. Given that the process of

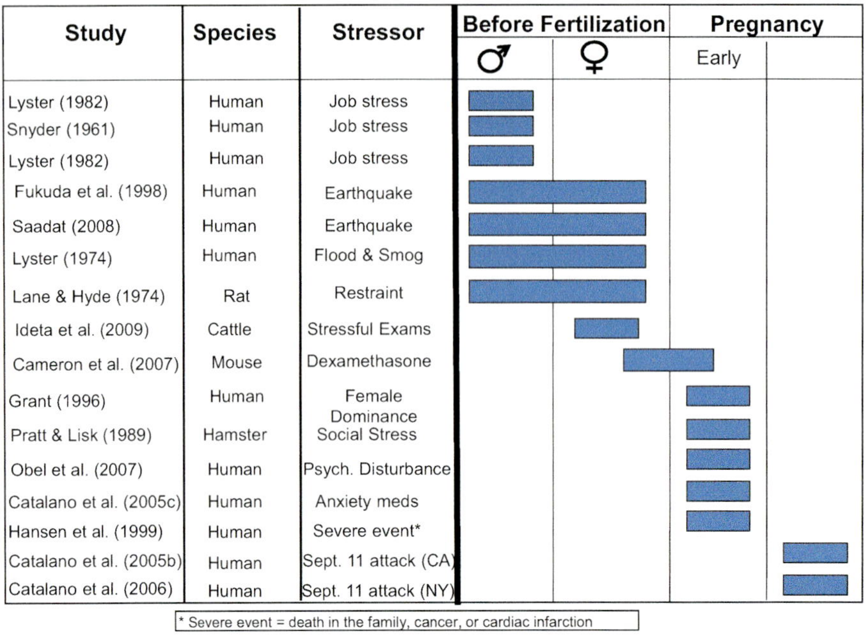

Study	Species	Stressor	Before Fertilization ♂	Before Fertilization ♀	Pregnancy Early	Pregnancy
Lyster (1982)	Human	Job stress	▬			
Snyder (1961)	Human	Job stress	▬			
Lyster (1982)	Human	Job stress	▬			
Fukuda et al. (1998)	Human	Earthquake	▬▬			
Saadat (2008)	Human	Earthquake	▬▬			
Lyster (1974)	Human	Flood & Smog	▬▬			
Lane & Hyde (1974)	Rat	Restraint	▬▬			
Ideta et al. (2009)	Cattle	Stressful Exams		▬		
Cameron et al. (2007)	Mouse	Dexamethasone		▬▬▬		
Grant (1996)	Human	Female Dominance			▬	
Pratt & Lisk (1989)	Hamster	Social Stress			▬	
Obel et al. (2007)	Human	Psych. Disturbance			▬	
Catalano et al. (2005c)	Human	Anxiety meds			▬	
Hansen et al. (1999)	Human	Severe event*			▬	
Catalano et al. (2005b)	Human	Sept. 11 attack (CA)				▬
Catalano et al. (2006)	Human	Sept. 11 attack (NY)				▬

* Severe event = death in the family, cancer, or cardiac infarction

Fig. 4.4 Summary of studies in humans showing significant influences of stress on natal sex ratios, as well as when during development and in which parent the stress was experienced. Reproduced from Navara (2010)

reproduction is especially costly for female mammals, it would be expected that control of sex ratios in humans would lie with women. Indeed, the majority of evidence appears to support the idea that events controlled within the female reproductive tract are responsible for skewing sex ratios, and even in results that show relationships between, for example, the man's occupation and the sex ratio of offspring he produces, there was not a direct link drawn between the sex ratio and the proportion or survival rate of Y-bearing sperm. Thus, it appears that future studies should be directed towards examining whether sex ratios are adjusted after fertilization, or immediately prior as a result of discriminatory mechanisms within the oviduct and uterus.

4.8 Can We Harness These Mechanisms to Artificially Control Offspring Sex?

As we've discussed in Chaps. 2 and 3, there are great needs for technology that would allow us to purposefully skew sex ratios produced by both humans and domestic animals. For as long as we have human records, humans have been trying to find ways to purposefully alter whether a woman produces a girl or boy baby. The ancient Greeks believed that human had two uteruses, one for producing boy babies

4.8 Can We Harness These Mechanisms to Artificially Control Offspring Sex?

Fig. 4.5 Common old wives' tales about how to choose the sex of a baby

and the other for producing girl babies. Aristotle believed that both men and women produced semen and that the baby was determined by which semen won the competition. Shettles and Rorvik (2006) do an excellent job of summarizing the extensive, and often misguided, historical attempts to alter the sexes of babies. It wasn't until the late 1800s when it was determined that sperm and egg fusion resulted in a baby and that sperm cells had one mismatched set of chromosomes that, in 1902, were dubbed the X and Y sex chromosomes (McClung 1902). After this point, more plausible methods of sex ratio manipulation were explored.

Today, many old wives' tales about how to select the sex of a baby remain (Fig. 4.5), but there are also methods that have been designed based on the science of the reproductive process. The Shettles method of sex selection was designed based on the idea, described above, that X- and Y-bearing sperm are functionally different and that Y-bearing sperm are faster but also die faster than X-bearing sperm. Shettles and Rorvik (2006) suggest that optimizing the timing of intercourse can influence the sex of the baby that is produced. They suggest that, to produce a boy, intercourse should be timed as close to ovulation as possible because Y-bearing sperm will swim quickly to the waiting egg and outcompete the X-bearing sperm. To conceive a girl, intercourse should be timed 2–3 days before ovulation and then should be avoided immediately prior to ovulation. Since X-bearing sperm theoretically live longer, they would be better able to hang on during the wait for the egg, outcompeting the Y-bearing sperm and producing a female baby. Shettles and Rorvik (2006) summarize several studies supporting the idea that sperm differ in speed and hardiness and also support studies reporting success using the Shettles method (Benendo 1970; Seguy 1974). However, more recent studies call this method into question. (Guerrero 1974) found that the chances of conceiving a boy were higher when intercourse took place 6 days before ovulation and suggested that his method was superior to the Shettles method because data supporting the Shettles method were based on artificial inseminations, which did not take into account many natural influences. Indeed, through a review of the literature, Zarutskie et al. (1989) concluded that more *females* were produced when intercourse took place closer to ovulation, and some studies showed no relationship between timing of intercourse and the sex of the baby (e.g., Wilcox et al. 1995). Further, the Shettles method does not take into account the timing of changes in the viscosity of cervical mucus, which, as described above, appears to cause a U-shaped influence on sex ratio relative to the timing of intercourse. Elizabeth Whalen Sc.D. developed an alternative method based on the findings of Guerrero (1974) described above. She suggests that, for a boy, intercourse should be timed 5–6 days prior to the rise in basal body temperature that indicates ovulation, while for a girl, intercourse should occur 2–3 days before this temperature rise (Whelan 1977). The efficacy of both of these methods remains controversial.

Still, the observations that Y-bearing sperm are weaker and swim faster have led to some other recommendations about how to select for a particular sex (Carson 1988). For example, adjusting sexual position to allow rear entry allows for deposition of sperm closest to the cervix, minimizing the distance that Y-bearing sperm have to travel. Some suggest that men wear boxers rather than briefs to prevent the degradation of the "more fragile" Y-bearing sperm. Some have recommended that dietary changes can be used to regulate vaginal pH and thus promote the survival of sex containing a particular sex chromosome. This idea is based on work by Stolkowski and Choukroun (1981), showing that the ionic balance, produced by the ratio of sodium and potassium:calcium and magnesium, may influence the sex ratio at conception. The recommendations are to consume high levels of potassium and sodium and low levels of magnesium and calcium to increase the chances that the Y-bearing sperm will fertilize the egg. On the more extreme side, some even suggest that purposefully adjusting the vaginal pH through douching with an alkaline solution may aid in the survival of Y-bearing sperm.

To date, it is unclear whether any of these proposed methods work to help alter the sexes of babies produced. While each of these individual factors may influence the probability that a particular sperm fertilizes the egg, these factors do not act alone, and thus the results are complicated by the interactions of many factors. As a result, for any one method to reliably and repeatedly result in the production of a particular sex, the effect must be extremely strong, and it seems we have not yet found a natural method strong enough to produce such reliable results. However, there are three patented and effective methods by which offspring sex can be determined in mammals (Fig. 4.6). The first was popular as of the 1970s and was developed by Ronald Ericsson (1982) (Fig. 4.6a). This method takes advantage of the differential motility of sperm. Sperm are suspended in an isotonic vehicle solution which is overlaid onto an aqueous fluid that makes migration more difficult. Sperm are allowed to migrate until a significant number of motile sperm have migrated and then the layer containing those sperm is collected. Because Y-bearing sperm move faster, the layer contains approximately 56–60% Y-bearing sperm. This process is repeated 2–3 more times to obtain 70–90% Y-bearing sperm

4.8 Can We Harness These Mechanisms to Artificially Control Offspring Sex?

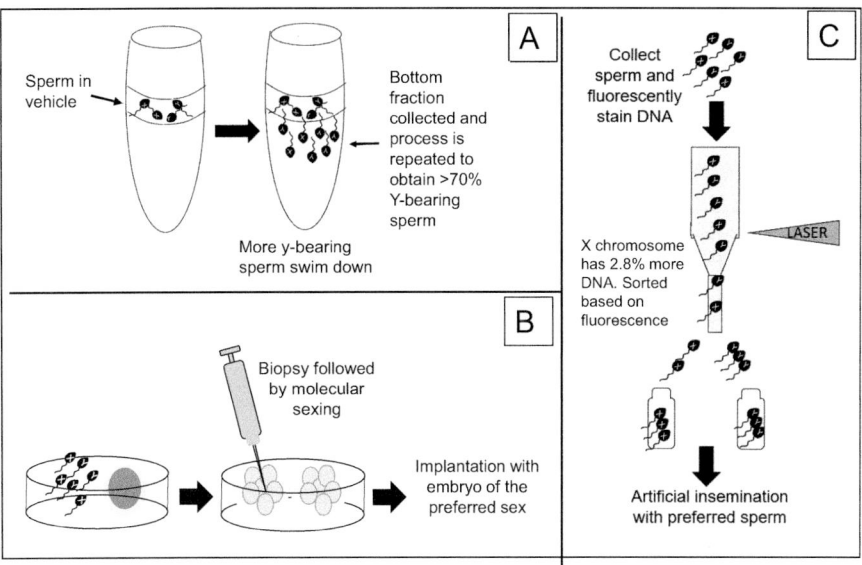

Fig. 4.6 Three common and effective methods of sex selection: (**a**) the Ericsson method using sperm swim methods, (**b**) in vitro fertilization with preimplantation genetic diagnosis, and (**c**) sperm sorting using flow cytometry

that are functional and motile. These can be used during artificial insemination to produce an offspring with a higher likelihood of being male; however, the results cannot be guaranteed.

Two other methods exist with a near-perfect success rate. One of them requires the method of in vitro fertilization (Fig. 4.6b). Briefly, sperm and eggs are collected from the parents and combined in vitro to form embryos. Those embryos are then biopsied, and sex is determined via genetic sexing using a single cell. The embryo of the desired sex is then implanted into the recipient. While this method has a 100% chance of producing the desired sex, the process of in vitro fertilization itself has a much lower success rate (20% or lower depending on the case). In addition, the biopsy itself comes with additional risk.

Finally, the most popular and successful method now used is sperm sorting through the use of flow cytometry (Fig. 4.6c). Sperm collected from the male are stained with a fluorescent DNA marker and are then loaded into a flow cytometer. Inside the flow cytometer, a laser excites the fluorescent label causing emission of light. This light can be quantified and used to sort cells with different DNA content. This works for sorting sperm because X-bearing sperm contain 2.8% more DNA than Y-bearing sperm, thus producing a different emission pattern that can be used to sort the sperm by the sex chromosomes they carry. Sperm carrying the preferred sex chromosome are then used for artificial insemination. For humans, a company called Microsort© is the leader in this technology, though this method is also routinely used for domestic animals.

Preselecting offspring of agricultural species has been a goal of animal breeders for quite some time because obtaining offspring of a preferred sex could substantially increase profits in agricultural industries. In addition, as discussed in Chap. 2, technologies that allow humans to select a baby of a particular sex could prevent the practice of sex-biased abortion that is a known practice in some Asian cultures. However, the methods currently available for humans are still quite expensive and require additional development before they will be commercially available to the mainstream. In addition, this technology now generates questions in the moral realm. Should we have the power to choose the sexes of our babies? What happens if this technology causes artificial biases in the sex ratios of our populations? As these technologies become more widely used, these questions will need to be discussed.

References

Aitken RJ, Krausz C (2001) Oxidative stress, DNA damage and the Y chromosome. Reproduction 122(4):497–506

Almiñana C, Caballero I, Heath PR, Maleki-Dizaji S, Parrilla I, Cuello C, Gil MA, Vazquez JL, Vazquez JM, Roca J (2014) The battle of the sexes starts in the oviduct: modulation of oviductal transcriptome by X and Y-bearing spermatozoa. BMC Genomics 15(1):1

Benendo F (1970) The problem of sex determination in the light of personal observations. Polish Endocrinol 21(200):1954

Carson SA (1988) Sex selection: the ultimate in family planning. Fertil Steril 50(1):16

Catalano R, Bruckner T, Smith KR (2008) Ambient temperature predicts sex ratios and male longevity. Proc Natl Acad Sci USA 105(6):2244–2247

Catalano R, Bruckner T, Marks AR, Eskenazi B (2006) Exogenous shocks to the human sex ratio: the case of September 11, 2001 in New York City. Hum Reprod 21(12):3127–3131

Chaudhury I, Jain M, Halder A (2014) Sperm sex ratio (X: Y Ratio) and its variations. Austin J Reprod Med Infertil 1(1):7

Cui K-H (1997) Size differences between human X and Y spermatozoa and prefertilization diagnosis. Mol Hum Reprod 3(1):61–67

Diasio RB, Glass R (1971) Effects of pH on the migration of X and Y sperm. Fertil Steril 22(5):303–305

Ericsson RJ (1982) Method of increasing the incidence of female offspring. Google Patents

Ericsson RJ (1994) Sex selection: sex selection via albumin columns: 20 years of results. Hum Reprod 9(10):1787–1788

Flint APF, Albon SD, Jafar SI (1997) Blastocyst development and conceptus sex selection in red deer Cervus elaphus: studies of a free-living population on the Isle of Rum. Gen Comp Endocrinol 106(3):374–383

Fox C, Meldrum S, Watson B (1973) Continuous measurement by radio-telemetry of vaginal pH during human coitus. J Reprod Fertil 33(1):69–75

García-Herreros M, Bermejo-Álvarez P, Rizos D, Gutiérrez-Adán A, Fahey AG, Lonergan P (2010) Intrafollicular testosterone concentration and sex ratio in individually cultured bovine embryos. Reprod Fertil Dev 22(3):533–538

Grant VJ, Chamley LW (2010) Can mammalian mothers influence the sex of their offspring periconceptually? Reproduction 140(3):425–433

Grant VJ, Irwin R (2005) Follicular fluid steroid levels and subsequent sex of bovine embryos. J Exp Zool A Comp Exp Biol 303(12):1120–1125

Graves JAM (1995) The origin and function of the mammalian Y chromosome and Y-borne genes–an evolving understanding. Bioessays 17(4):311–320

Guerrero R (1974) Association of the type and time of insemination within the menstrual cycle with the human sex ratio at birth. N Engl J Med 291(20):1056–1059

Guerrero R (1975) Type and time of insemination within the menstrual cycle and the human sex ratio at birth. Stud Fam Plan 6(10):367–371

Gutiérrez-Adán A, Oter M, Martínez-Madrid B, Pintado B, De La Fuente J (2000) Differential expression of two genes located on the X chromosome between male and female in vitro–produced bovine embryos at the blastocyst stage. Mol Reprod Dev 55(2):146–151

Gutiérrez-Adán A, Perez-Crespo M, Fernandez-Gonzalez R, Ramirez M, Moreira P, Pintado B, Lonergan P, Rizos D (2006) Developmental consequences of sexual dimorphism during pre-implantation embryonic development. Reprod Domest Anim 41(s2):54–62

Helle S, Laaksonen T, Adamsson A, Paranko J, Huitu O (2008) Female field voles with high testosterone and glucose levels produce male-biased litters. Anim Behav 75(3):1031–1039

Holt W, Fazeli A (2016) Sperm selection in the female mammalian reproductive tract. Focus on the oviduct: Hypotheses, mechanisms, and new opportunities. Theriogenology 85(1):105–112

Ideta A, Hayama K, Kawashima C, Urakawa M, Miyamoto A, Aoyagi Y (2009) Subjecting holstein heifers to stress during the follicular phase following superovulatory treatment may increase the female sex ratio of embryos. J Reprod Dev 55(5):529–533

Ishijima S, Okuno M, Mohri H (1991) Zeta potential of human X-and Y-bearing sperm. Int J Androl 14(5):340–347

James WH (2000) Analysing data on the sex ratio of human births by cycle day of conception. Hum Reprod 15(5):1206–1208

Jongbloet P (2003) The male disadvantage and the seasonal rhythm of sex ratio at the time of conception. Hum Reprod 18(11):2491–2492

Kobayashi S, Isotani A, Mise N, Yamamoto M, Fujihara Y, Kaseda K, Nakanishi T, Ikawa M, Hamada H, Abe K (2006) Comparison of gene expression in male and female mouse blastocysts revealed imprinting of the X-linked gene, Rhox5/Pem, at preimplantation stages. Curr Biol 16(2):166–172

Larson MA, Kimura K, Kubisch HM, Roberts RM (2001) Sexual dimorphism among bovine embryos in their ability to make the transition to expanded blastocyst and in the expression of the signaling molecule IFN-τ. Proc Natl Acad Sci USA 98(17):9677–9682

Macaulay AD, Hamilton CK, King WA, Bartlewski PM (2013) Influence of physiological concentrations of androgens on the developmental competence and sex ratio of in vitro produced bovine embryos. Reprod Biol 13(1):41–50

Martin JF (1997) Length of the follicular phase, time of insemination, coital rate and the sex of offspring. Hum Reprod 12(3):611–616

Masters W (1960) Influence of male ejaculate on vaginal acidity. In: Endocrine dysfunction and infertility, pp 76–78

McClung CE (1902) The accessory chromosome—sex determinant? Biol Bull 3(1–2):43–84

Moein-Vaziri N, Phillips I, Smith S, Almiñana C, Maside C, Gil MA, Roca J, Martinez EA, Holt WV, Pockley AG (2014) Heat-shock protein A8 restores sperm membrane integrity by increasing plasma membrane fluidity. Reproduction 147(5):719–732

Moghissi KS (1966) Cyclic changes of cervical mucus in normal and progestin-treated women. Fertil Steril 17(5):663–675

Muschat M (1926) The effect of variation of hydrogen-ion concentration on the motility of human spermatozoa. Surg Gynecol Obstet 42:778–781

Navara KJ (2009) Humans at tropical latitudes produce more females. Biol Lett 5(4):524–527

Navara KJ (2010) Programming of offspring sex ratios by maternal stress in humans: assessment of physiological mechanisms using a comparative approach. J Comp Physiol B 180(6):785–796

Orzack SH, Stubblefield JW, Akmaev VR, Colls P, Munné S, Scholl T, Steinsaltz D, Zuckerman JE (2015) The human sex ratio from conception to birth. Proc Natl Acad Sci USA 112(16):E2102–E2111

Pérez-Crespo M, Ramirez M, Fernández-González R, Rizos D, Lonergan P, Pintado B, Gutiérrez-Adán A (2005) Differential sensitivity of male and female mouse embryos to oxidative induced heat-stress is mediated by glucose-6-phosphate dehydrogenase gene expression. Mol Reprod Dev 72(4):502–510

Pyrzak R (1994) Sex selection: separation of X-and Y-bearing human spermatozoa using albumin gradients. Hum Reprod 9(10):1788–1790

Ray P, Conaghan J, Winston R, Handyside A (1995) Increased number of cells and metabolic activity in male human preimplantation embryos following in vitro fertilization. J Reprod Fertil 104(1):165–171

Robbins WA, Wei F, Elashoff DA, Wu G, Xun L, Jia J (2008) Y: X sperm ratio in boron-exposed men. J Androl 29(1):115–121

Rohde W, Porstmann T, Doerner G (1973) Migration of Y-bearing human spermatozoa in cervical mucus. J Reprod Fertil 33(1):167–169

Ross MT, Grafham DV, Coffey AJ, Scherer S, McLay K, Muzny D, Platzer M, Howell GR, Burrows C, Bird CP (2005) The DNA sequence of the human X chromosome. Nature 434(7031):325–337

Saragusty J, Hermes R, Hofer H, Bouts T, Göritz F, Hildebrandt TB (2012) Male pygmy hippopotamus influence offspring sex ratio. Nat Commun 3:697

Seguy B (1974) [Methods of natural and voluntary sex selection. Value for the prevention of sex-linked malformations and of certain recurrent abortions]. J Gynecol Obstet Biol Reprod 4(1):145–149

Shettles LB (1960a) Human spermatozoa shape in relation to sex ratios. Fertil Steril 12:502–508

Shettles LB (1960b) Nuclear morphology of human spermatozoa. Obstet Gynecol 16(1):10

Shettles LB, Rorvik DM (2006) How to choose the sex of your baby: the method best supported by scientific evidence. Harmony Books, New York

Stolkowski J, Choukroun J (1981) Preconception selection of sex in man. Isr J Med Sci 17(11):1061–1067

Tiido T, Rignell-Hydbom A, Jönsson B, Giwercman YL, Rylander L, Hagmar L, Giwercman A (2005) Exposure to persistent organochlorine pollutants associates with human sperm Y: X chromosome ratio. Hum Reprod 20(7):1903–1909

Trivers RL, Willard DE (1973) Natural selection of parental ability to vary the sex ratio of offspring. Science 179(4068):90–92

Unterberger F (1930) Das problem der willkürlichen Beeinflussung des Geschlechts beim Menschen. Dtsch Med Wochenschr 56(08):304–307

Whelan EM (1977) Boy or girl? The sex selection technique that makes all others obsolete. Bobbs-Merrill, New York

Wilcox AJ, Weinberg CR, Baird DD (1995) Timing of sexual intercourse in relation to ovulation—effects on the probability of conception, survival of the pregnancy, and sex of the baby. N Engl J Med 333(23):1517–1521

Zarutskie PW, Muller CH, Magone M, Soules MR (1989) The clinical relevance of sex selection techniques. Fertil Steril 52(6):891–905

The Bees Do It, but What About the Birds? Evidence for Sex Ratio Adjustment in Birds

5

> *At our zoo, to adjust sex ratios, we must currently perform in ovo sexing and avoid incubating eggs of the undesired sex, since we do not euthanize birds post hatch for management. This solution is not ideal if the species is threatened/ endangered or genetically valuable*
> Tom Jensen, Senior Scientist at the San Diego Zoo

In the egg laying industry, more than 200 million chicks are killed each year because they are the "wrong sex" and are not profitable. After all, only females can lay eggs. If there was a way to purposefully skew chicken sex ratios, millions of dollars and hundreds of millions of animal lives would be saved each year. Luckily, nature may have actually solved this problem for us, because it appears that birds have an inherent ability to skew the sex ratios of the offspring they produce, and these skews occur prior to hatch. Since the early 1900s, researchers have documented skews in the offspring sex ratios produced by wild birds in response to changing environmental and social variables. If we focus on primary sex ratios alone, there have been at least 130 studies since the 1970s showing evidence for facultative adjustment of offspring sex ratios. Not only are these results exciting in an adaptive context, but they indicate that there may be mechanism that can be harnessed for use, not only in the poultry industry but also in conservation efforts.

In 1997, Donald (2007) did an extensive assessment of adult sex ratios in avian species designated within four threat categories: not globally threatened, near threatened or vulnerable, endangered, or critically endangered. He found that as threat level increased, the adult sex ratio became more and more male-biased. He concluded that this was the result of higher levels of female mortality rather than the initial production of more males; however, this highlights the need for an understanding of whether birds can adjust sex ratios purposefully and whether we can harness this potential for use in conservation efforts.

5.1 Case Study: Extreme Sex Ratio Biases in Eclectus Parrots

Eclectus parrots are native to Australia and its surrounding islands, though you'd more likely find them these days in a pet store. They are coveted due to their remarkably extreme sexual dimorphism in plumage coloration; males are bright emerald green while females are bright red with some purple/blue feathers around the wings and on the chest and back of the neck. However, they are also remarkable birds in terms of sex ratio adjustment as well (Fig. 5.1). Using breeding records from aviculturists in five Australian states, Heinsohn et al. (1997) examined sex ratios of 209 fledglings produced by 12 females. What they found was that females would produce long strings of the same sex before switching to the other sex. One female produced 20 males before switching and producing 13 females. Another female produced 11 males before switching and producing 9 females. It wasn't always males that were produced first; one female produced 11 female fledglings before she produced a male. While the overall sex ratio produced was an even 46%, these long runs of each sex defied expectations of the sex combinations that would have been produced based on random chance. Further, in the 41 cases where there were two successful young in the broods, 18 were female–female pairs, 17 were male–male pairs, and only 6 were male–female. Using a simulation model, the authors determined that even if sex-specific nestling mortality were occurring, the observed patterns of sexes are still extremely improbably if the primary sex ratio is 50:50. This strongly supports the idea that eclectus parrots can bias the sexes of offspring at fertilization. The reasons behind the long strings of particular sexes remain unknown; however, this work highlights the ability of an avian species to modulate the sexes of offspring produced both within and among clutches.

Bird ID	Sexes of Offspring
1	f,f,f,f,f,f,f,f,f,f,f,m,f,ff,f,f,ff,m,f,m,f,m,f,f,f,f,ff,m,m,m,m,f,f,f,ff
2	m,m,mm,m,mm,m,mm,m,m,mm,m,m,m,m,m,m,f,f,f,f,f,f,f,f,f,f,f,ff
3	m,mm,mm,m,mm,m,m,m,f,f,f,f,f,f,f,f,f
4	ff,ff,f,ff,mf,mf,m,m,m,m,mm,ff,ff,ff
5	ff,f,f,mf,mm,mm,m,m,mm,mm,m,mm,mm

Fig. 5.1 Sexes of offspring produced by five individual eclectus parrot pairs. Figure adapted from Heinsohn et al. (1997)

5.2 Can Birds Facultatively Adjust Offspring Sex Ratios?

To date, there is clear evidence that primary sex ratios often differ from an expected binomial distribution. However, whether birds have the ability to *facultatively* bias sex ratios to maximize fitness remains controversial. The difficulty of answering such a question lies in the fact that the ecological dynamics of avian species varies so widely, as do the potential motivational forces that would drive facultative sex allocation and the variety of possible mechanisms underlying the sex ratio variation. In their review focusing on the complexities of adaptive sex allocation in birds, Komdeur and Pen (2002) point out three main problems associated with applying classical theories of sex allocation, most of which were developed for invertebrates, to avian systems. (1) Classic sex ratio theory assumes that sex allocation comes at no cost to the parents. In avian systems, there are clear costs associated with secondary sex ratio manipulation, as it involves death of offspring after a significant level of investment in egg production, incubation, and feeding of offspring has already been made. At the level of primary adjustment, the potential costs are less clear, and would be determined by the precise mechanism responsible for the adjustment. (2) Classic theory assumes that generations do not overlap and that there is a fixed amount of resources available for reproduction. In reality, birds produce multiple offspring per clutch, multiple clutches per season, and usually reproduce over multiple seasons, resulting in complex dynamics involving the costs and benefits of producing a male or female, as well as the influences of coexisting with those offspring in the future. (3) Classic theory assumes a short burst of uniparental care, while birds often exhibit biparental care, and at varying levels. Predicting which sex would provide the greatest fitness benefits becomes much more difficult, and predictions may even differ between the sexes within a single pair of birds.

Still, a variety of avian studies available have attempted to examine variation in offspring sex ratios against the framework of these classic sex allocation theories, and this has led to a confusing compilation of results that do not allow us to draw a final conclusion about whether facultative sex allocation occurs in *any* species. Because of the diversity of life histories in birds, the answer to this question will likely never apply across all birds and must be assessed at the species level. In addition, to fully understand how sex allocation is functioning in an adaptive context in each individual bird species, we must first answer the following questions:

1. Which sex is more costly? This will depend on if there is sexual dimorphism in offspring sizes and growth rates, and which sex is more likely to remain in the environment and become competition for limited resources.
2. Which sex is more likely to be beneficial? This will depend on whether either sex helps in future reproductive attempts and provides an adaptive benefit in doing so, which sex is more likely to gain access to mates when resources are limited, whether traits associated with attractiveness are heritable, and whether one sex is more likely to reproduce when produced early or late in the season, and early or late in the brood.

3. What mating system do the birds exhibit, and in the case of monogamy, what is the rate of extra-pair copulations? The amount of time and energy the male devotes to a partner and care of his offspring can determine how beneficial it is to produce a male in a given environment.
4. What is the level of intra-brood competition? This will depend on whether offspring hatch synchronously, brood sizes in relation to the availability of resources, and whether siblicide is a common behavior exhibited by the species in general.
5. What are the details of the breeding environment? The results of studies conducted in the same species often differ, and this could be because those studies are done in different locations where environmental and social pressures are different, or during different years with differing weather patterns, social dynamics, and resource availability.

To my knowledge, for no species has all of this information been compiled. However, given the breadth of information available on the potential drivers of sex allocation, it may be possible to draw some conclusions regarding the potential for adaptive sex allocation to occur. Here, I discuss results of studies that examine seven potential drivers of sex allocation: dispersal and philopatry, the presence/absence of helpers, male quality/attractiveness, female quality/condition, food availability, seasonal effects, and variation across the laying order.

5.3 Case Study: Seasonal Variation in Offspring Sex Ratios in American Kestrels

While limits of time, funding, and sample sizes often don't allow for testing of more than one sex allocation theory at a time, instances where sex ratio adjustment has been tested in the same species in relation to a range of variables can be invaluable to our understanding of what factors might influence patterns of sex allocation. The large body of work done on the American Kestrel is a perfect example (Fig. 5.2). American Kestrels, like many other raptors, exhibit reversed-sexual dimorphism where the female is the larger sex. Interestingly, the eggs that produce male kestrels are larger than those that produce females, and it has been suggested that this may be a way for kestrel mother to help control the competitive dynamic in the nest (Anderson et al. 1997). This, however, would suggest that mothers either have a way of knowing which follicles will produce male and female offspring to stimulate production of a larger egg, or the production of a larger egg stimulates the production of a male embryo. These birds also exhibit hatching asynchrony, which results in a size hierarchy within the brood. There is no evidence, however, that females adjust sex ratios of nestlings according to laying order in this species (Sockman and Schwabl 2000).

The most interesting part of the sex ratio story in American Kestrels comes in relation to parental condition and seasonal progression. Wiebe and Bortolotti (1992) examined sex ratios in a population of kestrels in Canada and showed that

5.3 Case Study: Seasonal Variation in Offspring Sex Ratios in American Kestrels

Fig. 5.2 A summary of studies examining sex ratios in American Kestrels in relation to a variety of cues

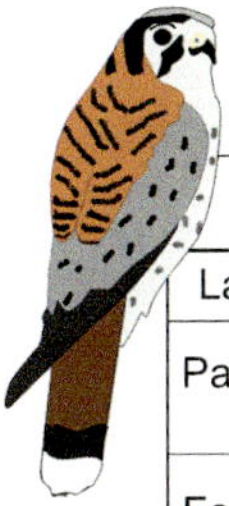

Cue	Effect on Sex Ratio
Laying Order	None
Parental Condition	Better condition of both parents → more females
Food availability	As food availability declines → more males
Seasonal Progression	Canada – no effect. Wisconsin & Florida – more males as season progresses

parents in better condition produced more female offspring, and that the proportion of males increased as the amount of food available declined. It is unclear whether this effect was driven by the effects of food availability on female condition, or whether it was a reaction of the female to a low condition male; however, further analysis showed that if pre-laying condition of males decreased, clutches sired by those males were more likely to be female-biased. Females all increased in their pre-laying condition and produce a mix of male and female-biased clutches. Thus, it appears likely that male condition may play a bigger role than the condition of the female herself. More testing is needed to truly flush this out, though.

The story becomes even more complex when comparing seasonal patterns of sex ratio adjustment among populations. Wiebe and Bortolotti (1992) found no evidence of a seasonal pattern in offspring sex ratios in their Canadian population. However, in a Wisconsin (Griggio et al. 2002) and a Florida (Smallwood 1998) population, the proportion of male offspring produced decreased as the season progressed. If sex ratios were directly linked to maternal condition, we would expect to see an increase in the proportion of males across the season (especially given that female condition was negatively related to the proportion of male offspring), because while female condition across the season has not to my knowledge been documented in American Kestrels, female condition generally decreases across the season in most species studied. Perhaps sex ratios are actually driven by female characteristics.

Alternatively, while Wiebe and Bortolotti (1992) posited that their patterns of sex ratios in kestrels follow the Trivers–Willard hypothesis of sex allocation (Trivers and Willard 1973), Smallwood (1998) suggested that sex allocation in the Florida population may be driven according to a different adaptive hypothesis called the "Early Bird hypothesis," whereby, in nonmigratory species where nest sites are limited, males fledged early in the season are at a competitive advantage to those fledged later. In this scenario, it would be expected that more males would be produced early in the season, as was seen in the nonmigratory Florida population, but there would be no seasonal trend in a migratory population where competition for nest sites does not begin until the following breeding season. This would explain the lack of a seasonal trend in the migratory Canadian population.

But what about the seasonal decrease in the proportion of males observed in the migratory Wisconsin population? Griggio et al. (2002) suggest that this may result from the fact that the Wisconsin population is on the geographical limit between the migratory and nonmigratory populations, and point out that the variation in the sex ratios of the Wisconsin population were much lower (41–53% males) compared to the Florida population (36–58% males). This could indicate a gradual latitudinal change in the pattern. However, it could also relate to the competitive pressures related to nest site limitation. In the Florida and Wisconsin populations, nest sites were far more limited than in the Canadian population. This still wouldn't explain the significant pattern in the migratory Wisconsin population, though. Perhaps instead, there is a hybrid explanation for the patterns seen in American Kestrels. Total circulating protein content, a measure of condition that has been shown to be positively correlated with the classic measure of condition (the residual of size divided my mass), actually increases in males during incubation (Dawson and Bortolotti 1997). It is unclear whether males are able to continue increasing in condition as the season progresses, particularly since he is the chief provider for the offspring (Wiehn 1997). However, our understanding of the adaptive significance of sex ratio adjustment in this species would benefit from measurements of male condition across the breeding season and correlations of those condition measures with offspring sex ratios produced by those males.

Despite the complexity of this system and the sex ratio patterns observed here, the observations of multiple populations, and assessment of several potential driving forces generate more substantial speculation about whether and how females are adaptively allocating offspring sex in this species compared to other species in which sex ratios are regressed against only one or two parameters. These studies in kestrels highlight the variety of variables that could influence sex allocation. These will be addressed individually below.

5.4 Seasonal Variation in Sex Ratios of Other Avian Species

American Kestrels join a group of raptor species that alter sex ratios with the progression of the breeding season. The direction of these skews, however, varies among raptors. For example, goshawks, marsh harriers, and sparrowhawks increase the proportion of male offspring as the season progresses, whereas lesser kestrels and European kestrels decrease the proportion of males across the season [reviewed in Daan et al. (1996)], much like American Kestrels did (Griggio et al. 2002; Smallwood 1998). Daan et al. (1996) proposed a new explanation for this variation in seasonal trends; they suggested that the direction of sex ratio adjustment across the season varies based sex-specific differences in maturation time, and that the sex whose maturation time is most significantly reduced by an early hatch should be produced at a higher proportion early in the season. In European kestrels, for example, males hatched early in the season have a higher probability of breeding as yearlings (Dijkstra et al. 1990), which could explain why female kestrels produce a higher proportion of males earlier in the season. For marsh harriers, on the other hand, males do not generally generate adult plumage in time to breed, whereas females have been known to breed as yearlings. Their seasonal pattern of sex ratio adjustment complies with the predictions of Daan et al. (1996); they produce a higher proportion of *females* early in the season.

While this hypothetical model has been applied only to raptor species, seasonal trends in offspring sex ratios are certainly not restricted to raptors. Spotless starlings, for example, increase the proportion of male offspring as the season advances (Cordero et al. 2001). This pattern also complies with the model described above, as male spotless starlings rarely breed as yearlings while females do have the potential to breed that first year if hatched early enough. European shags also seem to follow this pattern. Male shags hatched early in the season are more likely to be successful reproductively, and these birds produce a higher proportion of male offspring early in the season (Velando et al. 2002). Lincoln's sparrows showed a dramatic decrease in the proportion of males produced, from 80 to 40% males over the short 19 days season. Given the short duration of the season, the maturation hypothesis, as it is described in raptors, does not likely apply here. However, the authors suggest that the decrease in the proportion of males across the season in Lincoln's sparrows may be because faster bill development later in the season results in suboptimal bills for singing, leading females to produce fewer males at this time (Graham et al. 2011). Thus, the idea that sex ratios are designed to maximize the reproductive potential of offspring still applies.

However, while a majority of raptors studied exhibit seasonal variation in sex ratios, work in passerines is more mixed. Despite the seasonal skews documented in the passerine species discussed above, dark-eyed juncos, red-winged blackbirds, western bluebirds, yellowhammers, white-throated magpie-jays, and wild zebra finches showed no seasonal trend in the sex ratios of their offspring. Barn swallows and common grackles both change the proportion of males produced across the season (in barn swallows it increases and in grackles it decreases), but for both, this pattern only emerges for large clutches. Results of studies in great tits and house sparrows are mixed; some show seasonal variation in sex ratios while others don't. The directions of the patterns, when present, are equally variable. In three species, increases in the proportion of male offspring were observed, while in four species, decreases were observed. A study following up on the maturation model proposed by Daan et al. (1996) in passerines may shed light onto whether birds in this family are exercising similar strategies of sex allocation when they bias sex ratios across the season.

If the maturation model of sex ratio adjustment is, in fact, the main driver of seasonal sex ratio adjustment, perhaps it is not surprising that gull species, which generally take longer to mature, show little to no variation in sex ratios across the season. Studies conducted in black-headed gulls, lesser black-backed gulls, western gulls, and yellow-legged gulls have failed to show variation in sex ratios across the season. In fact, only two species of gulls have shown signs of seasonal variation in sex ratios; Ryder (1983) showed some evidence that female ring-billed gulls that started mating particularly early in the season during a year of high food availability produced more males early in the season, while females mating late in a year of lower food availability produced more females. However, the statistical analysis of sex ratios with seasons was not shown in this study. Genovart et al. (2003) showed a seasonal pattern of sex ratios in Audouin's gull, but it was U-shaped; more males were produced at the middle of the season.

Overall, it is still unclear why some species vary the sex ratios of their offspring across the season while others don't. The maturation hypothesis has much potential towards explaining the adaptive significance of seasonal patterns seen in raptors and some passerines. An updated meta-analysis examining how life history traits may contribute to these seasonal patterns and how the patterns fit into each of the suggested adaptive hypotheses may help to disentangle the complexities of seasonal sex ratio adjustment.

5.5 Variation in Sex Ratios Across the Laying Order

Some of the confusion regarding patterns of sex allocation across the season and in response to other variables may result from sex allocation strategies that are specific to a particular egg in the laying sequence. For example, in marsh harriers, there was a seasonal trend in the proportion of male offspring, but this was most pronounced when looking at first and second-hatched offspring in the brood (Zijlstra et al. 1992). The trend described above for European shags, where more males were produced early in the season, was completely driven by eggs in the first clutch position; in early broods, 77% of first-hatched chicks were male compared with 30% in late broods (Velando 2002). For crimson rosellas, sexes of first eggs followed a seasonal pattern while sexes of middle and late eggs did not (Krebs et al. 2002). Variation within the clutch may cloud interpretation of sex allocation in response to other cues as well. For example, when bald eagles experience a year of low food availability, sex ratios are male-biased, but only in the first clutch position (Dzus et al. 1996). So, in some cases, if clutch positions are not examined separately, significant trends may be missed. Using a simulation model, Rosivall and colleagues (2008) show that this is indeed the case. The detection of sex ratio manipulation at the clutch level is extremely difficult if the target of manipulation is only the first egg in the sequence, because random segregation of sex chromosomes in later eggs can mask the maternal manipulation in the first. This would result in conflicting results much like we are currently seeing in how sex ratios respond to many variables. It is important in future studies to examine how the sex of each egg in the sequence responds to the cue being examined to rule out the possibility of Type II statistical error.

But what about laying order as its own cue leading to sex allocation? There are some key reasons that it may benefit females to vary sex ratios of offspring across the laying sequence. For example, if males and females are not equally affected by competition within the nest, or if one sex is better at competing than the other, then the sex composition of the brood can significantly influence the overall

5.5 Variation in Sex Ratios Across the Laying Order

reproductive output (Bortolotti 1986b). As a result, parents may manipulate the sex ratio within the brood to minimize the potential for offspring mortality that may result from competition over resources in the nest or siblicide. This may be particularly important in species that exhibit hatching asynchrony, where incubation begins before the last egg is laid. This can result in a dramatic size hierarchy among offspring, where the last-hatched chicks are significantly smaller and unable to compete with larger siblings in the nest. Bald eagles are a good example of a species in which this occurs (Bortolotti 1986a). Female bald eagles are larger than males, and eggs within the brood hatch asynchronously, often resulting in brood reduction. Bortolotti (1986b) examined sex ratios of eggs within bald eagle broods and found that first eggs were 63% female while second eggs were 31% female, so bald eagles produce more females early in the brood. The author hypothesized that this was because males are larger and better able to compete, which would further stimulate brood reduction if the size effects of hatching asynchrony compounded the already smaller sizes of females. Indeed, when broods contained a first hatched male and a second-hatched female, there was a much higher rate of brood reduction than with other hatching combinations. This suggests that allocation of sexes to a particular place in the laying sequence may be adaptive. Interestingly, however, Dzus et al. (1996) showed that when food availability was low, bald eagles reversed the pattern of sex allocation across the laying order, producing more males in the first clutch position instead. This suggests that laying order can interact with other factors to influence sex ratios.

House finches are a particularly applicable species for testing how multiple factors, including hatching asynchrony and environmental factors, drive adaptive sex allocation because there are two recently established populations in which selection on juvenile morphology and thus growth rates of different sexes across the laying order have been demonstrated to be very different. In the Montana population of house finches, the growth rate increases with hatch order for females, but decreases with hatch order for males, while in the Alabama population, growth rates decrease with hatch order for females and are highest in the middle of the hatching order for males. Thus, we would expect that sex allocation patterns across the laying order would also differ, and in fact, Baydaev and colleagues showed that they do (Badyaev et al. 2002). In Montana, first-laid eggs are mostly female and last-laid eggs are mostly male, while in Alabama, this pattern is reversed. Further, in both populations, survival often depended on whether the correct sex hatched in the correct position. For example, males hatching in male-biased hatching positions had a better chance of survival than males hatching in female-biased hatching positions. This suggests that sex allocation across the laying order may be adaptive in this species as well.

It seems that sometimes, birds can use patterns of sex allocation across the laying sequence to adjust sex ratios in a larger context. European kestrels produce a higher proportion of males early in the sequence, but only early in the breeding season. Later in the breeding season, chicks hatching from early clutch positions are more likely to be female (Dijkstra et al. 1990). In general, resource availability is lower for these birds later in the breeding season, and late broods suffer higher rates of mortality. Late-hatched males that do survive are generally unable to breed during the first year. Thus, males become less profitable as the breeding season progresses (which complies with the maturation hypothesis described above). The authors suggest that the method by which European kestrels may alter which sex is produced early and late in the season is by allocating sexes in a particular way across the laying sequence, because follicles ovulated early in the sequence are more likely to produce chicks that hatch and survive compared to those ovulated late in the season. So, ovulating male eggs early in the sequence allows more males to survive at a time when they will be better able to reproduce during the first year.

5.6 Case Study: Male Attractiveness and Sex Allocation in Blue Tits

It is also possible that seasonal variation in sex ratios may be related to the quality of the pair-male. For example, in species where males compete for mates and/or territories, the best quality males secure their mates earlier in the season. If the traits leading to male success are heritable, females mated to good quality males should produce more sons (Male Attractiveness hypothesis—see below for more detailed discussion). As a result, seasonal variation may only be a by-product of adjustments based on male quality.

Since the mid-1990s, blue tits have been particularly popular study animals for testing the hypothesis that offspring sex ratios relate to measures of male quality (Fig. 5.3). There is a large body of literature on these species because they are among the first avian species in which the UV aspect of their plumage color has been assessed in a sexual selection context. It was thought that they were nearly

Study	Result
Svensson & Nilsson (1996)	Males that produced more offspring more likely to survive the following winter
Griffith et al. (2003)	Positive correlation between male UV chroma and proportion of males produced
Sheldon et al. (1999)	Controls: positive correlation b/t UV chroma and proportion male offspring Blocking UV reversed this relationship
Korsten et al. (2006)	No effect of UV blocking on sex ratios
Delhey et al. 2007)	Reduced UV within physiological range. No effect on sex ratios

Fig. 5.3 A summary of studies examining sex ratios in relation to UV coloration in blue tits

5.6 Case Study: Male Attractiveness and Sex Allocation in Blue Tits

monomorphic in coloration, until it was discovered that the UV reflectance of male crown feathers was dimorphic (Hunt et al. 1998). It is now known the females choose mates based on the UV aspect of their crown feathers and that males and females likely even mate assortatively based on this characteristic (Hunt et al. 1999). In 1996, Svensson and Nilsson (1996) showed that males producing a higher proportion of male offspring were more likely to survive the following winter, indicating that these males were of better quality. Given emerging links between condition and UV reflectance of the crown feathers, several researchers set out to test whether sex ratios were related to UV reflectance in crown feathers of the male. Griffith et al. (2003) found the clearest effect in a population of blue tits on the island of Gotland in Scotland. There was a strong positive correlation between the UV chroma of the male's crown feathers and the proportion of males he and his mate produced in 2 of the 3 years studied, and this correlation held when all 3 years were analyzed together. Still, these results show that there can be substantial variation in these effects based on the year of study, likely caused by unknown environmental and social variables that differed between years.

Using a population in the same area, Sheldon et al. (1999) experimentally reduced UV coloration of the crown and found that in control birds, UV reflectance was indeed positively related to both the proportion of males produced, and masking the UV reflectance reversed this relationship. They did not, however, show whether there were overall treatment differences in the proportion of males produced between the two treatment groups, and it is unclear why the relationship between UV color and sex ratios would reverse after sunblock was applied. Korsten et al. (2006) conducted a similar experimental study in the Netherlands and this time did report on overall differences between the treatment groups; the proportion of males produced by birds whose UV coloration was blocked was not different from controls. However, here the story gets more complicated. In one of the 2 years during which the study was conducted, there was an interaction between the pretreatment UV chroma of the crown feathers and the UV treatment; there was a positive relationship between the *pre*-manipulation UV chroma and proportion of sons produced by the control group, the same relationship seen by Sheldon et al. (1999), but no such relationship was present in the UV-blocked group. When trying to interpret this result, it is important to remember that the clutch was produced after the females had the chance to assess both the pre-treatment UV chroma and the change in UV chroma, so it is likely that the change, itself, somehow influenced how females would normally vary sex ratios in response to UV chroma. This might suggest that UV blocking interfered with the female's ability to assess her mate's quality and prevent her from making adjustments in sex allocation as a result. Indeed, the manipulations of plumage color in these two studies decreased UV in the males far below what would be seen in the dullest male in nature. The relationship between posttreatment UV chroma and sex ratios in control birds was unfortunately not reported.

In 2007, Delhey et al. (2007) worked with a population in Austria to further address the question of whether females allocate sex according to the male's UV coloration. This time, they used manipulations that either increased or decreased the

UV coloration within the physiological range using "T-shirt marker pens." Like Korsten et al. (2006), they found no effect of treatment on the sex ratios produced by the pairs with UV-blocked males. But there was again an interaction with pre-manipulation UV chroma. This time, though, the trend was only evident in the group with *reduced* UV chroma; males with higher UV chroma before manipulation produced fewer sons. The authors suggested that perhaps if males started with a higher UV chroma, they experienced a greater reduction in UV chroma after treatment, causing females to allocate fewer sons. The same explanation could actually explain what was seen in the Korsten et al. (2006) study; perhaps the dramatic decrease in the males that started with higher UV chroma prevented females from producing the higher proportion of males seen in the control group.

So, in three of the studies described above, there was a positive correlation between UV chroma of the male's crown feathers and the sex ratio of offspring he produced. These patterns were all found using unmanipulated feathers, including never-treated feathers (Griffith et al. 2003) or pre-treated feathers (Korsten et al. 2006; Sheldon et al. 1999). The relationships of offspring sex ratios with UV chroma of the crown feathers *after* treatment in studies where treatment occurred were either insignificant (Delhey et al. 2007) or were not shown (Korsten et al. 2006; Sheldon et al. 1999). To add to these correlational analyses, a French study showed no relationship between UV coloration and offspring sex ratio (Dreiss et al. 2006); however, this group took measures on collected feathers rather than live birds, which could have affected results. In terms of experimental studies, it appears that the degree of UV reduction may have influenced the sex allocation by females. If males started out with a high UV chroma, blocking that coloration reduced the proportion of males more than it did if a male started out with a low UV chroma. Thus despite the complexity of the results in this species, these studies point to a potentially adaptive manipulation of offspring sex in response to this UV-based ornament in blue tits.

5.7 Other Studies Relating Sex Ratios with Male Quality

Studies testing for a relationship between male quality and offspring sex ratios are certainly not restricted to blue tits. In fact, male quality is the best studied driver of sex allocation in a variety of avian species. A recent meta-analysis including species from all taxa showed a small but significant relationship between offspring sex ratios and male attractiveness; a higher proportion of male offspring were produced when males were more attractive (Booksmyth et al. 2015). In birds, studies examining the relationship of sex ratios with the expression of male ornaments in multiple species has produced mixed results (Table 5.1) yet in all studies in which a significant relationship was found, the relationship was a positive one; the more elaborate the ornament, the more male offspring were produced. Taken together, these studies indicate that there are species-level differences in whether sexually selected ornaments are used as criteria in strategies of sex allocation, but in the species that do use it, most if not all appear to bias sex ratios towards males.

5.7 Other Studies Relating Sex Ratios with Male Quality

Table 5.1 Studies examining offspring sex ratios in relation to a male ornament

Species	Ornament	Effect on sex ratio	References
House Sparrow	Size of throat patch	No	Husby et al. (2006)
House Sparrow	Size of throat patch	No	Westneat et al. (2002)
Coal Tit	Bib saturation/size	No	Dietrich-Boschoff et al. (2006)
Great Tit	Breast stripe size	No	Radford and Blakey (2000)
Common yellowthroat	Black mask size	No	Abroe et al. (2007)
Common yellowthroat	Black mask size	Male-bias (only when mated to young male)	Taff et al. (2011)
Collared flycatcher	Forehead patch size	Male-bias	Ellegren et al. (1996)
Collared flycatcher	Forehead patch size	No	Rosivall et al. (2004)
Canary	Song complexity	No	Leitner et al. (2006)
Peacock	Eyespot # on train	Male-bias	Pike and Petrie (2005b)
Blue tit	Length of strophe bout	Male-bias	Dreiss et al. (2006)
Blue tit	UV crown color	Male-bias	Delhey et al. (2007)
Blue tit	UV crown color	No	Dreiss et al. (2006)
Blue tit	UV crown color	No	Korsten et al. (2006)
Blue tit	UV crown color	Male-bias	Griffith et al. (2003)
Blue tit	UV crown color	Male-bias	Sheldon et al. (1999)

There are two main adaptive hypotheses that may be applied to patterns of sex ratio adjustment and male quality. The first is the Male Attractiveness Hypothesis, which requires that a particular trait linked to reproductive success is heritable and posits that females should overproduce sons when mating with a male that is attractive or of otherwise high quality. All of the studies identified above that link sex ratios with ornament quality follow the predictions of the Mate Attractiveness Hypothesis in that more males were produced when ornaments were most elaborate. This could result from heritable components to those ornaments. Indeed, in great tits, collared flycatchers, peacocks, and blue tits, there is evidence that these ornaments have a heritable component (Johnsen et al. 2003; Norris 1993; Petrie et al. 2009; Qvarnström et al. 2006).

The second hypothesis requires that there is a link between a male's condition and his territory quality and posits that the sex allocation patterns we see in relation to male attractiveness are actually occurring on the basis of territory quality; in a better territory with more resources, females could, in theory, produce more of the larger and costlier sex. In all of the species above that show a positive correlation between ornament expression and offspring sex ratios, males do establish and defend territories (great tit—Krebs 1971, collared flycatcher—Pärt and Qvarnström 1997, barn swallow—Møller 1990, peacock—Loyau et al. 2005, and blue tit—Alonso-Alvarez et al. 2004). However, in some of these species, there is little evidence that the ornament in question functions directly in territoriality. For example, the breast stripes on great tit males and the tail length of male barn swallows appear to function mostly in female choice (Møller 1990; Norris 1990). The function of the UV component of the crown in blue tits in male–male competition is controversial (Korsten et al. 2007). Thus, based on the evidence to date, there is no overwhelming support for this hypothesis, though in a direct test of this in house wrens, addition of boxes to a male's territory stimulated his mate to produce a higher proportion of males (Dubois et al. 2006), so perhaps additional experimental tests of this hypothesis are necessary.

It is also possible that ornament quality is directly linked to another aspect of male quality that is, itself, responsible for sex allocation decisions made by the female. For example, great tits exhibit black breast stripes as ornaments. Males with larger breast stripes are preferred by females as mates (Norris 1990). In this species, both the male breast stripe and male tarsus length are positively correlated with offspring sex ratios, and the sizes of the breast stripe and the tarsus are both heritable in this species (Gebhardt-Henrich and Van Noordwijk 1991; Norris 1990). Yet when tarsus length is controlled for, the relationship between breast strip size and sex ratios disappears (Kölliker et al. 1999), suggesting that the link with tarsus length is the stronger one. In fact, another study in great tits showed no relationship between breast stripe size and offspring sex ratios (Radford and Blakey 2000). Similarly, in peacocks, longer trains were associated with acquisition of better territories, but so were longer tarsi (Loyau et al. 2005). Thus, it is important to examine the relationships of offspring sex ratios not only with expression of ornamental traits but also with related measures of quality as well.

The most widespread assessment of a male bird's condition is the use of body mass corrected for skeletal size. Male condition shows potential to be a driver of sex allocation decisions. Out of five studies examining sex allocation in relation to this classic measure of body condition (calculated by regressing mass to skeletal size and using the residuals) in males, four showed a significant relationship; however, the direction of these relationships varied. In great tits and white-winged fairy wrens, females mated to males in good condition produced a higher proportion of female offspring. However, wandering albatrosses and American kestrels exhibit the opposite pattern, where females mated to males in good condition produce a higher proportion of *female* offspring. Why might the directions differ among species? If this behavior is adaptive, then the costs and benefits of producing males and females must differ among these species. But how do we assess the relative costs and benefits of producing a particular sex?

Wandering albatrosses, white-winged fairy wrens, and great tits all show classic sexual dimorphism, with males being larger and more ornamented. It is generally accepted that raising the larger sex is more costly for the female and requires a more optimal situation in terms of resources and perhaps mate assistance. Thus, for the white-winged fairy wren and the great tit, it makes sense for a female to produce more males when her mate is in better condition and can perhaps provide more parental care or better genes. In support of this idea, great tit males provide a large amount of parental care to offspring, and females make mate choice decisions based on ornaments that appear to indicate the future level of parental care (Norris 1990) White-winged fairy wren males also provision offspring of their social mates (Rathburn and Montgomerie 2004), so a mate in good condition could allow females to produce more costly male offspring. American Kestrel males also help provision offspring; however, this species shows reverse size dimorphism, which could explain why female kestrels mated to males in better condition produce more *female* offspring. Thus, all three species could be making sex allocation decisions by producing the "costlier" sex when mated to good quality males.

These are relatively large inferences to make, however, and not all species follow the predicted patterns of sex allocation. Wandering albatrosses, for example, show very high levels of male parental care and take part in the longest rearing period of any bird species. Males are larger than females, but the single egg in each clutch was more likely to be female, the smaller sex, when the pair-mate was in good condition. This relationship was weaker than the relationship found between the mother's structural size and offspring sex ratio; larger mothers produced more sons. The authors report no assortative mating based on any measure of size or condition, and it would be surprising if large females preferentially mated with males in low condition. However, it is possible that the link between male condition and offspring sex ratios was somehow indirect, and this study highlights the need to take into account both male and female traits when testing for adaptive patterns of sex allocation.

5.8 Case Study: Sex Ratio Adjustment and Food Supplementation in Kakapos

Kakapos are large, nocturnal parrots that can be found in New Zealand, and they are among the most critically endangered bird species. In the 1970s and 1980s, kakapo numbers were critically low and were declining due to predation by feral cats, so in 1987, all living kakapos were rounded up and moved to offshore island bird sanctuaries that were free of introduced predators. As of 2000, there were only 62 individuals remaining, and breeding was still difficult because rats were killing offspring and the sex ratio of surviving adults was biased strongly in favor of males. Because females only mate when they reach a critical mass, supplementary food was provided to many of the females in the sanctuaries to encourage breeding. However, with this supplemental feeding came an undesired effect: sex ratios of offspring were dramatically biased towards males. Thirteen male and five female fledglings were produced by supplemented females, while 4 male and 11 female offspring were produced by non-supplemented females (Clout et al. 2002) (Fig. 5.4). This led to changes in the food supplementation practices, and the population totaled 126 individuals as of 2015 (White et al. 2015).

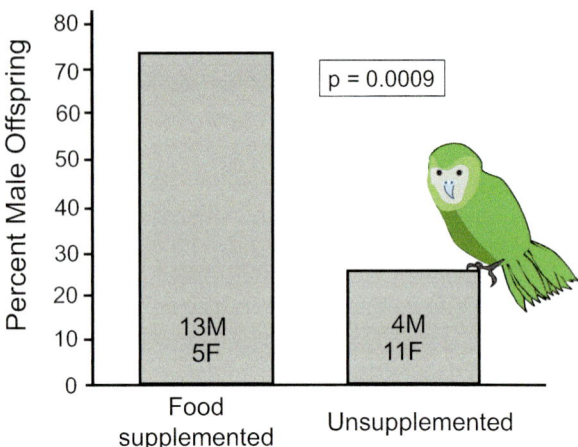

Fig. 5.4 Influences of food supplementation of sex ratios produced by kakapos. Data are summarized from Clout et al. (2002)

5.9 Sex Ratio Adjustment Based on Female Quality and Food Availability

This now classic conservation example highlights the potential for using maternal condition and resource availability to predict strategies of adaptive sex allocation in birds. Out of the 23 studies I found that examined offspring sex ratios in relation to some measure of food availability, only 4 showed no significant effect. Likewise, in 19 studies I found examining sex ratios in relation to female body condition or weight, only 6 showed no effect (Table 5.2). This small number of papers showing no effect could very well be due to publication bias towards significant results. Yet the number of studies showing a significant effect suggests that it is important to further explore the relationships between resource availability, maternal condition, and sex ratios. Interestingly, aside from a few studies, a majority show no relationship of sex ratios with female age or skeletal size. Thus, it appears that weight gain, in particular, is an important factor underlying sex allocation decisions in many avian species.

Nager et al. (1999) performed an elegant study in which they experimentally altered the condition of female lesser black-backed gulls and examined the resulting sex ratios produced by those females. To decrease female condition, the authors removed eggs as they were laid, driving females to compensate for those lost eggs by laying more, a manipulation that has been shown to adversely affect the condition of the female over time. Half of these females were supplemented with nightly boluses of baked hen egg, a treatment that has been shown to increase female condition in this species. Sex ratios in both groups started out as 50:50, but females forced to lay extended clutches that were not supplemented with hen egg produced significantly fewer males across the clutch, reaching on 25% males by the 12th egg in the sequence. Females that were supplemented maintained a 50:50 sex ratio all the way up to the 12th egg in the sequence. Eggs produced by supplemented

5.9 Sex Ratio Adjustment Based on Female Quality and Food Availability

Table 5.2 Studies examining offspring sex ratios in relation to food availability

Species	Effect induced by limitation/low quality	Effect	References
American Kestrel	Male-bias	Proportion of males increased as food availability declined	Wiebe and Bortolotti (1992)
Bald Eagle	Male-bias	Male biased sex ratio during years with low vole density	Dzus et al. (1996)
Bell Miner	Female-bias	Low food availability → female-biased sex ratios. Opposite for high food availability	Ewen (2003)
Black-legged kittiwake	Female-bias	Supplemental feeding led to production of more males. Unfed parents had female-biases.	Merkling et al. (2012)
Blue-footed booby	Male-bias	More males produced in poorer years when less food available	Torres and Drummond (1999)
Eurasian tree creeper	Female-bias	Female-bias in pine-dominated territories with low food availability	Suorsa et al. (2003)
Great skua	Female-bias	Experimentally induced energy expenditure decreased proportion of males. Food supplementation prevented the effect	Kalmbach et al. (2001)
Kakapo	Female-bias	Food supplementation produced more males. Low food → female bias	Clout et al. (2002)
Lesser black-backed gull	Male-bias	Male-bias as female condition deteriorates. Food supplementation prevents the effect	Nager et al. (1999)
Montagu's harrier	None	No effect of food supplementation	Leroux and Bretagnolle (1996)
Parrot finch	Male-bias	Male-bias on low quality diet	Pryke and Rollins (2012)
Pigeon	Female-bias	Reduction of food stimulated female-bias in 1st and 2nd eggs	Pike (2005)
Tengmalm's Owl	Female-bias	Sex ratio declined as food declined over 3-year period	Hipkiss and Hornfeldt (2004)
Ural owl	Male-bias	Male-biased sex ratios when main vole food source disappeared from 1999 to 2001	Brommer et al. (2003)
Yellow-legged gull	None	No effect of food supplementation	Saino et al. (2010)
Zebra finch	Mixed	Females on low quality diet produced male-bias in large clutches and female-bias in large clutches	Arnold et al. (2003)
Zebra finch	Male-bias	Females on low quality diet produced more males	Bradbury and Blakey (1998)

(continued)

Table 5.2 (continued)

Species	Effect induced by limitation/low quality	Effect	References
Zebra finch	Female-bias	Clutches more male-biased when food was in excess	Kilner (1998)
Zebra finch	Female-bias	Females on high quality diet produced more males	Okekpe (2009)
Zebra finch	Male-bias	More males produced on low quality diet	Rutstein et al. (2004)

females were larger across the clutch than those produced by unsupplemented females, suggesting that these females were, in fact, in better condition than their unsupplemented counterparts (Table 5.3).

I talked above about the potential for biasing sex ratios according to the relative costs of each sex, and the same hypotheses could apply here. If a female experiences limited food availability, it makes sense that she would preferentially produce the sex that requires the least investment of resources for survival. However, it is also possible to predict patterns of sex allocation using the potential benefit of offspring in terms of reproductive success. The Trivers–Willard hypothesis of sex allocation (discussed in detail in Chaps. 1 and 2) posits that females in good condition should produce more of the sex that has the greatest variance in reproductive success (Trivers and Willard 1973). In the caribou on which this hypothesis was based, the predicted bias for high condition females was towards males, because males must produce large costly antlers to win a mate. It appears that maternal condition does drive adaptive sex allocation in some way in birds. However, it becomes difficult to predict the direction of sex ratio biases and to fit these biases with the model in an exact sense because many bird species do not fit the assumptions of the model. For example, the Trivers–Willard model assumes that the condition of the mother will be highly correlated with the condition of the resulting offspring. In birds, however, we must also often factor in a high level of parental investment from the father. The second prediction is that differences in condition during parental investment endure into adulthood; however, there are examples in birds of compensatory growth that eliminates size differences among nestlings. Finally, it is assumed that slight differences in condition have disproportionately large effects on male reproductive success, as is true in most systems where males provide low levels of investment. However, >90% of birds are monogamous, and a majority exhibit moderate to high levels of paternal care of offspring. As a result, it becomes unclear whether males in these situations have the ability to benefit more from slight differences in maternal condition. This makes it difficult to predict the direction of avian sex ratio skews in this context, and in the studies examining the link between maternal condition or food availability and sex ratios in birds, the direction of the significant effects varied. Whether these patterns are adaptive and if so, why, is unclear.

Table 5.3 Studies examining offspring sex ratios in relation to maternal condition

Species	Relationship	References
Peafowl	Positive	Pike and Petrie (2005a)
Ruff	Negative	Thuman et al. (2003)
Blue-footed booby	Negative	Velando (2002)
Ural owl	None	Brommer et al. (2003)
Tree swallow	Positive	Delmore et al. (2008) and Whittingham and Dunn (2000)
Western bluebird	None	Koenig and Dickinson (1996)
Fairy Martin	None	Magrath et al. (2002)
Lesser black-backed gull	Positive	Nager et al. (1999)
Savi's warbler	Positive	Neto et al. (2011)
Great tit	None	Radford and Blakey (2000)
House sparrow	None	Westneat et al. (2002)
Superb starling	Negative	Rubenstein (2007)
House wren	Positive	Whittingham et al. (2002)
American Kestrel	Positive	Wiebe and Bortolotti (1992)

5.10 Case Study: Seychelles Warblers Adjust Sex in Response to Helpers at the Nest

One of the most dramatic sex ratio skews found in an avian species was observed in Seychelles warblers, a species in which females are philopatric and, during times of high breeding density, frequently serve as helpers. When breeding on low quality territories, female Seychelles warblers produced 77% sons, and those on high quality territories produced 13% sons (Komdeur et al. 1997). These were all "unhelped" females, but when females in the high quality territory had two or more helpers, they shifted the sex ratios of their offspring towards males (Fig. 5.5). The suggested reasoning for this extreme shift in sex ratios is that females that remain on the same breeding territories of their parents are costly because they deplete prey. The authors followed up on this by examining sex ratios in populations that had been moved from a low quality habitat on Cousin Island to high quality habitats on Aride and Cousine Islands in the Seychelles. Before translocation, females produced male-biased sex ratios, but after translocation, they flipped offspring sex ratios towards females. In one of these groups, helpers were experimentally removed from six experimental units. Before helper removal, sexes produced by all six were male. After removal, these units produced one male and five females. To date, this study remains the best evidence in birds for adaptive sex allocation based on the presence of helpers; however, sample sizes were very small compared to other studies examining sex allocation.

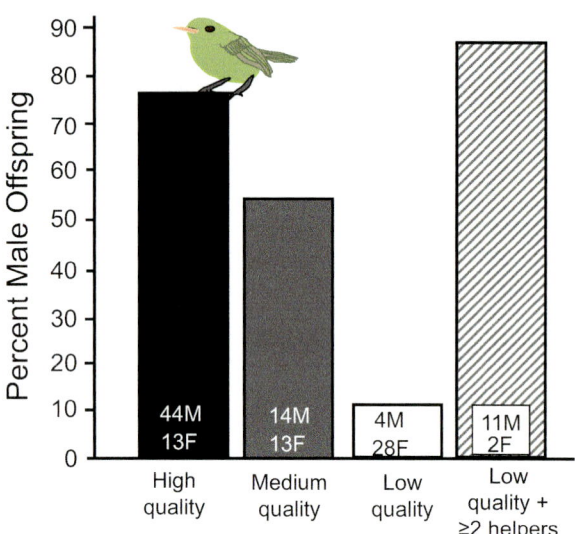

Fig. 5.5 Influences of territory quality and the presence of two or more helpers on sex ratios produced by Seychelles warblers. Data are summarized from Komdeur et al. (2007). The first three bars represent average sex ratios produced over a 3-year period. Sex ratio data from low-quality territories with the presence of helpers were collected during a separate year; however, similar comparisons all within the same year are available in the original reference

5.11 Influences of Helpers on Sex Ratio Adjustment in Other Species

Since the work in Seychelles warblers, the possibility of sex allocation according to the presence of helpers has been studied in more than 20 other species. In some, evidence for a sex ratio skew relative to the helping sex was observed at the population level. Evidence for a population-level skew was found in four of eight species studied. Half showed sex ratios biased towards that of the helping sex and the other half showed the opposite. In the remaining species, sex ratios were examined at both the population and individual levels. In most cases, population-level biases were not found, and out of these 16 studies in which individual sex ratios were compared to the number of helpers, only 5 showed a significant effect. The remaining studies showed no relationship between the number of helpers and offspring sex ratios. Among the studies that showed a significant effect is one conducted in laughing kookaburras, in which groups with helpers, especially when all helpers were female, produced male-biased sex ratios, while pairs without helpers or with male helpers produced female-biased sex ratios. This result, along with the findings in the Seychelles warblers above, follows the predictions of the Local Resource Competition theory (LRC), which states that females should preferentially produce the sex that competes least for limiting resources (see Chap. 3 for a more detailed discussion of this idea). Under this theory, cooperative breeders should bias sex ratios towards the sex that disperses rather than the sex that remains to help. Another species that appears to follow this pattern is the red-backed fairy wren, in which males are the helping sex while females disperse. The overall population sex ratio is biased towards females in this species, and females with helpers produce female-biased broods while those without produce unbiased broods.

There are, however, species that do not follow the predictions of the LRC. For example, for sociable weavers, in which pairs with helpers produced more males, the philopatric helper sex, than pairs without helpers. Further, for Harris's hawks and bell miners, the population sex ratio was found to be biased towards the philopatric sex, in these cases males. These cases match the predictions of an alternative hypothesis, the Local Resource Enhancement theory (LRE), which suggests that when the non-dispersing sex helps, rather than competes, with the mother, sex ratios should be biased towards that helper sex. Griffin et al. (2005) approached the question of when the LRE should apply in avian species and showed using a meta-analysis that there is greater adjustment of sex ratios according to this theory when the helpers bring larger benefits. Indeed, in all cases where sex ratios are biased toward the helping sex, helper-induced increases in reproductive success have been documented.

5.12 How Do We Determine the Adaptive Significance of Avian Sex Ratio Adjustment?

Despite the many examples of skewed sex ratios that appear to occur in response to environmental and social variables, it still remains unclear whether these biases represent examples of facultative manipulation and whether they function adaptively. For every study that has shown a significant sex ratio skew, it is possible that there are many more studies that remain unpublished due to negative results. Palmer (2000) showed that the number of significant results showing sex ratio manipulation was highly dependent on sample sizes such that there were fewer significant results in studies with low sample sizes. He suggested that this may indicate a publication bias, which would explain the large number of significant published results. However, because Palmer (2000) only examined data at the population level, results at smaller sample sizes may have been neutralized by individual variation. West and Sheldon (2002) used data at the individual level to show that publication bias in sex ratio studies does not exist, and mechanisms of sex determination in birds may, in fact, allow for the evolution of facultative sex ratio adjustment. Cassey et al. (2006) examined the heterogeneity of effect sizes in sex ratio studies and showed significant deviation from zero particularly in relation to male quality, laying order, and season. These results suggest that facultative control of avian sex ratios likely exists in response to key factors, though effects are weak. However, there has been an explosion of studies on avian sex ratios in response to a variety of factors since these three key studies have been done. Given a now much larger sample size, there is now a need for an even larger study exploring the potential for facultative manipulation of avian sex ratios.

In addition, rather than focusing on testing a particular adaptive hypothesis, future studies should collect as much data as possible regarding the life history of the species as well as the conditions surrounding breeding pairs at the time of the study and consider combinations of adaptive hypotheses. For example, Wild and West (2007) constructed a model that incorporated both the Local Resource Competition and Condition-Dependent ideas together. The importance of this approach is evidenced by studies in which sex ratios vary according to multiple factors, such as the wandering albatross study discussed above; offspring sex ratios were related to both condition of males and structural size of females (Blanchard et al. 2007). Because many of the variables that appear to influence offspring sex ratios are often correlated with one another (e.g. male and female condition, female condition and season, etc.), focusing on only one variable may lead to false conclusions regarding the adaptive significance of sex ratio skews in a particular species.

Multiyear studies are also particularly helpful, for example, the finding that bald eagles bias sex ratios of first eggs towards males only in years with lower productivity (Dzus et al. 1996) may have been completely missed if the study focused on only 1 year. Looking for consistency across years can also give us an idea of the strength of the pattern. Kentish plovers, for example, decreased the proportion of male offspring as the season progressed, and this effect was the same in all 3 years examined (Székely et al. 2004).This suggests that there is a strong drive in this species to vary sex ratios seasonally. We also need to consider the variety of environmental conditions surrounding the study species. Collecting data from multiple populations can help tease out adaptive patterns, such as with the population differences in sex allocation across the laying order in house finches (Badyaev et al. 2002), and the different seasonal patterns observed in American Kestrels in Florida, Wisconsin, and California (Griggio et al. 2002; Smallwood 1998; Wiebe and Bortolotti 1992).

Finally, in their review of sex allocation in birds, Komdeur and Pen (2002) emphasized that experimental studies are critical to our understanding of how and if female birds facultatively bias sex ratios. Work in captive or domesticated species, in particular, allows control of environmental conditions while adjusting the variable of choice. Zebra finches were among the first species in which experimental studies were used to test the potential for sex allocation in birds. Burley (1981, 1986) manipulated male attractiveness using colored leg bands and showed that more males are produced when the pair-male is wearing a more attractive red leg band. Since then, this species has been used extensively to test the idea that food quality and availability influences how sex is allocated, though results are still mixed (Arnold et al. 2003; Bradbury and Blakey 1998; Okekpe 2009; Rutkowska and Cichoń 2002; Rutstein et al. 2004) perhaps due to a high level of domestication. Peafowl in captivity were used to demonstrate that experimental removal of eyespots from a male's train significantly influences the sex ratios produced by the female he mates with (Pike and Petrie 2005a, b), and some excellent work has been conducted in Gouldian finches showing that pairing females with genetically incompatible morphs produce more than 80% male offspring (Pryke and Griffith 2009). There have also been several elegant experimental studies conducted in the wild. The studies described above, including the Nager et al. (1999) study that altered maternal condition of lesser black-backed gulls, the Komdeur et al. (1997)

study that transferred Seychelles warblers to a better quality environment and removed helpers at the nest, and the multiple studies in which the level of UV coloration in the crown of blue tits was experimentally blocked (Delhey et al. 2007; Korsten et al. 2006; Sheldon et al. 1999) are excellent examples. More experimental studies would benefit our understanding of how and if female birds adaptively skew sex ratios.

References

Abroe B, Garvin JC, Pedersen MC, Whittingham LA, Dunn PO, Lank D (2007) Brood sex ratios are related to male size but not to attractiveness in common yellowthroats (Geothlypis trichas). Auk 124(1):176–184

Alonso-Alvarez C, Doutrelant C, Sorci G (2004) Ultraviolet reflectance affects male-male interactions in the blue tit (Parus caeruleus ultramarinus). Behav Ecol 15(5):805–809

Anderson D, Reeve J, Bird D (1997) Sexually dimorphic eggs, nestling growth and sibling competition in American kestrels Falco sparverius. Funct Ecol 11(3):331–335

Arnold KE, Griffiths R, Stevens DJ, Orr KJ, Adam A, Houston DC (2003) Subtle manipulation of egg sex ratio in birds. Proc R Soc Lond B Biol Sci 270(Suppl 2):S216–S219

Badyaev AV, Hill GE, Beck ML, Dervan AA, Duckworth RA, McGraw KJ, Nolan PM, Whittingham LA (2002) Sex-biased hatching order and adaptive population divergence in a passerine bird. Science 295(5553):316–318

Blanchard P, Hanuise N, Dano S, Weimerskirch H (2007) Offspring sex ratio in relation to parental structural size and body condition in the long-lived wandering albatross (Diomedea exulans). Behav Ecol Sociobiol 61(5):767–773

Booksmyth I, Mautz B, Davis J, Nakagawa S, Jennions MD (2015) Facultative adjustment of the offspring sex ratio and male attractiveness: a systematic review and meta-analysis. Biol Rev 92:108–134

Bortolotti GR (1986a) Evolution of growth rates in eagles: sibling competition vs. energy considerations. Ecology 67(1):182–194

Bortolotti GR (1986b) Influence of sibling competition on nestling sex ratios of sexually dimorphic birds. Am Nat:495–507

Bradbury R, Blakey J (1998) Diet, maternal condition, and offspring sex ratio in the zebra finch, Poephila guttata. Proc R Soc Lond B Biol Sci 265(1399):895–899

Brommer JE, Karell P, Pihlaja T, Painter JN, Primmer CR, Pietiäinen H (2003) Ural owl sex allocation and parental investment under poor food conditions. Oecologia 137(1):140–147

Burley N (1981) Sex ratio manipulation and selection for attractiveness. Science 211(4483):721–722

Burley N (1986) Sex-ratio manipulation in color-banded populations of zebra finches. Evolution:1191–1206

Cassey P, Ewen JG, Møller AP (2006) Revised evidence for facultative sex ratio adjustment in birds: a correction. Proc R Soc Lond B Biol Sci 273(1605):3129–3130

Clout MN, Elliott GP, Robertson BC (2002) Effects of supplementary feeding on the offspring sex ratio of kakapo: a dilemma for the conservation of a polygynous parrot. Biol Conserv 107(1):13–18

Cordero P, Vinuela J, Aparicio J, Veiga J (2001) Seasonal variation in sex ratio and sexual egg dimorphism favouring daughters in first clutches of the spotless starling. J Evol Biol 14(5):829–834

Daan S, Dijkstra C, Weissing FJ (1996) An evolutionary explanation for seasonal trends in avian sex ratios. Behav Ecol 7(4):426–430

Dawson RD, Bortolotti GR (1997) Total plasma protein level as an indicator of condition in wild American kestrels (Falco sparverius). Can J Zool 75(5):680–686

Delhey K, Peters A, Johnsen A, Kempenaers B (2007) Brood sex ratio and male UV ornamentation in blue tits (Cyanistes caeruleus): correlational evidence and an experimental test. Behav Ecol Sociobiol 61(6):853–862

Delmore KE, Kleven O, Laskemoen T, Crowe SA, Lifjeld JT, Robertson RJ (2008) Sex allocation and parental quality in tree swallows. Behav Ecol 19(6):1243–1249

Dietrich-Bischoff V, Schmoll T, Winkel W, Krackow S, Lubjuhn T (2006) Extra-pair paternity, offspring mortality and offspring sex ratio in the socially monogamous coal tit (Parus ater). Behav Ecol Sociobiol 60(4):563–571

Dijkstra C, Daan S, Buker J (1990) Adaptive seasonal variation in the sex ratio of kestrel broods. Funct Ecol 4:143–147

Donald PF (2007) Adult sex ratios in wild bird populations. Ibis 149(4):671–692

Dreiss A, Richard M, Moyen F, White J, Møller A, Danchin E (2006) Sex ratio and male sexual characters in a population of blue tits, Parus caeruleus. Behav Ecol 17(1):13–19

Dubois NS, Kennedy ED, Getty T (2006) Surplus nest boxes and the potential for polygyny affect clutch size and offspring sex ratio in house wrens. Proc R Soc Lond B Biol Sci 273(1595): 1751–1757

Dzus EH, Bortolotti GR, Gerrard JM (1996) Does sex-biased hatching order in bald eagles vary with food resources? Ecoscience 3(3):252–258

Ellegren H, Gustafsson L, Sheldon BC (1996) Sex ratio adjustment in relation to paternal attractiveness in a wild bird population. Proc Natl Acad Sci 93(21):11723–11728

Ewen JG (2003) Facultative control of offspring sex in the cooperatively breeding bell miner, Manorina melanophrys. Behav Ecol 14(2):157–164

Gebhardt-Henrich SG, Van Noordwijk AJ (1991) Nestling growth in the great tit I. Heritability estimates under different environmental conditions. J Evol Biol 4(3):341–362

Genovart M, Oro D, Ruiz X, Griffiths R, Monaghan P, Nager RG (2003) Seasonal changes in brood sex composition in Audouin's Gulls. Condor 105(4):783–790

Graham EB, Caro SP, Sockman KW (2011) Change in offspring sex ratio over a very short season in Lincoln's Sparrows: the potential role of bill development. J Field Ornithol 82(1):44–51

Griffin AS, Sheldon BC, West SA (2005) Cooperative breeders adjust offspring sex ratios to produce helpful helpers. Am Nat 166(5):628–632

Griffith S, Örnborg J, Russell A, Andersson S, Sheldon B (2003) Correlations between ultraviolet coloration, overwinter survival and offspring sex ratio in the blue tit. J Evol Biol 16(5): 1045–1054

Griggio M, Hamerstrom F, Rosenfield RN, Tavecchia G (2002) Seasonal variation in sex ratio of fledgling American Kestrels: a long term study. Wilson Bull 114(4):474–478

Heinsohn R, Legge S, Barry S (1997) Extreme bias in sex allocation in Eclectusparrots. Proc R Soc Lond B Biol Sci 264(1386):1325–1329

Hipkiss T, Hörnfeldt B (2004) High interannual variation in the hatching sex ratio of Tengmalm's owl broods during a vole cycle. Popul Ecol 46(3):263–268

Hunt S, Bennett AT, Cuthill IC, Griffiths R (1998) Blue tits are ultraviolet tits. Proc R Soc Lond B Biol Sci 265(1395):451–455

Hunt S, Cuthill IC, Bennett AT, Griffiths R (1999) Preferences for ultraviolet partners in the blue tit. Anim Behav 58(4):809–815

Husby A, Sæther BE, Jensen H, Ringsby TH (2006) Causes and consequences of adaptive seasonal sex ratio variation in house sparrows. J Anim Ecol 75(5):1128–1139

Johnsen A, Delhey K, Andersson S, Kempenaers B (2003) Plumage colour in nestling blue tits: sexual dichromatism, condition dependence and genetic effects. Proc R Soc Lond B Biol Sci 270(1521):1263–1270

Kalmbach E, Nager RG, Griffiths R, Furness RW (2001) Increased reproductive effort results in male-biased offspring sex ratio: an experimental study in a species with reversed sexual size dimorphism. Proc R Soc Lond B Biol Sci 268(1481):2175–2179

Kilner R (1998) Primary and secondary sex ratio manipulation by zebra finches. Anim Behav 56(1):155–164

Koenig WD, Dickinson JL (1996) Nestling sex-ratio variation in Western Bluebirds. Auk:902–910

Kölliker M, Heeb P, Werner I, Mateman A, Lessells C, Richner H (1999) Offspring sex ratio is related to male body size in the great tit (Parus major). Behav Ecol 10(1):68–72

Komdeur J, Pen I (2002) Adaptive sex allocation in birds: the complexities of linking theory and practice. Phil Trans R Soc Lond B Biol Sci 357(1419):373–380

Komdeur J, Daan S, Tinbergen J, Mateman C (1997) Extreme adaptive modification in sex ratio of the Seychelles warbler's eggs. Nature 385(6616):522–525

Korsten P, Lessells CKM, Mateman AC, Van der Velde M, Komdeur J (2006) Primary sex ratio adjustment to experimentally reduced male UV attractiveness in blue tits. Behav Ecol 17(4): 539–546

Korsten P, Dijkstra TH, Komdeur J (2007) Is UV signalling involved in male-male territorial conflict in the blue tit (Cyanistes caeruleus)? A new experimental approach. Behaviour 144(4): 447–470

Krebs JR (1971) Territory and breeding density in the Great Tit, Parus major L. Ecology 52(1): 2–22

Krebs EA, Green DJ, Double MC, Griffiths R (2002) Laying date and laying sequence influence the sex ratio of crimson rosella broods. Behav Ecol Sociobiol 51(5):447–454

Leitner S, Marshall RC, Leisler B, Catchpole CK (2006) Male song quality, egg size and offspring sex in captive canaries (Serinus canaria). Ethology 112(6):554–563

Leroux A, Bretagnolle V (1996) Sex ratio variations in broods of Montagu's Harriers Circus pygargus. J Avian Biol:63–69

Loyau A, Jalme MS, Sorci G (2005) Intra-and intersexual selection for multiple traits in the peacock (Pavo cristatus). Ethology 111(9):810–820

Magrath MJ, Green DJ, Komdeur J (2002) Sex allocation in the sexually monomorphic fairy martin. J Avian Biol 33(3):260–268

Merkling T, Leclaire S, Danchin E, Lhuillier E, Wagner RH, White J, Hatch SA, Blanchard P (2012) Food availability and offspring sex in a monogamous seabird: insights from an experimental approach. Behav Ecol 23:751–758

Møller AP (1990) Male tail length and female mate choice in the monogamous swallow Hirundo rustica. Anim Behav 39(3):458–465

Nager R, Monaghan P, Griffiths R, Houston D, Dawson R (1999) Experimental demonstration that offspring sex ratio varies with maternal condition. Proc Natl Acad Sci 96(2):570–573

Neto JM, Hansson B, Hasselquist D (2011) Sex allocation in Savi's warblers Locustella luscinioides: multiple factors affect seasonal trends in brood sex ratios. Behav Ecol Sociobiol 65(2):297–304

Norris K (1990) Female choice and the quality of parental care in the great tit Parus major. Behav Ecol Sociobiol 27(4):275–281

Norris K (1993) Heritable variation in a plumage indicator of viability in male great tits Parus major. Nature 362(6420):537–539

Okekpe CC (2009) Evidence that maternal diet alters steroid levels and primary offspring sex ratio in the zebra finch. Thesis, Auburn University, AL.

Palmer AR (2000) Quasireplication and the contract of error: lessons from sex ratios, heritabilities and fluctuating asymmetry. Annu Rev Ecol Syst 31:441–480

Pärt T, Qvarnström A (1997) Badge size in collared flycatchers predicts outcome of male competition over territories. Anim Behav 54(4):893–899

Petrie M, Cotgreave P, Pike TW (2009) Variation in the peacock's train shows a genetic component. Genetica 135(1):7–11

Pike TW (2005) Sex ratio manipulation in response to maternal condition in pigeons: evidence for pre-ovulatory follicle selection. Behav Ecol Sociobiol 58(4):407–413

Pike TW, Petrie M (2005a) Maternal body condition and plasma hormones affect offspring sex ratio in peafowl. Anim Behav 70(4):745–751

Pike TW, Petrie M (2005b) Offspring sex ratio is related to paternal train elaboration and yolk corticosterone in peafowl. Biol Lett 1(2):204–207

Pryke SR, Griffith SC (2009) Genetic incompatibility drives sex allocation and maternal investment in a polymorphic finch. Science 323(5921):1605–1607

Pryke SR, Rollins LA (2012) Mothers adjust offspring sex to match the quality of the rearing environment. Proc Biol Sci 279(1744):4051–4057

Qvarnström A, Brommer JE, Gustafsson L (2006) Testing the genetics underlying the co-evolution of mate choice and ornament in the wild. Nature 441(7089):84–86

Radford A, Blakey J (2000) Is variation in brood sex ratios adaptive in the great tit (Parus major)? Behav Ecol 11(3):294–298

Rathburn MK, Montgomerie R (2004) Breeding biology and social structure of White-winged Fairy-wrens (Malurus leucopterus): comparison between island and mainland subspecies having different plumage phenotypes. Emu 103(4):295–306

Rosivall B (2008) Contradictory results in sex ratio studies: populations do not necessarily differ. Behav Ecol Sociobiol 62(6):1037–1042

Rosivall B, Török J, Hasselquist D, Bensch S (2004) Brood sex ratio adjustment in collared flycatchers (Ficedula albicollis): results differ between populations. Behav Ecol Sociobiol 56(4):346–351

Rubenstein DR (2007) Temporal but not spatial environmental variation drives adaptive offspring sex allocation in a plural cooperative breeder. Am Nat 170(1):155–165

Rutkowska J, Cichoń M (2002) Maternal investment during egg laying and offspring sex: an experimental study of zebra finches. Anim Behav 64(5):817–822

Rutstein A, Slater P, Graves J (2004) Diet quality and resource allocation in the zebra finch. Proc R Soc Lond B Biol Sci 271(Suppl 5):S286–S289

Ryder JP (1983) Sex ratio and egg sequence in ring-billed gulls. Auk 100(3):726–728

Saino N, Romano M, Caprioli M, Ambrosini R, Rubolini D, Fasola M (2010) Sex allocation in yellow-legged gulls (Larus michahellis) depends on nutritional constraints on production of large last eggs. Proc R Soc Lond B Biol Sci 277(1685):1203–1208

Sheldon BC, Andersson S, Griffith SC, Örnborg J, Sendecka J (1999) Ultraviolet colour variation influences blue tit sex ratios. Nature 402(6764):874–877

Smallwood PD (1998) Seasonal shifts in sex ratios of fledgling American kestrels (Falco sparverius paulus): the early bird hypothesis. Evol Ecol 12(7):839–853

Sockman KW, Schwabl H (2000) Yolk androgens reduce offspring survival. Proc R Soc Lond B Biol Sci 267(1451):1451–1456

Suorsa P, Helle H, Huhta E, Jäntti A, Nikula A, Hakkarainen H (2003) Forest fragmentation is associated with primary brood sex ratio in the treecreeper (Certhia familiaris). Proc R Soc Lond B Biol Sci 270(1530):2215–2222

Svensson E, Nilsson J-A (1996) Mate quality affects offspring sex ratio in blue tits. Proc R Soc Lond B Biol Sci 263(1368):357–361

Székely T, Cuthill IC, Yezerinac S, Griffiths R, Kis J (2004) Brood sex ratio in the Kentish plover. Behav Ecol 15(1):58–62

Taff CC, Freeman-Gallant CR, Dunn PO, Whittingham LA (2011) Relationship between brood sex ratio and male ornaments depends on male age in a warbler. Anim Behav 81(3):619–625

Thuman KA, Widemo F, Griffith SC (2003) Condition-dependent sex allocation in a lek-breeding wader, the ruff (Philomachus pugnax). Mol Ecol 12(1):213–218

Torres R, Drummond H (1999) Variably male-biased sex ratio in a marine bird with females larger than males. Oecologia 118(1):16–22

Trivers RL, Willard DE (1973) Natural selection of parental ability to vary the sex ratio of offspring. Science 179(4068):90–92

Velando A (2002) Experimental manipulation of maternal effort produces differential effects in sons and daughters: implications for adaptive sex ratios in the blue-footed booby. Behav Ecol 13(4):443–449

Velando A, Graves J, Ortega-Ruano JE (2002) Sex ratio in relation to timing of breeding, and laying sequence in a dimorphic seabird. Ibis 144(1):9–16

West SA, Sheldon BC (2002) Constraints in the evolution of sex ratio adjustment. Science 295(5560):1685–1688

Westneat DF, Stewart IR, Woeste EH, Gipson J, Abdulkadir L, Poston JP (2002) Patterns of sex ratio variation in house sparrows. Condor 104(3):598–609

White K, Eason D, Jamieson I, Robertson B (2015) Evidence of inbreeding depression in the critically endangered parrot, the kakapo. Anim Conserv 18(4):341–347

Whittingham LA, Dunn PO (2000) Offspring sex ratios in tree swallows: females in better condition produce more sons. Mol Ecol 9(8):1123–1129

Whittingham LA, Valkenaar SM, Poirier NE, Dunn PO, Blem C (2002) Maternal condition and nestling sex ratio in house wrens. Auk 119(1):125–131

Wiebe KL, Bortolotti GR (1992) Facultative sex ratio manipulation in American kestrels. Behav Ecol Sociobiol 30(6):379–386

Wiehn J (1997) Plumage characteristics as an indicator of male parental quality in the American kestrel. J Avian Biol:47–55

Wild G, West SA (2007) A sex allocation theory for vertebrates: combining local resource competition and condition-dependent allocation. Am Nat 170(5):E112–E128

Zijlstra M, Daan S, Bruinenberg-Rinsma J (1992) Seasonal variation in the sex ratio of marsh harrier Circus aeruginosus broods. Funct Ecol:553–559

Potential Mechanisms of Sex Ratio Adjustment in Birds

6

A hen is only an egg's way of making another egg.
Samuel Butler

In many ways, female birds appear to function much like machines. They manufacture large balls of nutrients and energy that ultimately nurture their offspring through the first weeks of life. In addition, we now know that female birds can control not only the nutritional environment that surrounds the embryos, they can also use hormones to alter the genetic composition of those offspring and trigger post-inheritance modifications of inherited DNA to permanently program the phenotypic traits that the offspring will express throughout their lives. So the hen is a perfectly designed vessel for producing optimal offspring that will produce more eggs down the line. In Chap. 5, I presented abundant evidence that female birds alter the sex ratios of offspring produced in response to surrounding environmental and social conditions. Despite decades of research on this phenomenon, however, the mechanism responsible still remains unknown. In this chapter, I will outline the potential target times and mechanisms throughout female gametogenesis and offspring development during which sex ratio adjustment in birds may take place.

6.1 How Is Sex Determined in Birds?

Oocytes are present within the avian embryo from early incubation and reach approximately 480,000 in number by the time of hatch, when oogenesis stops (Johnson 2015) (Fig. 6.1). These oocytes are initially organized into primordial follicles that are surrounded by a perivitelline membrane layer and a single layer of granulosa cells. When the female reaches sexual maturity, the development of an additional layer of cells, the theca cells, marks the progression of primordial follicles into primary follicles, which then begin to project from the surface of the ovary. White yolk that is high in lipoproteins is deposited into the oocyte, and the

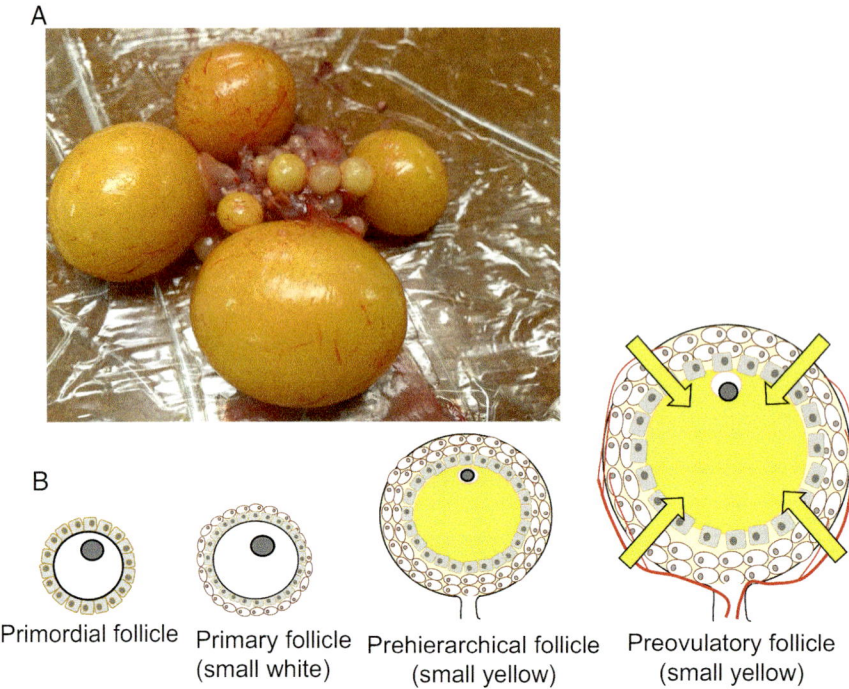

Fig. 6.1 (**a**) shows a photograph of a chicken ovary, in which small white primary follicles, small yellow prehierarchical follicles, and large yellow pre-ovulatory follicles are visible. The pre-ovulatory follicles are marked in order of size (F1 = largest, F5 = smallest), and the follicles ovulate in this order with the largest ovulating first. (**b**) shows the structure of ovarian follicles at different developmental stages. Primordial follicles are made up of the nuclear material inside a germinal vesicle, which is surrounded by a perivitelline membrane and a single layer of granulosa cells. Progression to the primary follicle stage is marked by the development of a layer of theca cells that surround the granulosa layer. Prehierarchical follicles have developed two layers of theca cells (the theca interna and theca externa) and a small quantity of yolk precursors have passed into the oocyte generating a yellow appearance. Finally, once selected into the pre-ovulatory hierarchy, the granulosa cells move further apart to allow a large flood of yolk precursors to enter the oocyte, increasing its size up to fivefold

follicle begins a slow growth. After this, a smaller pool (approximately 8–12) of these white follicles are selected to transition into prehierarchical follicles, during which the theca layer differentiates into a theca externa and a theca interna, and the follicle accumulates yellow yolk, growing to sizes between 6 and 8 mm. It is from this pool of follicles that the final pre-ovulatory hierarchy will be selected at a rate of one follicle per day. Once selected into this pre-ovulatory hierarchy, the likelihood of atresia drops dramatically, and there is >90% chance that the follicle will ovulate. After selection, these follicles begin the process of rapid yolk deposition, during which the granulosa cells change shape to allow more yolk precursors to pass from the bloodstream to the oocyte. Those precursors enter the oocyte through a receptor-mediated process; they are then further processed into the components found in egg yolk, such as triglycerides, cholesterol, and phospholipids, to name a few. During the stage of rapid yolk deposition, the follicle grows up to five times its size in a period of days and then stops incorporating yolk approximately 24 h prior to ovulation (Johnson 2015). The ovary of a mature hen contains follicles of all sizes ranging from small white primary follicles to large yellow pre-ovulatory follicles (Fig. 6.1).

The genetic information carried by the follicles is located in the germinal vesicle, which is located within a small white germinal disk at the periphery of the oocyte. Also contained therein are the sex chromosomes that will ultimately determine whether the resulting embryo becomes a male or a female. Birds demonstrate a ZW system of genetic sex determination, meaning that the female determines the sexes of offspring, donating either a male-producing Z chromosome or a female-producing W chromosome. Figure 6.2 illustrates the steps of meiosis I, leading to the segregation of sex chromosomes, which will be described in detail in the following sections. Oocytes begin the process of meiosis at day 15.5 of incubation, progressing through the leptotene, ziplotene, and pachytene phases of prophase I, but this process arrests in the diplotene phase of meiotic prophase I at the time of hatch. This means that every oocyte in the ovary contains both Z and W chromosomes and theoretically has the potential to ultimately produce either a male or female offspring. Oocytes remain arrested in meiotic prophase I until just hours before ovulation when meiosis then resumes, and the sex chromosomes segregate. In the chicken, this process occurs 3–5 h prior to ovulation, while in the quail, it occurs approximately 2 h prior to ovulation. It is at this point, just hours before ovulation, that the offspring sex is irrevocably determined, because when pairs of homologous chromosomes separate during anaphase I, only one set is retained in the oocyte and destined to produce the embryo, while the other set ends up in an unfertilizable polar body that eventually degrades (see Fig. 6.2 for an overview of meiosis I in the avian oocyte).

Ultimately, the important factors in the determination of which sex chromosome is retained in the oocyte are (1) the position of the chromosomes at the time of meiotic spindle formation and (2) the direction in which the spindle pulls those chromosomes. This is because it is the location within the germinal vesicle that determines whether a chromosome is retained or expelled into the unfertilizable polar body. Until approximately 24 h prior to ovulation, the germinal vesicle is located in the center of the germinal disk; however, as the time of ovulation approaches, the germinal vesicle migrates to the periphery of the germinal disk, with one side flattened against that peripheral edge. It is at this time, approximately 6–12 h prior to ovulation, that the homologous chromosome pairs line up along a horizontal meiotic plate with metaphase spindle fibers attached and oriented perpendicular to the upper edge of the germinal disk (see Fig. 6.2). Given this orientation, the ultimate effect is that when the fibers pull the homologous chromosome pairs apart during anaphase I, one set is pulled towards the periphery of the germinal disk, and one is pulled towards the center. The chromosome set that is pulled towards the center will be retained in the oocyte, while the chromosome set that is pulled to the periphery will bud off via cytokinesis to be consumed by a small

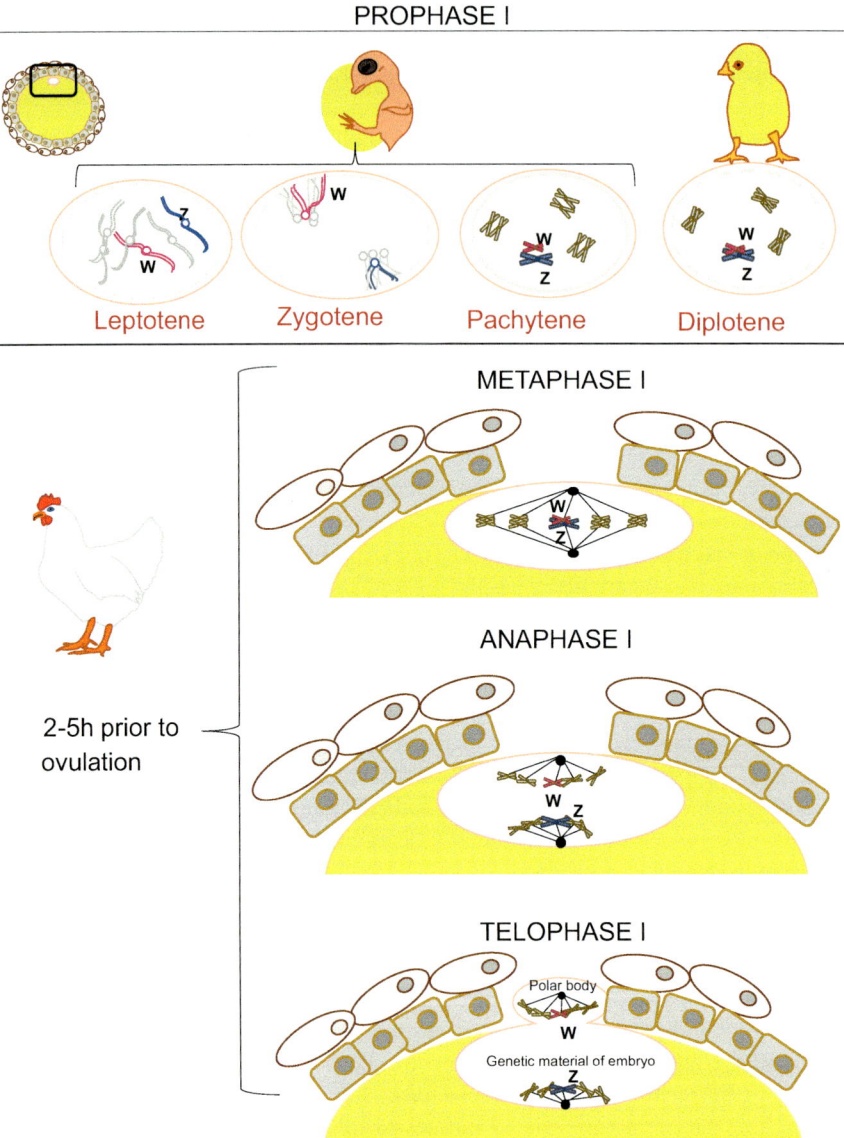

Fig. 6.2 A visual representation of the process leading to sex chromosome segregation (telophase I) in the avian ovary. Meiosis begins at approximately day 15.5 of embryonic development and progresses through the initial stages of prophase. During the leptotene stage, chromosomes are scattered within the germinal vesicle and exist in a paired and lampbrush form. During the zygotene stage, chromosomes form "bouquets" due to attachment of their telomeres to specific points on the nuclear envelope of the germinal vesicle. During the pachytene phase, the chromosomes condense into tight pairs via cohesin proteins. During the diplotene phase, chromosomes synapse, forming chiasmata, which are traded segments of DNA that are then responsible for holding the chromosome pairs together. Oocytes remain arrested in this stage until hours prior to ovulation when the germinal vesicle disappears, and the

polar body (reviewed in Rutkowska and Badyaev 2008). The ultimate fate of this polar body is currently not known; however, we do know that the chromosome set retained in the oocyte is the set that will go on in the ovulated egg to serve as half of the genetic material for the resulting embryo when the egg is fertilized. The following sections outline the potential time points at which manipulation of sex ratio adjustment may occur, and how the processes of follicular maturation and/or meiosis may be manipulated to achieve it.

6.2 Do Oocytes Have a Predetermined Sex?

It has generally been believed that all oocytes have equal potential to retain the W and Z chromosomes and thus have equal potential to produce male and female offspring. Badyaev et al. (2006) challenged this idea with their finding that in two populations of house finches, oocytes that ultimately produced males and females were both spatially clustered within the ovary during follicle development and were temporally clustered such that males or females hatched at the optimal positions within the clutch sequence. The authors suggest that oocytes may be predestined to retain a particular sex chromosome and that females may be able to select these predestined oocytes into the ovulatory hierarchy such that offspring of a particular sex hatch at the optimal time in the clutch sequence. They posit that this process may explain variation in sex ratios produced at different clutch positions.

Despite the fact that the process of sex chromosome segregation does not occur until hours prior to ovulation, it may be possible to set up which sex chromosome an oocyte will retain much earlier, during embryonic development. It is at this time that the early portion of meiosis occurs, whereby DNA is replicated (interphase), chromosomes begin to condense, and homologous chromosomes begin to associate with one another (leptotene stage of prophase I). Next, connections between the homologous chromosomes become very strong via the appearance of synaptonemal complexes, or protein structures that form between homologous chromosomes (zygotene stage of prophase I), and trading of genetic content occurs between the bivalent arms of the chromosomes, forming chiasmata, which thereafter hold the homologous chromosomes together (pachytene phase of prophase I) (see Fig. 6.2). Even during these early phases, there are several forces controlling where in the germinal vesicle the chromosomes are located. For example, during the zygotene phase, the chromosomes are bunched in "bouquets" that occur due

Fig. 6.2 (continued) chromosome pairs align along the meiotic plate, which is oriented perpendicular to the upper oocyte plasma membrane. The spindles pull the homologous chromosomes apart during anaphase I, and cytokinesis during telophase I results in the formation of one polar body with no fertilization potential and one oocyte containing a diploid set of chromosomes. The sex chromosome that is pulled up into the polar body is lost, and the one that is pulled down into the oocyte determines the sex of the future offspring

to the fact that the telomeres on the ends of the chromosomes link to the nuclear envelope and then form clusters. By the diplotene phase of prophase I, these clusters disperse, and protein bodies form that gather the chromosomes in a karyosphere, which is a conglomeration of protein bodies, chromosomes, and the nucleolus. The sizes and numbers of protein bodies are known to vary among species, and importantly, between the W and Z chromosomes. The formation of these protein bodies and the ways in which they associate with one another as well as the initial position within the telomere-driven "bouquets" could potentially determine where on the karyosphere, and thus where in the germinal vesicle, the chromosomes are located when the spindle fibers seek attachment to them. This idea could serve as a mechanism underlying a predetermination of whether the oocyte will retain the W or Z chromosome from this point forward. If, then, the female had a mechanism of selecting only follicles in which the Z or W chromosome was in a particular position within the germinal vesicle, she could skew sex ratios as a result.

If the location of the chromosome "bouquets" predetermines which sex chromosome the oocyte will retain, how, then, could information about this predetermination be used to skew sex ratios? If follicles that will retain Z chromosomes congregate in one special segment of the ovary while those that will retain the W congregate in another, then microclimates within the ovary could drive selection of follicles from one region versus another into the hierarchy (Fig. 6.3). It is believed that follicle-stimulating hormone (FSH) as well as several other factors expressed in the granulosa layer, such as antimullerian hormone (AMH), BMP-15, and the c-Kit/Kit ligand, influences which follicle is selected into the pre-ovulatory hierarchy (reviewed in Johnson 2012). If one segment of the ovary were exposed to a

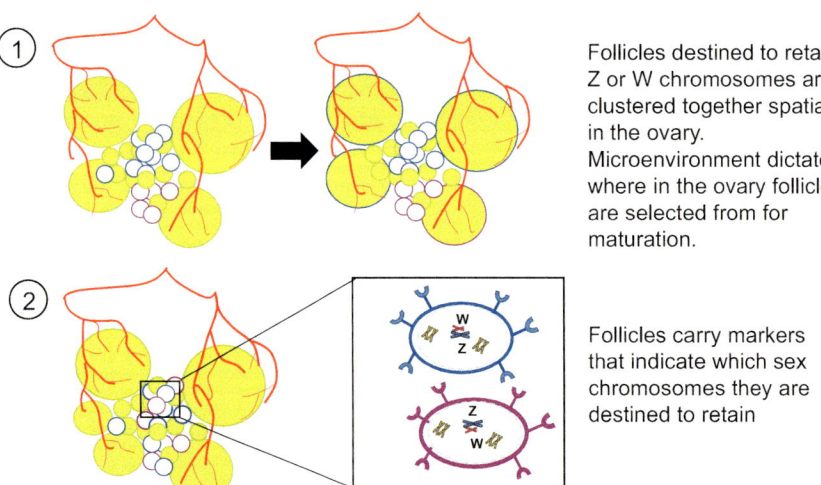

Fig. 6.3 Two potential mechanisms by which follicles that are predetermined to retain the W (pink) or Z (blue) chromosomes may be selected to alter offspring sex ratios

higher influx of FSH, perhaps that would stimulate preferential selection of follicles from that side of the ovary. Alternatively, if follicles in a particular region expressed higher levels of FSH receptors, this could also result in region-specific selection into the pre-ovulatory hierarchy. Such a mechanism might explain the production of clusters of same-sex offspring in the absence of a clear cue known to skew sex ratios, such as patterns observed within the clutch sequence in at least 10 avian species (reviewed in Krackow 1995). Heinsohn et al. (1997) showed a striking example of sex clustering within the laying sequence in eclectus parrots, in which females produced long strings of the same sex; one female produced 20 male offspring before switching and producing 13 female offspring.

Alternatively, it may not be necessary that follicles predetermined to carry a particular sex chromosome congregate together within the ovary. If the chromosome orientation within the oocyte could somehow influence how that follicle responds to cues that would select it into the pre-ovulatory hierarchy, and if physiological cues that are responsive to environmental or social changes could alter how orientation affects the follicular responsiveness, this could be another mechanism by which oocytes of a particular sex are selected (Fig. 6.3). The existence of such a mechanism would mean, however, that the process of skewing avian sex ratios is much more complicated than previously thought. Additional studies are needed to test whether ovarian follicles in the avian ovary are preset to retain a particular chromosome, and perhaps whether oocytes recruited from specific parts of the ovary are more likely to produce a male or female offspring. Further, more recent work shows that treatment with hormones (see below) just hours prior to ovulation can successfully skew avian sex ratios. This treatment occurred well after the follicles had been selected into the pre-ovulatory hierarchy. This does not preclude the possibility that follicles have a predetermined tendency to retain a specific sex chromosome. It is possible that the hormone treatment simply overrode such a predetermination.

6.3 Do Factors During Rapid Yolk Deposition Influence Which Sex Chromosome Is Retained?

While the findings by Badyaev et al. (2006), Heinsohn et al. (1997), and others could provide support for the idea that oocytes are predetermined to retain a specific sex chromosome, there is another explanation that could explain temporal and/or spatial clustering of oocytes and offspring with a specific sex chromosome. Once selected into the hierarchy, ovarian follicles grow very rapidly over a period of about 5–7 days. This process involves accumulation of large quantities of yolk precursors, as well as hormones, nutrients, and other physiologically active molecules along with them. It is well known that environmental changes, such as stress or territorial intrusion, influence the quantities of hormones, immunological molecules, and other physiologically relevant resources that females deposit into the yolks of their eggs. While the germinal disk experiences less of this yolk transport than other follicular regions (hence the reason why it appears as a white

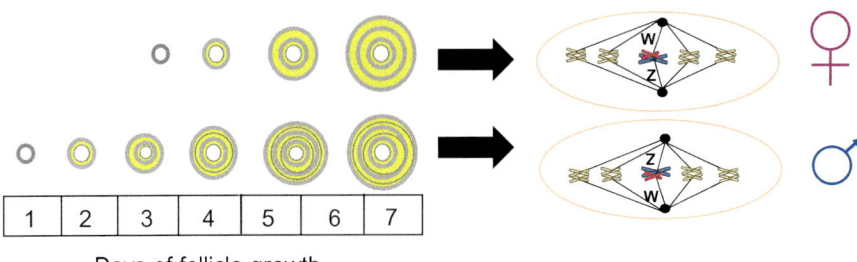

Fig. 6.4 An illustration of the theoretical mechanism in which follicle growth rates determine which sex chromosome the oocyte retains, determining offspring sex

disk), it still lies on top of the accumulating yolk on its posterior side, and on its anterior side, it is also exposed to granulosa cells which are known to actively communicate with the oocyte (Johnson et al. 2005; Yao and Bahr 2001). Hence, it is possible that follicles developing a particular region of the ovary, or during a particular time period, may accumulate more of the mediator that controls sex ratios (Fig. 6.4). Badyaev et al. (2006) showed that eggs that produced male offspring had accumulated larger quantities of testosterone and carotenoids compared to those that produced females. Further, Young and Badyaev (2004) measured rates of yolk deposition in house finch eggs by staining the lipid rings, and they showed that the rates at which the yolk was deposited were related to the sex of the offspring produced by the follicle; follicles that grew faster were more likely to produce male offspring. Whether this was the result of larger and more frequent incorporation of a chemical mediator that exerted an influence on the ultimate outcome of meiosis, or the physical process of oocyte expansion exerted an effect, is yet unclear. This possibility still does not explain the finding that sex ratio skews can be induced just hours prior to ovulation, given that yolk incorporation into the follicle is thought to cease 24 h prior to ovulation. However, this possibility could be tested by exposing birds to cues known to skew sex ratios only during the period of follicle growth and testing for subsequent effects on offspring sex.

6.4 Can Sex Ratios Be Altered via Direct Manipulation of Meiotic Segregation?

There are a select few ways that the process of meiosis could be manipulated to preferentially retain a particular sex chromosome. Rutkowska and Badyaev (2008) provide an excellent review of the possibilities, and I will summarize some of their ideas along with some additional possibilities here. After months to years of remaining arrested in the diplotene stage of meiotic prophase, meiosis suddenly resumes just hours prior to ovulation, and the process progresses into metaphase I. There, the homologous chromosomes undergo congression, aligning along a horizontal meiotic plate. Given that one set of these chromosomes will be pulled down into the center of

6.4 Can Sex Ratios Be Altered via Direct Manipulation of Meiotic Segregation?

the oocyte, and the other half will be pulled up into the unfertilizable polar body, the position of the chromosomes relative to the upper membrane of the oocyte and to their partner chromosomes may determine their ultimate fate. Understanding how congression and segregation of the homologous chromosomes occurs may offer some insight into what parts of the process could be manipulated to select a particular sex chromosome.

Little to nothing is known about how congression of chromosome pairs and localization along the meiotic plate occurs in birds. In mammalian systems, actin networks appear to control the localization of chromosome "bouquets" during the zygotene phase of prophase I described earlier (Koszul et al. 2008), and work in starfish suggests that actin also aids in the process of chromosome congression during metaphase (Lénárt et al. 2005; Rutkowska and Badyaev 2008). However mammalian work instead indicates that a complex interaction of microtubules and associated proteins underlies the very precise movements of the chromosome pairs. Before congression can occur, chromosomes must be co-oriented such that homologous chromosomes stay together, and pairs of sister chromatids are linked to the same spindle pole via microtubules. This is a complex process that requires precise timing of the spindle connections. Microtubules connect to a protein structure found at the centromere of chromatids called the kinetochore. During this time, spindle microtubules attach and release chromosomes until both kinetochores from sister chromatids are attached to the same spindle (Sakuno and Watanabe 2009). This is very important because both sister chromatids must move together towards the same pole during anaphase.

It is thought that congression results from the activity of kinesins, which make up part of a family of motor proteins that move along microtubules, and the process by which this may occur is summarized below and in Fig. 6.5. Kinesins are located both at the kinetochore of the chromosome (the protein complex that links the chromosomes to the microtubules of the meiotic spindle) and also on chromosome arms. Kinesins located at the kinetochore are termed "minus-directed motors" and are responsible for pulling chromosomes towards the spindle poles. There is another group of kinesins that are located on the chromosomes arms, and these are termed "plus-directed motors" that act to push the chromosomes away from the spindle poll (Levesque and Compton 2001; Carpenter 1991; Fuller 1995). This force is also called the polar ejection force or the polar wind. Through a delicate balance of these two mechanisms, the chromosomes oscillate within the oocyte until they move to the center of the oocyte. The cell also contains a set of lateral microtubules that provide tracks for chromosome movements and appear to be critical for meiotic alignment in oocytes (Wignall and Villeneuve 2009). The connections between the homologous chromosomes are then released and the chromosomes are pulled towards opposite poles. The microtubules holding them have the ability to shorten or elongate without detaching from the kinetochore on the chromosome or the spindle poll, suggesting that they can actively shed and acquire segments to move chromosomes within the cell and cytoplasmic kinesins which help facilitate this (Mitchison and Kirschner 1985). It is currently unknown how homologous chromosomes recognize and pair with one another, what

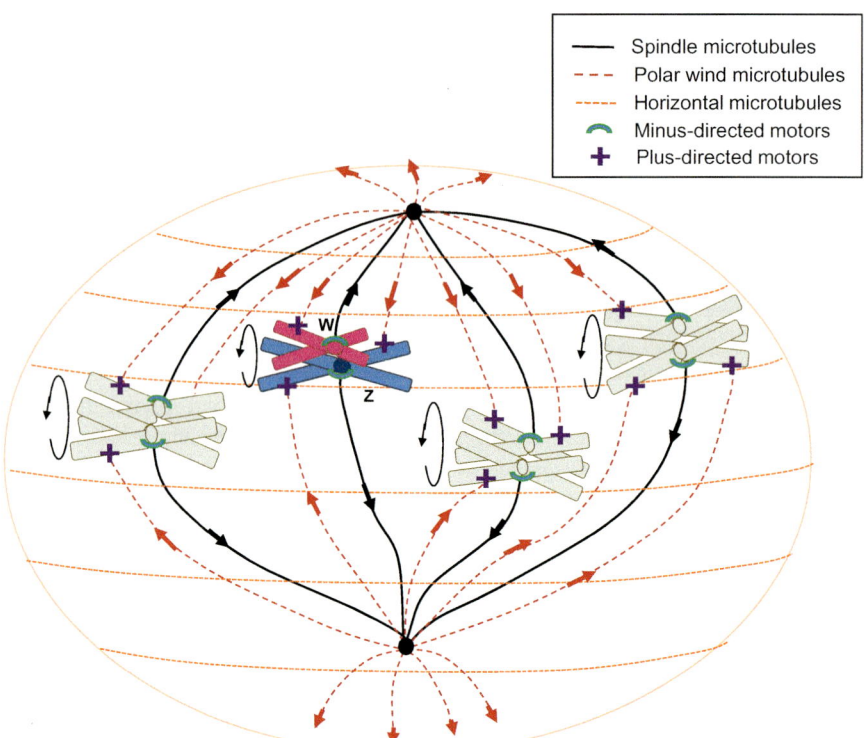

Fig. 6.5 An illustration of the hypothesized mechanisms responsible for chromosome orientation and congression at the meiotic plate. Chromosome movement is driven by the action of kinesins, which are proteins that "walk" along a microtubule network within the cell. Kinesins located on the kinetochore (the protein center that attaches the chromosome center to the spindle microtubules) are called minus-directed motors because they connect to the chromosomes at their minus ends and pull the centromere towards the spindle pole. Kinesins located on the arms of the chromosomes are called plus-directed motors because they are attached to the chromosomes at their plus ends and push the chromosomes *away* from the spindle poles. This force has been called the polar wind, or the polar ejection force. These antagonistic forces cause the chromosomes to continuously oscillate, and horizontal microtubules form a lattice that keeps the chromosomes in the correct orientation to align them along the meiotic plate. The congression event is driven through a combination of push and pull forces (spindle vs. polar wind forces) in addition to the actions of many additional protein helpers not included in this figure

determines which homologous chromosomes is oriented towards the periphery and which is oriented towards the center of the oocyte, and if chromosomes located closer to the center of the oocyte when on the meiotic plate can then be pulled out to the polar body despite their initial positions on the meiotic plate.

What we do know is that the avian W and Z chromosomes differ from one another in many ways. Most obviously, they differ in size; the W chromosome is substantially smaller than the Z chromosome, especially when in the lampbrush form (Solovei et al. 1993; Solari 1993), and it has been shown that chromosomes

6.4 Can Sex Ratios Be Altered via Direct Manipulation of Meiotic Segregation?

with smaller sizes are more likely to mis-segregate (Spence et al. 2006), suggesting that chromosome size can influence the process of chromosome movement. In addition, because the polar ejection force results from the activities of motor proteins on the chromosome arms, the strength of the polar ejection force is smaller for smaller chromosomes (Carpenter 1991), perhaps because they have fewer motor proteins to drive it. Because chromosome size likely affects movement within the cell, Rutkowska and Badyaev (2008) suggest that bird species with greater dimorphism in sizes of the sex chromosomes should exhibit a greater tendency to bias offspring sex ratios. To date, this idea has not yet been tested; however, the adjustment of chromosome lengths may represent a mechanism by which sex ratios could be altered (Fig. 6.6a). For sex ratio adjustment to occur in response to environmental and social stimuli, however, there must be a physiological transducer that not only responds to these stimuli but also has the capacity to alter chromosome lengths. One possibility is via the adjustment of telomerase activity (see below for more detail about this idea).

Size differences between the sex chromosomes also correlate with large differences in content; the Z chromosome in the chicken has over 200% more DNA than the W (Mendonça et al. 2010). Because DNA has a negative charge, this may also indicate that there is a charge difference between the two sex chromosomes. Given that microtubule movements and the activities of the kinesins are based on charge, it is possible that charge differences between the sex chromosomes could control their movements within the cell. A potential underlying mechanism could involve the ion content of the oocyte at critical times during meiosis. For example, in frog oocytes, liberation of intracellular calcium stores via second messenger-mediated processes results in the formation of "calcium puffs" within the oocyte (Parker and Yao 1995). During these events, the calcium moves in waves throughout the oocyte. It is unknown whether other ions are released through avian oocytes and whether the calcium ion gradients are found in within them, though there is some evidence that such charge gradients occur in human oocytes (Tesarik and Mendoza 1995). If charged ions are, in fact, released in waves in avian oocytes, a resulting change in the oocyte charge balance at a critical time during meiosis could influence to which sides the Z and W chromosomes segregate (Fig. 6.6e).

Another key difference between the avian sex chromosomes that may provide a control over their movement is the sizes and positions of their centromeres. Noted in the Rutkowska and Badyaev review are examples in fruit flies and voles in which the chromosome with the most centrally located centromere preferentially migrates either to the gamete or to the polar body, depending on the species (Novitski 1967; Gileva and Rakitin 2006). If similar effects are at play in birds and if centromeres are more centrally located on the Z versus the W chromosome, then this could represent a mechanism by which meiotic drive could occur. However, for sex ratios to be altered in response to environmental and social conditions, there would need to be a plasticity in either the location of the centromere on the two sex chromosomes or the directional orientation of the chromosome with the most centrally located centromere towards either the oocyte or the polar body. How could this potentially occur? One possibility is via alterations in telomere lengths. The avian W chromosome has a mega-telomere,

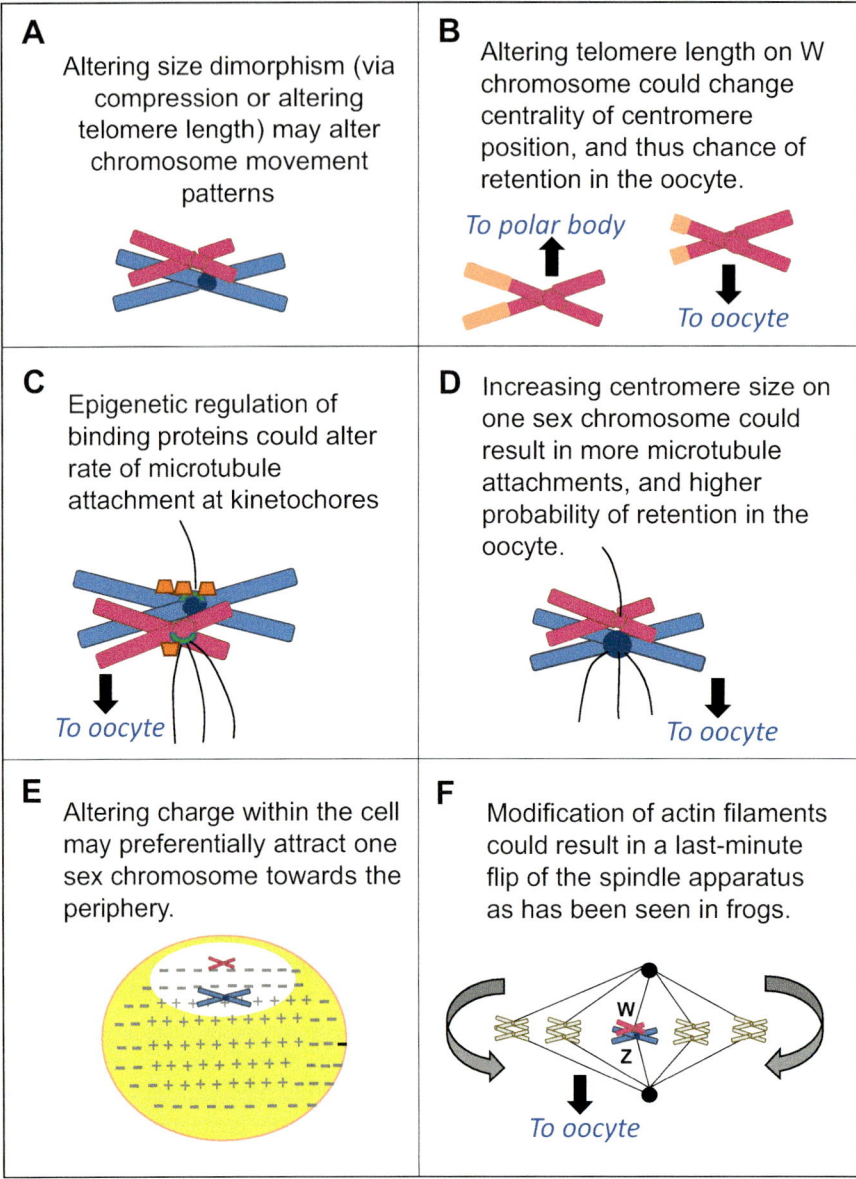

Fig. 6.6 Depiction of the six potential mechanisms by which environmental and social influence may ultimately influence the process of meiotic segregation to influence which sex chromosome is retained in the oocyte

the longest of all the chromosomes found in chickens. In addition, adjacent to the mega-telomere, and closer to the centromere, lies a telomere of a length more typical based on the telomere lengths found in the other chromosomes. Given the position of

6.4 Can Sex Ratios Be Altered via Direct Manipulation of Meiotic Segregation?

the mega-telomere, it is likely not susceptible to recombination (Rodrigue et al. 2005). In addition, Rodrigue et al. (2005) further suggest that telomere lengths may be unstable, and regulated by a chromosome-specific mechanism, given the large variety in telomere lengths among the chromosomes found in chickens. If the telomere length on the W chromosome was reduced, the centromere would not be as centralized on the chromosome, which may alter the potential of that chromosome to move into the oocyte as described in other species above. Telomerase is an enzyme responsible for elongation of telomeres and, in mammals, is highly expressed in oocytes during metaphase of the first meiotic division (Bekaert et al. 2004). If the activity of this enzyme on the W chromosome was altered, this could affect both telomere length and centromere position (Fig. 6.6b).

The relative sizes of the centromeres could also determine which sex chromosome is pulled into the oocyte, as it has been found in other systems that chromosomes attached to more microtubules tend to migrate preferentially into the oocyte (a mechanism termed "centromere drive") (reviewed in Rutkowska and Badyaev 2008) (Fig. 6.6c). This is because centromere size is correlated to kinetochore size, and larger kinetochores have greater interactions with microtubules (Malik and Bayes 2006). It is currently unknown if avian sex chromosomes differ in centromere size, or if centromere size can be controlled at any point during meiosis. We do know, however, that there are proteins associated with the kinetochore that are responsible for blocking attachments to microtubules (Cheeseman et al. 2002). If expression or function of these proteins could be regulated in a manner specific to the W or Z sex chromosome, this could also result in a skewed sex ratio. In order for the localization and/or function of these proteins to respond to environmental or social stimuli, there would likely need to be epigenetic regulation of the DNA sequences that produce the proteins, or of the DNA sequences on the chromatin that attracts the proteins. DNA methylation can change the level of compaction in particular regions of chromosomes (Gupta et al. 2006), and there is evidence of hypermethylation of the Z chromosome during embryonic development in chickens (Teranishi et al. 2001), so it is possible that epigenetic modification of the sex chromosomes could influence the binding of associated proteins that then influence microtubule attachment.

Finally, there is one final way in which the directional movement of sex chromosomes could potentially be controlled. In some species, such as horses, a marsupial species, and a frog species, the spindle apparatus rotates just prior to anaphase I to determine the ultimate fate of the chromosomes (Merry et al. 1995; Gard et al. 1995; Tremoleda et al. 2001). In the case of the frog, an inhibitor of actin assembly inhibited the rotation (Gard et al. 1995), and work in mammals also supports F-actin networks as key regulators of spindle positioning (Almonacid et al. 2014). Rutkowska and Badyaev (2008) suggest that hormonal or ionic gradients within the oocyte could influence how the actin filaments position the spindle (Fig. 6.6f). How this would result in spindle movement in such a way that orients a particular sex chromosome towards the periphery of the oocyte remains unknown.

Overall, given that treatment with hormones just hours prior to ovulation can skew sex ratios and that treatment with steroid hormones and endocrine disruptors has been shown to disrupt meiotic segregation (see below for details), it is possible that the target for sex ratio skews lies in the process of chromosome segregation itself. It is also possible that there is a combination of events, perhaps involving a change in follicular growth rates for example, which ultimately lead to altered segregation of the sex chromosomes. The first step would be to expose birds to a stimulus or treatment known to stimulate sex ratio skews and then collect oocyte at various stages of development to view whether the Z or W chromosome preferentially segregates during the meiotic process and when this may happen. In addition, visualization of the spindle apparatus could provide clues to whether microtubule connections may differ between the Z and W chromosome in response to the stimulus or treatment. Finally, we should measure levels of telomerase and quantify the lengths of the telomeres on the W chromosomes to determine whether the treatment or stimulus alters telomere lengths and perhaps centromere positions.

6.5 Can Sex Ratio Adjustment Occur via Disruption of the Normal Meiotic Process?

Recently, Tagirov and Rutkowska (2013) put forth a new idea towards the mechanism responsible for sex ratio biases in birds. They suggested that sex ratio biases may not, in fact, result from selecting *or* discarding a particular sex chromosome, but may instead result from retaining *both* sex chromosomes; the competition among cells containing those sex chromosomes distinguishes whether the embryo differentiates into a male or female. The beginning of this process would actually occur prior to ovulation during meiotic segregation, by blocking the extrusion of the first polar body. Normally, at the completion of meiosis I, the oocyte contains one diploid set of chromosomes while the remaining diploid set of chromosomes are shuttled to the polar body and ultimately undergo degradation. If polar body extrusion is blocked, however, the oocyte would retain two individual sets of diploid chromosomes.

The second meiotic division completes at the time of fertilization, during which the remaining sister chromatids separate, and another polar body, containing half of the chromatids, is extruded and degrades. One sperm is usually allowed to fertilize the egg, resulting in a new diploid set of chromosomes containing the sex chromosome donated by the female and the Z chromosome donated by the male. But what if the double set of diploid chromosomes resulting from a failed polar body extrusion allows more than one sperm to fertilize the egg? Unlike in mammalian oocytes, in avian oocytes, polyspermy, in which more than one sperm penetrates the perivitelline membrane surrounding the oocyte, is common. Figure 6.7 shows a microscopic view of a perivitelline membrane isolated from an egg after artificial insemination with sperm is common; in fact, it is possible to observe upwards of 60 holes in the membrane, each indicating penetration by an individual spermatozoa. It is thought that only one spermatozoa penetrating at the center of the germinal disk accomplishes fertilization (Johnson 2015); however, it is possible that the presence of three to four pronuclei (depending on whether the second polar body extrusion at meiosis II was

6.5 Can Sex Ratio Adjustment Occur via Disruption of the Normal Meiotic Process?

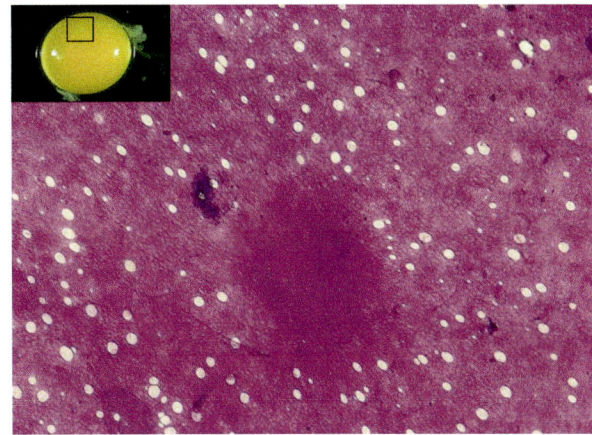

Fig. 6.7 Photograph of the germinal disk region of the perivitelline membrane of a chicken egg stained with Schiff's reagent. Each tiny hole was made by a single sperm that penetrated the membrane, showing that multiple sperm penetrate this layer of the avian egg during fertilization

also disrupted) could allow for sperm that are already present at the site to fertilize (Fig. 6.8). This would result in the formation of up to four diploid blastomeres, some of which are WZ (female) and some of which are ZZ (male). Tagirov and Rutkowska (2013) suggest that environment surrounding those blastomeres (such as the hormonal environment, for example) could differentially affect the proliferation of ZZ versus WZ blastomeres, and eventually, cells of one sex win the competition and determine the sex of the embryo.

How would failure of polar body extrusion occur? Lee and Sydeman (2009) provide an excellent overview of the normal processes involved in cytokinesis in mouse oocytes. Normally, remodeling actin filaments allows for the movement and the anchoring of the spindle carrying the chromosomes in the area of the cortex. The cleavage furrow then forms around the spindle, and cytokinesis occurs to create the polar body. Oocytes are unique in that their contents separate asymmetrically, such that the majority of the contents stay in the segment that will be ovulated so it can be used to nourish the resulting embryo, and only a small amount of the contents is retained in the polar body. In mouse oocytes, a protein called anillin forms a ring around the spindle fibers, and recruits and attaches to several regulators, including RhoA GTP-ase, which controls actin polymerization and myosin contraction, and an actomyosin contractile ring to the plasma membrane that will ultimately constrict, resulting in complete cytokinesis (Lee et al. 2016). Interference with any of these steps would inhibit cytokinesis.

To bias sex ratios in this way, the process of cytokinesis would have to be disrupted as a result of exposure to particular environmental or social stimuli, and this disruption would likely have to occur during *both* polar body extrusion events (telophase I and II). If only one polar body extrusion was disrupted, there would be three pronuclei in the oocyte ready for fusion with sperm DNA, two containing one sex chromosome and only one containing the other. As a result, there would not be equal potential for the blastocyst to develop as male or female. Telophase I occurs while the follicle is still in the ovary, prior to ovulation, while telophase II occurs

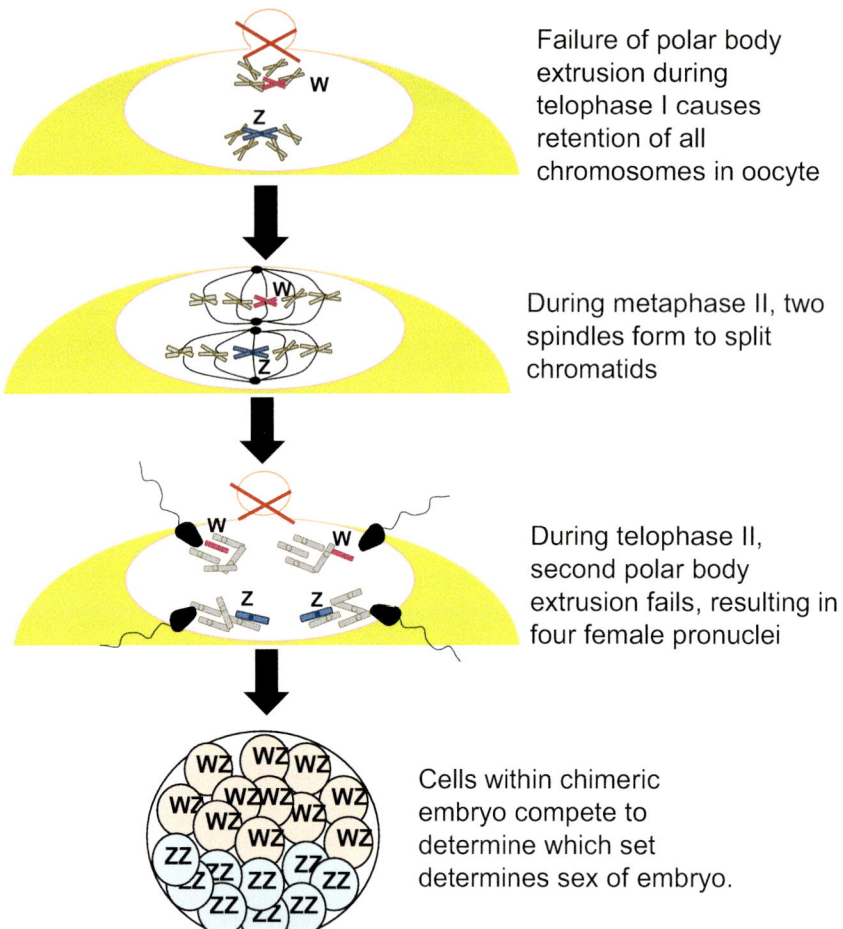

Fig. 6.8 Visualization of the way which sex ratios could be adjusted via the formation of chimeric embryos. If cytokinesis during telophase I fails, the oocyte would retain both sets of homologous chromosomes. Both sets would then undergo meiosis II, aligning along a meiotic plate and segregating to opposite sides of the spindle. If telophase II is successful and a polar body is formed, three sets of haploid chromosomes would remain in the oocyte; however, if cytokinesis failed again, four sets of diploid chromosomes would remain. If four sperm were to enter the egg and fertilize the four pronuclei that formed from these sets of chromosomes, the blastocyst would then have equal numbers of WZ and ZZ cells that would then compete to determine the sex of the offspring

after ovulation when the oocyte meets the sperm in the infundibulum of the oviduct. Thus, if a physiological mediator triggered failure of both polar body extrusion events and the early competitive environment of the blastocyst cells, it would have to be present in both the ovary and the oviduct.

6.6 Can Sex Ratios Be Skewed After Meiotic Segregation?

After the sex chromosomes have segregated, there are still additional ways to alter the sexes of the offspring that ultimately end up in the egg and hatch. The first is via sex-specific follicle atresia. Atretic follicles are those that may initially begin the growth process but then fail to reach ovulation. Once the follicle becomes unviable, the yolk becomes reabsorbed and the follicle disintegrates. Generally, atresia occurs as a result of a loss of gonadotropin support or a destruction of the germinal disk region (reviewed in Johnson 2015). Atresia is the normal fate of most small follicles that do not make it into the pre-ovulatory hierarchy (Gilbert et al. 1983). It is less common in large yellow follicles that have already been recruited into the hierarchy (Waddington et al. 1985), but has been induced in chickens via fasting (Proszkowiec-Węglarz et al. 2005) and corticosterone treatment (Etches et al. 1984). In addition, the prevalence of atresia in hierarchical, pre-ovulatory follicles differs among strains of chickens (Hocking et al. 1987), with broiler breeders showing higher rates of follicle atresia than white leghorns. It is possible that rates of atresia are also higher in species where sex ratio skews have been observed in response to environmental and social variables. This is currently unknown. When prehierarchical follicles become atretic, it is easy to see, as the follicle becomes misshapen and the color of the yolk becomes brighter (Fig. 6.9a). If atresia was triggered in follicles containing the "wrong" sex chromosomes before they ovulated, this could result in a sex ratio skewed towards one sex. Given that the segregation of sex chromosomes completes approximately two hours prior to ovulation, this leaves very little time for the process of atresia to be initiated. Further, there would need to be a mechanism by which the sex chromosome within the oocyte would be detected and/or involved in the process of triggering atresia. It is unknown whether mechanisms exist to support this.

The second mechanism by which sex may be selected after meiosis I has completed is via sex-specific losses of ovulated oocytes, also called internal ovulation. Approximately 1 h prior to ovulation, the infundibulum, which contains nerves, begins to show electrical activity (Shimada and Tanabe 1982) and then

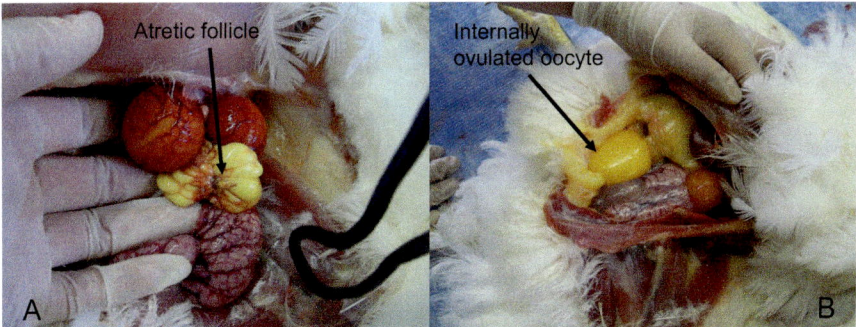

Fig. 6.9 Photographs of (**a**) an atretic follicle in the hen ovary and (**b**) an oocyte that ovulated into the abdominal cavity instead of the infundibulum of the oviduct

rotates to actively engulf the follicle that is about to ovulate. The control of this process is not well understood; however, it is known that the fimbria of the infundibulum engorge with blood, the muscle fibers within those fimbria contract (Aitken 1971), and the wrinkled fimbrial lip, composed of a highly vascularized, double-layered mucosa, grows to surround the oocyte that is being ovulated (Fujii 1981). The infundibulum will not pick up inanimate objects, so this process is likely coordinated by a cross talk with the oocyte (reviewed in Gilbert 1968). After the infundibulum surrounds the oocyte, cilia within the infundibulum beat to draw the oocyte inside (Fujii 1981; Mohammadpour and Keshtmandi 2008). If the sex chromosome carried by the oocyte interrupts the ovary–infundibulum cross talk in some way, or interferes with the beating of the cilia, this would inhibit the proper connection between the follicle and the infundibulum, and the oocyte would instead ovulate into the abdominal cavity (Fig. 6.9b).

Finally, there is one more post-ovulatory mechanism that may act to influence the sexes of offspring that make it to hatch: blocks on fertilization of oocytes carrying a specific sex chromosome. Fertilization is a choreographed group of events that involves the binding of sperm to the zona pellucida, or the extracellular glycoprotein coat, of the egg. Located on the zona pellucida are a group of proteins called zona pellucida, or ZP, proteins. It is currently unclear whether these proteins are produced by the oocyte, the follicular cells, or both, but they play several roles in the process of fertilization. To successfully fertilize an egg, sperm must initially bind to a specific type of ZP protein that allows attachment of the sperm plasma membrane to the zona pellucida. After this point, protein–protein interactions trigger the acrosome reaction, during which the acrosomal contents within the sperm exocytose to fuse with the egg. In mammals, the protein responsible for both steps is called ZPC (Bleil and Wassarman 1980, 1983). If the expression of these critical proteins was blocked when the oocyte carried a particular sex chromosome, this would result in the production of offspring of only one sex. It is hard to think of a mechanism by which environmental and social stimuli would trigger changes in the way W- versus Z-bearing oocytes express ZP proteins, but environmental stimuli that make the *sperm* more or less effective at binding could ultimately alter sex ratios. This is one of the few cases in which the experiences of the *male* would exert the control over the sex ratio adjustment process.

Because birds generally ovulate very regularly, and often at a rate of one egg per day, sex ratio biases caused by any of these three mechanisms would likely result in obvious changes in the rate of egg laying or the fertility rates within eggs. In a majority of studies in which sex ratio biases have been observed in bird, there do not appear to be corresponding changes in the rates of missing or infertile eggs; however, missing eggs (resulting in laying gaps) due to disruptions of ovulation or oviposition *are* known to occur routinely in response to stressful conditions and territorial intrusions in some avian species (Nilsson and Svensson 1993; Wiebe and Martin 1995; Low 2008). Thus, there are mechanisms that link exposure to environmental and social stimuli to the mechanisms responsible for the successful production of an egg. In addition, there are ways that the loss of oocytes could go unnoticed by researchers counting eggs in the laying sequence. This would be particularly true for species that lay only 1–2 eggs per clutch, such as pigeons. In fact, Goerlich et al. (2009) showed that testosterone treatment of pigeon females biases sex ratios towards males but only in first eggs of the clutch. This could indicate that oocytes carrying the W chromosome were discarded, and ovulation was delayed until an oocyte containing the Z chromosome was ready for ovulation. Given that this occurred in the first egg of the sequence, missing eggs may not be detected, particularly if one-egg clutches are relatively common. Further, Pike (2005) suggested that another way that internal ovulation events might go undetected is if two follicles ready for ovulation are maintained in the ovary at the same time. He formed this idea to fit experimental results showing that pigeons with reduced body condition caused by repeated egg laying and limited food intake produced female-biased sex ratios. He did not directly observe the follicles in the ovary or the presence of internal ovulations, however. Broiler breeder hens have been observed to have two similar sized follicles ready for ovulation at a given time, and this can result in the ovulation of more than one oocyte at a time (Hocking et al. 1987). Broiler breeders are known for their many reproductive dysfunctions, however, and it is unclear whether this possibility would apply in a wild bird where sex ratio skews are seen. In fact, a direct study of the ovaries of pigeons that were adjusting sex ratios in response to body mass showed no evidence for sex-specific follicle atresia or internal ovulations (Goerlich et al. 2010).

6.7 Can Females Control Sex After Eggs Are Laid?

In most cases of sex ratio bias in birds, the possibility that sex ratio adjustment occurs via sex-specific death of conceptuses, either during the blastocyst or embryo stage, is often discarded because evidence of this would be immediately obvious; there would be more eggs with little to no embryonic development present when sex ratios are biased, and such patterns are generally not found. However, there are some cases in which sex-specific susceptibility to factors during incubation leads to biased sex ratios at hatch. For example, excitement was generated when Göth and Booth (2005) showed the first evidence of temperature-dependent sex ratio biases in a bird. Australian brush turkeys build large mounds in which eggs are laid by females, and males attending those nests routinely assess the temperatures within the nests using a special gland on the beak, and then adjust temperatures within an optimal range by adding or taking away nesting material. Göth and Booth (2005) artificially incubated brush turkey eggs under cool (31 °C), average (34 °C), or hot (36 °C) conditions and showed that nearly 80% males hatched from eggs incubated at the cool temperature while about 70% females hatched from eggs incubated at the hot temperature. Those incubated at the average temperature hatched approximately 50% of each sex. Later work confirmed that temperature-driven sex ratio biases in this species are caused by sex-specific patterns of death during the early stages of embryonic development; females are more susceptible at high temperatures, while males are more susceptible at low temperatures.

After this study in brush turkeys was published, the potential for temperature-dependent influences on embryo mortality in the two sexes was studied in other avian species. A study of Japanese quail, in which eggs were incubated at five temperatures ranging from 36.7 to 38.7 °C, showed that over 60% males hatched from eggs incubated at the lowest two temperatures, while only 42–44% males hatched from eggs incubated at the higher three temperatures (Yılmaz et al. 2011). Work in wood ducks also showed that eggs incubated at a lower temperature (35 °C) were more likely to produce male offspring (63%) compared to those incubated at two higher temperatures (35.9 °C: 44% male, 37 °C: 47% male) (DuRant et al. 2016). Thus, both quail and wood ducks show similar patterns of sex-specific mortality to those seen in brush turkeys; males are more susceptible to temperatures that are too high and females are more susceptible to temperatures that are too low.

What might be the mechanism for a sex-specific sensitivity to high or low temperatures early in embryonic development? To my knowledge, there are no studies that have examined the different metabolic needs of male versus female avian embryos at various stages of development, but work in mammals shows that metabolic rates of male and female embryos differ substantially from very early on in development (Tiffin et al. 1991). It has been shown in some birds that male and female embryos have sex-specific responses to androgen exposure (Sockman et al. 2008), suggesting that male and female embryos differ in their physiological responses from even 1–2 days of incubation, when androgen concentrations are still high in the egg yolk (Elf and Fivizzani 2002). How such a physiological difference might affect susceptibility to suboptimal temperatures, however, remains unknown. Regardless, manipulation of temperatures during incubation appears to be a mechanism behind sex ratio biases in some exclusive cases.

6.8 Which Would Be the Optimal Mechanism of Sex Ratio Adjustment?

It is critical that we understand the mechanism(s) that underlie sex ratio adjustment in birds, because without this knowledge, we cannot hope to understand whether and how birds can alter sex ratios in an adaptive manner. Each mechanism outlined above brings with it a cost. For example, brush turkeys lay large eggs and supply very little postnatal care to offspring. Any mechanism involving the stimulation of embryonic death represents a large loss in terms of the energy allocated to forming the eggs. So, while it is clearly possible to adjust sex ratios by adjusting incubation temperatures to the extremes of the physiological norms, it may not be adaptively beneficial. In theory, from an adaptive standpoint, any mechanism that allows a female to avoid a significant loss in reproductive effort would be the optimal mechanism to use over another method that requires she fully form an egg, allocate significant resources to that egg, and even spend time and energy incubating it before sexes can be adjusted. Many studies in which offspring sex is examined at very early embryonic ages suggest that female birds have the ability to adjust sex ratios either at or before the completion of meiosis, which allows those females to avoid the need to discard a fully formed ovarian follicle. However, we still do not know the energy requirements for stimulating such a sex ratio bias or other costs involved. For example, if females must elevate testosterone concentrations to achieve a male-biased sex ratio, this could have significant side effects, including increases in metabolic rate and aggression, for example. In addition, exposure of offspring to high levels of testosterone in eggs can ultimately program those offspring to be less immunocompetent (Navara et al. 2005; Andersson et al. 2004). If sex ratio biases occur as a result of altering follicle growth rates, the production of the sex with lower follicle growth rates could also ultimately result in less nutritional support for the developing embryo, which could exert fitness costs down the line. The costs and benefits of each mechanism will likely be species specific, depending on factors such as the amount of energy invested in egg production, the sensitivity of those females to the side effects of hormone elevations, and the level of variation in yolk sizes and follicle growth rates, which also varies significantly among species. Before we can hope to understand whether females can adaptively bias sex ratios, we need to first understand how they do it, and the costs involved in the process.

References

Aitken R (1971) The oviduct. In: Bell DJ, Freeman BM (eds) Physiology and biochemistry of the domestic fowl, vol 3. Academic, London, pp 1237–1289

Almonacid M, Terret M-É, Verlhac M-H (2014) Actin-based spindle positioning: new insights from female gametes. J Cell Sci 127(3):477–483

Andersson S, Uller T, Lõhmus M, Sundström F (2004) Effects of egg yolk testosterone on growth and immunity in a precocial bird. J Evol Biol 17(3):501–505

Badyaev A, Acevedo Seaman D, Navara K, Hill G, Mendonca M (2006) Evolution of sex-biased maternal effects in birds: III. Adjustment of ovulation order can enable sex-specific allocation of hormones, carotenoids, and vitamins. J Evol Biol 19(4):1044–1057

Bekaert S, Derradji H, Baatout S (2004) Telomere biology in mammalian germ cells and during development. Dev Biol 274(1):15–30

Bleil JD, Wassarman PM (1980) Mammalian sperm-egg interaction: identification of a glycoprotein in mouse egg zonae pellucidae possessing receptor activity for sperm. Cell 20(3):873–882

Bleil JD, Wassarman PM (1983) Sperm-egg interactions in the mouse: sequence of events and induction of the acrosome reaction by a zona pellucida glycoprotein. Dev Biol 95(2):317–324

Carpenter AT (1991) Distributive segregation: motors in the polar wind? Cell 64(5):885–890

Cheeseman IM, Drubin DG, Barnes G (2002) Simple centromere, complex kinetochore. J Cell Biol 157(2):199–203

DuRant SE, Hopkins WA, Carter AW, Kirkpatrick LT, Navara KJ, Hawley DM (2016) Incubation temperature causes skewed sex ratios in a precocial bird. J Exp Biol 219(13):1961–1964

Elf PK, Fivizzani AJ (2002) Changes in sex steroid levels in yolks of the leghorn chicken, Gallus domesticus, during embryonic development. J Exp Zool A Ecol Genet Physiol 293(6):594–600

Etches R, Williams J, Rzasa J (1984) Effects of corticosterone and dietary changes in the hen on ovarian function, plasma LH and steroids and the response to exogenous LH-RH. J Reprod Fertil 70(1):121–130

Fujii S (1981) Scanning electron microscopic observation on ciliated cells of the chicken oviduct in various functional stages. J Fac Appl Biol Sci Hiroshima Univ 20:1–11

Fuller MT (1995) Riding the polar winds: chromosomes motor down east. Cell 81(1):5–8

Gard DL, Cha B-J, Roeder AD (1995) F-actin is required for spindle anchoring and rotation in Xenopus oocytes: a re-examination of the effects of cytochalasin B on oocyte maturation. Zygote 3(01):17–26

Gilbert A (1968) An observation bearing on the ovarian-oviduct relationship in the domestic hen. Br Poult Sci 9:301–302

Gilbert A, Perry M, Waddington D, Hardie M (1983) Role of atresia in establishing the follicular hierarchy in the ovary of the domestic hen (Gallus domesticus). J Reprod Fertil 69(1):221–227

Gileva E, Rakitin S (2006) Factors of maintaining chromosome polymorphism in common vole Microtus arvalis Pallas, 1779: reduced fertility and meiotic drive. Russ J Genet 42(5):498–504

Goerlich VC, Dijkstra C, Schaafsma SM, Groothuis TG (2009) Testosterone has a long-term effect on primary sex ratio of first eggs in pigeons—in search of a mechanism. Gen Comp Endocrinol 163(1):184–192

Goerlich VC, Dijkstra C, Groothuis TG (2010) No evidence for selective follicle abortion underlying primary sex ratio adjustment in pigeons. Behav Ecol Sociobiol 64(4):599–606

Göth A, Booth DT (2005) Temperature-dependent sex ratio in a bird. Biol Lett 1(1):31–33

Gupta S, Pathak RU, Kanungo MS (2006) DNA methylation induced changes in chromatin conformation of the promoter of the vitellogenin II gene of Japanese quail during aging. Gene 377:159–168

Heinsohn R, Legge S, Barry S (1997) Extreme bias in sex allocation in Eclectusparrots. Proc R Soc Lond B Biol Sci 264(1386):1325–1329

Hocking P, Gilbert A, Walker M, Waddington D (1987) Ovarian follicular structure of White Leghorns fed ad libitum and dwarf and normal broiler breeders fed ad libitum or restricted until point of lay. Br Poult Sci 28(3):493–506

Johnson P (2012) Follicle selection in the avian ovary. Reprod Domest Anim 47(s4):283–287

Johnson AJ (2015) Reproduction in the female. In: Scanes CG (ed) Sturkie's avian physiology, 6th edn. Academic, Boston, pp 635–665

Johnson P, Dickens M, Kent T, Giles J (2005) Growth differentiation factor 9: an oocyte factor regulating ovarian follicle development. In: Dawson A, Sharp PJ (eds) Functional avian endocrinology. Alpha Science International, p 313

Koszul R, Kim K, Prentiss M, Kleckner N, Kameoka S (2008) Meiotic chromosomes move by linkage to dynamic actin cables with transduction of force through the nuclear envelope. Cell 133(7):1188–1201

Krackow S (1995) Potential mechanisms for sex ratio adjustment in mammals and birds. Biol Rev 70(2):225–241

Lee DE, Sydeman WJ (2009) North Pacific climate mediates offspring sex ratio in Northern elephant seals. J Mammal 90(1):1–8

Lee SR, Jo YJ, Namgoong S, Kim NH (2016) Anillin controls cleavage furrow formation in the course of asymmetric division during mouse oocyte maturation. Mol Reprod Dev 83(9):792–801

Lénárt P, Bacher CP, Daigle N, Hand AR, Eils R, Terasaki M, Ellenberg J (2005) A contractile nuclear actin network drives chromosome congression in oocytes. Nature 436(7052):812–818

Levesque AA, Compton DA (2001) The chromokinesin Kid is necessary for chromosome arm orientation and oscillation, but not congression, on mitotic spindles. J Cell Biol 154(6):1135–1146

Low M (2008) Laying gaps in the New Zealand stitchbird are correlated with female harassment by extra-pair males. Emu 108(1):28–34

Malik H, Bayes J (2006) Genetic conflicts during meiosis and the evolutionary origins of centromere complexity. Biochem Soc Trans 34:569–573

Mendonça MAC, Carvalho CR, Clarindo WR (2010) DNA content differences between male and female chicken (Gallus gallus domesticus) nuclei and Z and W chromosomes resolved by image cytometry. J Histochem Cytochem 58(3):229–235

Merry N, Johnson MH, Gehring C, Selwood L (1995) Cytoskeletal organization in the oocyte, zygote, and early cleaving embryo of the stripe-faced dunnart (Sminthopsis macroura). Mol Reprod Dev 41(2):212–224

Mitchison T, Kirschner M (1985) Properties of the kinetochore in vitro. II. Microtubule capture and ATP-dependent translocation. J Cell Biol 101(3):766–777

Mohammadpour A, Keshtmandi M (2008) Histomorphometrical study of infundibulum and magnum in turkey and pigeon. World J Zool 3:47–50

References

Navara KJ, Hill GE, Mendonça MT (2005) Variable effects of yolk androgens on growth, survival, and immunity in eastern bluebird nestlings. Physiol Biochem Zool 78(4):570–578

Nilsson J-Å, Svensson E (1993) The frequency and timing of laying gaps. Ornis Scand 24:122–126

Novitski E (1967) Nonrandom disjunction in Drosophila. Annu Rev Genet 1(1):71–86

Parker I, Yao Y (1995) Calcium puffs in xenopus oocytes. In: Ciba foundation symposium 188-Calcium waves, gradients and oscillations. Wiley Online Library, pp 50–65

Pike TW (2005) Sex ratio manipulation in response to maternal condition in pigeons: evidence for pre-ovulatory follicle selection. Behav Ecol Sociobiol 58(4):407–413

Proszkowiec-Węglarz M, Rząsa J, Słomczyńska M, Paczoska-Eliasiewicz H (2005) Steroidogenic activity of chicken ovary during pause in egg laying. Reprod Biol 5:205–225

Rodrigue K, May B, Famula T, Delany M (2005) Meiotic instability of chicken ultra-long telomeres and mapping of a 2.8 megabase array to the W-sex chromosome. Chromosom Res 13(6):581–591

Rutkowska J, Badyaev AV (2008) Meiotic drive and sex determination: molecular and cytological mechanisms of sex ratio adjustment in birds. Philos Trans R Soc Lond B: Biol Sci 363 (1497):1675–1686

Sakuno T, Watanabe Y (2009) Studies of meiosis disclose distinct roles of cohesion in the core centromere and pericentromeric regions. Chromosom Res 17(2):239–249

Shimada K, Tanabe Y (1982) Electrical activity of the infundibulum in relation to ovulation in the chicken. J Reprod Fertil 65(2):419–423

Sockman KW, Weiss J, Webster MS, Talbott V, Schwabl H (2008) Sex-specific effects of yolk-androgens on growth of nestling American kestrels. Behav Ecol Sociobiol 62(4):617–625

Solari AJ (1993) Sex chromosomes and sex determination in vertebrates. CRC Press, Boca Raton

Solovei I, Gaginskaya E, Hutchison N, Macgregor H (1993) Avian sex chromosomes in the lampbrush form: the ZW lampbrush bivalents from six species of bird. Chromosom Res 1(3):153–166

Spence JM, Mills W, Mann K, Huxley C, Farr CJ (2006) Increased missegregation and chromosome loss with decreasing chromosome size in vertebrate cells. Chromosoma 115(1):60–74

Tagirov M, Rutkowska J (2013) Chimeric embryos—potential mechanism of avian offspring sex manipulation. Behav Ecol art007

Teranishi M, Shimada Y, Hori T, Nakabayashi O, Kikuchi T, Macleod T, Pym R, Sheldon B, Solovei I, Macgregor H (2001) Transcripts of the MHM region on the chicken Z chromosome accumulate as non-coding RNA in the nucleus of female cells adjacent to the DMRT1 locus. Chromosom Res 9(2):147–165

Tesarik J, Mendoza C (1995) Nongenomic effects of 17 beta-estradiol on maturing human oocytes: relationship to oocyte developmental potential. J Clin Endocrinol Metab 80(4):1438–1443

Tiffin G, Rieger D, Betteridge K, Yadav B, King W (1991) Glucose and glutamine metabolism in pre-attachment cattle embryos in relation to sex and stage of development. J Reprod Fertil 93 (1):125–132

Tremoleda J, Schoevers E, Stout T, Colenbrander B, Bevers M (2001) Organisation of the cytoskeleton during in vitro maturation of horse oocytes. Mol Reprod Dev 60(2):260–269

Waddington D, Perry M, Gilbert A, Hardie M (1985) Follicular growth and atresia in the ovaries of hens (Gallus domesticus) with diminished egg production rates. J Reprod Fertil 74(2):399–405

Wiebe KL, Martin K (1995) Ecological and physiological effects on egg laying intervals in ptarmigan. Condor 97:708–717

Wignall SM, Villeneuve AM (2009) Lateral microtubule bundles promote chromosome alignment during acentrosomal oocyte meiosis. Nat Cell Biol 11(7):839–844

Yao HH, Bahr JM (2001) Germinal disc-derived epidermal growth factor: a paracrine factor to stimulate proliferation of granulosa cells 1. Biol Reprod 64(1):390–395

Yılmaz A, Tepeli C, Garip M, Çağlayan T (2011) The effects of incubation temperature on the sex of Japanese quail chicks. Poult Sci 90(10):2402–2406

Young R, Badyaev AV (2004) Evolution of sex-biased maternal effects in birds: I. Sex-specific resource allocation among simultaneously growing oocytes. J Evol Biol 17(6):1355–1366

Hormones Rule the Roost: Hormonal Influences on Sex Ratio Adjustment in Birds and Mammals

7

> *In view of the apparent lack of genetic variance in the sex ratio in many species, a hormonal mechanism mediated by environmental factors provides a plausible explanation of many trends*
>
> Clutton-Brock and Iason (1986)

> *The purpose of this note is to persuade endocrinologists that mammalian sex ratios merit their attention*
>
> James (2008)

There is now abundant evidence supporting the idea that mammals and birds can facultatively control the sexes of offspring in response to environmental and social conditions, and we also know that there are some likely targets within the reproductive system during the process of gamete production and offspring development that may be manipulated to bias the sex ratios of the offspring produced. What we are missing now are the physiological transducers that act to convert environmental and social information into physiological signals that then act on either the developing gametes or the growing embryos to alter offspring sex. These transducers would need to interact with and coordinate a suite of both regulatory and responsive body tissues, and there are three main body systems known to act in this manner: the nervous, immune, and endocrine systems. There is now mounting evidence that these three systems are intimately interconnected, even leading to the renaming of the three as the neuroendocrine immune system (Wilder 1995). Together, these three systems control perception of external stimuli that an animal may encounter, the translation of that perception into a chemical messenger, and the reaction of target tissues. It is no wonder, then, that some have hypothesized that activities of this system underlie the mechanism behind sex ratio adjustment in vertebrates. In this chapter, I will present mounting evidence that hormones, steroid hormones in particular, are key players in the mechanisms responsible for sex ratio adjustment in both birds and mammals.

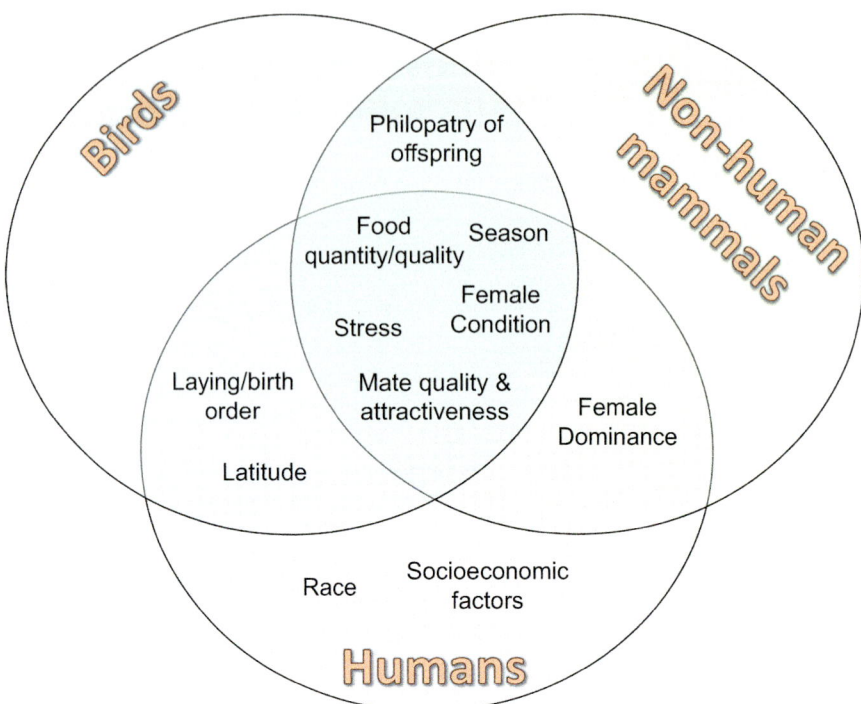

Fig. 7.1 A Venn diagram of the overlapping factors that have been shown to influence sex ratios in birds, humans, and nonhuman mammals

7.1 Links Between Hormones and Factors that Alter Sex Ratios

In Chaps. 2, 3, and 5, I outlined the factors that have been shown to influence sex ratios in humans, nonhuman mammals, and birds. Strikingly, despite the fact that birds and mammals utilize different modes of sex determination (WZ versus XY), nearly all factors that drive sex ratio biases are shared among the three groups (Fig. 7.1). These factors include food quality and quantity, female condition, progression through the breeding season, stress, and mate attractiveness and/or quality. Whether all of these factors have the capacity to trigger sex ratio biases on their own or whether a single factor that interrelates all of the others is responsible for the documented sex ratio skews in these systems is unknown. What *is* known, however, is that the physiological responses to a majority of these cues involve the same set of endocrine signals—steroid hormones. Figure 7.2 highlights the fact that while each of the cues known to trigger sex ratio biases influences physiological changes via a complex network of mediators, all of these mediators interact with the adrenal and reproductive steroid hormones, corticosterone in birds and cortisol in mammals, testosterone, estrogen, and progesterone.

7.1 Links Between Hormones and Factors that Alter Sex Ratios

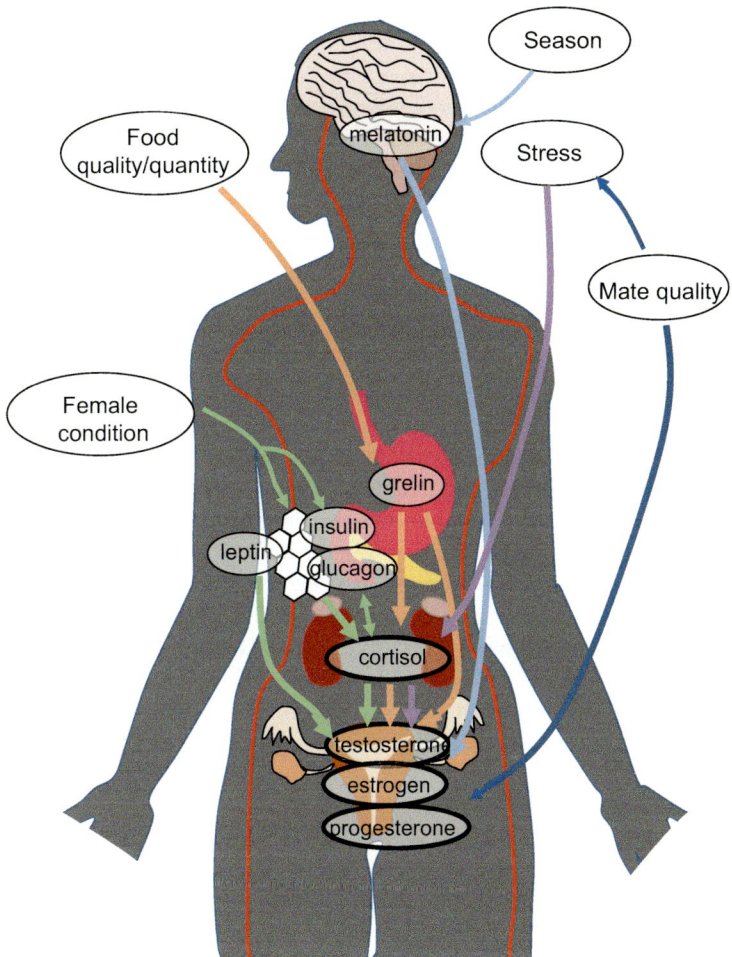

Fig. 7.2 A diagrammatic representation of how multiple factors that have been shown to influence avian and mammalian sex ratios ultimately interact through the same hormonal pathways involving actions of glucocorticoids and/or reproductive hormones

For example, the physiological response to a lack of food or a decrease in maternal body condition triggers the responses of mediators that directly interact in a reciprocal fashion with the adrenal glands, because a main action of the adrenal glucocorticoids is to stimulate the breakdown of energy reserves when the individual is in a stress state. This is true for both birds and mammals. We also know that many of these mediators directly interact with the reproductive system to stimulate or inhibit the function of the reproductive organs and the production of the sex steroids as a result. When looking at how these different cues collectively influence reproduction in both birds and mammals, it becomes clear that most of the responses funnel through the adrenal system, alter the production of glucocorticoids, and ultimately trigger changes in levels of sex steroids to influence the reproductive system. As a result, four prime candidates as mediators of sex ratio adjustment in both birds and mammals are cortisol/corticosterone, testosterone, progesterone, and estrogen. Below, I will present the evidence for and against these hormones as prime mediators of sex ratio adjustment in birds and mammals.

It is now well known that when the production of steroid hormones is triggered, developing gametes and offspring are exposed to those steroid hormones. First, several of these hormones are produced in the cells surrounding the developing gametes; in both birds and mammals, the granulosa and theca cell layers surrounding the oocyte produce androgens, estrogens, and progesterone. In addition, stress hormones (i.e., glucocorticoids such as cortisol in mammals or corticosterone in birds) are produced by the adrenal glands which lie in relatively close proximity to the ovary in both birds and mammals. Studies have documented accumulation of testosterone, estradiol, and corticosterone in the yolks of avian eggs (Schwabl 1993; Sockman and Schwabl 1999). Concentrations of these hormones can be influenced by conditions that the female experiences during rapid yolk deposition. For example, female Eastern bluebirds that experienced simulated territorial intrusions and/or high breeding densities deposited more testosterone into their eggs (Navara et al. 2006; Bentz et al. 2016b). Bluebirds are not the only species in which this occurs. Similar elevations occur in house sparrows, American coots, tree swallows, European starlings, and others (Schwabl 1997; Reed and Vleck 2001; Whittingham and Schwabl 2002; Bentz et al. 2013; Pilz and Smith 2004) reviewed in (Bentz et al. 2016a). Along the same lines, exposure to stressful experiences drives females to elevate concentrations of corticosterone in yolks. Barn swallows exposed to a predator during egg laying produced eggs with significantly higher concentrations of corticosterone in the yolks (Saino et al. 2002). Hayward and Wingfield (2004) showed in quail that when concentrations of corticosterone are elevated in circulating plasma, concentrations in egg yolks rise as well. These yolk hormones then go on to exert potent programming influences on offspring growing within the eggs (Groothuis and Schwabl 2008; Navara and Mendonça 2008; Hayward and Wingfield 2004). This paradigm is not restricted to birds. Abundant research suggests that hormones produced by mammalian mothers reach and influence offspring during development. For example, female spotted hyenas with higher dominance ranks produce higher concentrations of testosterone during gestation, and exposure to these higher hormone levels programs cubs to exhibit higher levels of aggression (Dloniak et al. 2006). It is also well known that stress-induced glucocorticoids reach and program developing mammalian embryos (e.g., Drake et al. 2005; Tangalakis et al. 1992). Thus, it is clear that there is a pathway by which steroid hormones can act to communicate information about the environment to the reproductive organs in a way that influences gamete and/or offspring growth and development. Given such a pathway, it is logical to hypothesize that these steroid hormones may also mediate changes in sex ratios in response to the same types of environmental and social conditions.

7.2 Evidence that Steroid Hormones Influence Avian Sex Ratios

7.2.1 Case Study: Peafowl

Some of the first solid evidence that steroid hormones play a role in the adjustment of avian sex ratios came from two studies on peafowl conducted by Thomas Pike and Marion Petrie in 2005. First, Pike found that when they manipulated the attractiveness of the peacocks that the peahens were mating with by removing eye spots on the train, females mating with these males deposited higher concentrations of corticosterone into egg yolks and also produced a significantly higher proportion of female offspring (Pike 2005). They then more directly examined the correlation between maternal hormone concentrations, body condition, and offspring sex ratios in peafowl; they found that females in lower body condition had higher concentrations of corticosterone in circulation, lower concentrations of testosterone, and they produced a significantly higher proportion of female offspring. A direct test of how the concentrations of the two steroid hormones related to sex ratios produced showed a significant positive correlation between sex ratio and testosterone concentrations and a significant negative correlation between sex ratios and corticosterone concentrations (Pike and Petrie 2005). Because corticosterone can influence levels of testosterone and vice versa, it was not clear whether both steroids participated in the determination of offspring sex, or whether one was related to sex ratios indirectly due to an interaction with the other. However, these two studies opened up a flood of subsequent experiments examining the roles of these two hormones in sex ratio adjustment.

7.2.2 Case Study: Japanese Quail

Following the peacock studies, the same authors continued examining the role of steroid hormones in adjustment of sex ratios, but this time in a different model—Japanese quail (Pike and Petrie 2006). To experimentally test whether elevated concentrations of corticosterone, estradiol, and testosterone influence sex ratios, they implanted quail with silastic implants containing one of the following: estradiol, fadrozole (an aromatase inhibitor that prevents production of estrogen), corticosterone, metyrapone (a corticosterone synthesis inhibitor), or testosterone. After implantation, they collected eggs for 10 days and compared sex ratios to pretreatment sex ratios. They found that females treated with corticosterone significantly reduced the proportion of male offspring they produced in comparison both to sex ratios of offspring that they produced prior to treatment and to sex ratios produced by females in the other treatment groups. None of the other treatment groups triggered a significant sex ratio bias compared to pretreatment sex ratios. It is interesting that elevating corticosterone concentrations exerted a significant effect on sex ratios, while inhibiting corticosterone production via treatment with metyrapone did not. Perhaps this indicates that the default sex ratio produced by quail females is always 50:50 unless a part of the sex determination process responsible for producing males is disrupted by exposure to high concentrations of corticosterone. Whether this is, in fact, true still remains unknown, and the effects of metyrapone have not been tested in any other additional species. It is interesting that testosterone did not exert an influence, given the positive relationship between testosterone and sex ratios produced by peahens. This could indicate that quail utilize a different mechanism of sex ratio adjustment or that the link between testosterone and sex ratios in peahens was due to an indirect linkage with corticosterone.

7.2.3 Influences of Corticosterone in Other Systems

Corticosterone shows great potential as a mediator of offspring sex ratios, not only because the results of the work in peafowl and quail strongly support corticosterone as a candidate mediator but also because the key function of this hormone is to respond to changes in surrounding stimuli and coordinate body systems to deal with those changes. As shown above (Fig. 7.2), corticosterone lies at the nexus of all of the responses to cues that have been shown to stimulate sex ratios in birds. In response to food restriction, corticosterone concentrations elevate and act to liberate energy reserves. When experiencing a stressful event, corticosterone is the primary responder that acts to help the animal maintain homeostasis in the face of that event. Corticosterone concentrations even change seasonally and could underlie the seasonal changes seen in avian sex ratios. A body of work now supports the idea that corticosterone plays a role in offspring sex ratios, though the story is not quite as simple as we might expect.

Following the quail study by Pike and Petrie (2006), Bonier et al. (2007) conducted a two-prong experiment in white-crowned sparrows. First, they measured corticosterone concentrations in naturally breeding females and showed that those with higher levels of corticosterone produced significantly higher proportions of female offspring. They then tested this relationship experimentally using time-release corticosterone pellets. Females implanted with corticosterone produced significantly more female offspring compared to controls, just as female quail did. What's more, a similar study in homing pigeons showed the same effect; females implanted with corticosterone produced more female offspring. Thus, elevating corticosterone over a period of a week or longer, either endogenously or via corticosterone implantation, stimulates the production of a higher proportion of female offspring in all four of these avian species. However, when during sex determination does corticosterone act to skew sex ratios, and does corticosterone act alone or in combination with other modulators?

Examining the timing of the treatments in these four studies may tell us a bit more about how and when corticosterone is acting (Fig. 7.3). In the white-crowned sparrow study, implants were given to birds after the first clutch, prior to re-nesting. The birds began their second clutches on average 10 days after implantation, which would mean that concentrations of corticosterone would have been elevated during

7.2 Evidence that Steroid Hormones Influence Avian Sex Ratios

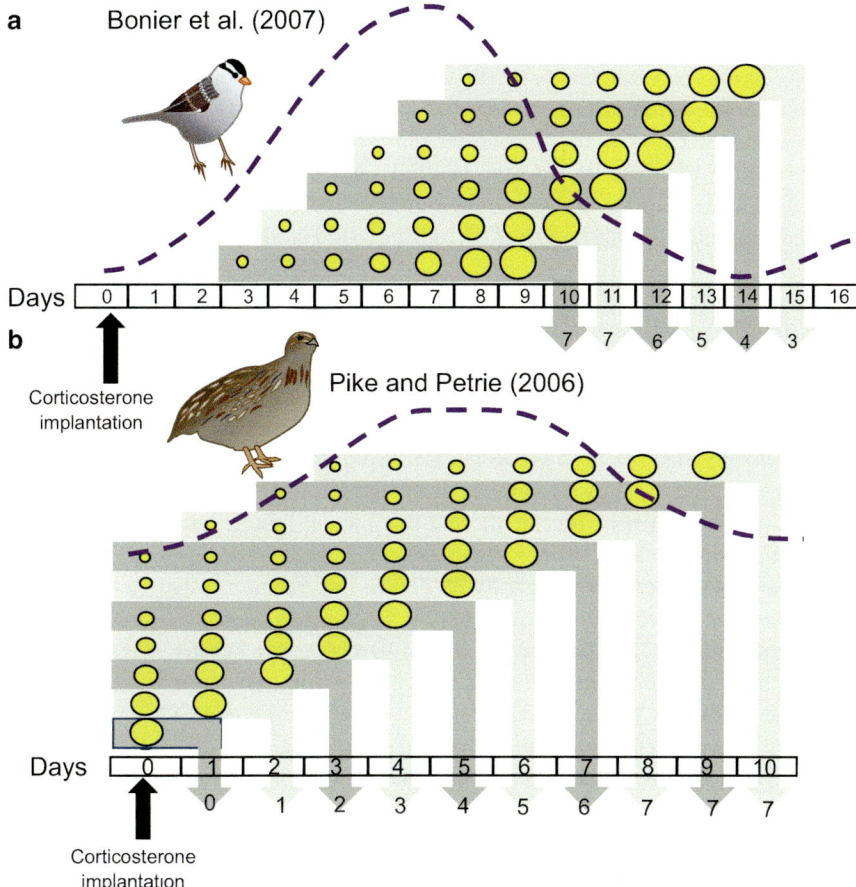

Fig. 7.3 Illustrations of the timing of corticosterone treatment relative to rapid yolk deposition within ovarian follicles in two key studies in which corticosterone implants were used (**a**) White-crowned sparrow: Bonier et al. (2007), (**b**) Japanese quail: Pike and Petrie (2006). In both cases, the dashed purple line indicates the estimated profile of corticosterone concentrations in circulation, downward pointing arrows indicate when the indicated follicle ovulates, and numbers within the arrows indicate the number of days that the developing follicle would likely have been affected by elevated concentrations of corticosterone while undergoing rapid yolk deposition. In the white-crowned sparrow study (**a**), the first egg of the second clutch was laid, on average, 10 days after implantation of corticosterone pellets, and the day of clutch completion was, on average, at 15 days after implantation. By day 15, corticosterone concentrations had returned to normal, which suggests that meiotic segregation and ovulation of the later few follicles occurred when corticosterone concentrations were low, yet resulting sex ratios were still female biased. In the quail study (**b**), eggs were collected for 10 days after implantation. As a result, corticosterone concentrations were likely elevated during meiotic segregation and ovulation for nearly all follicles, but because egg collections started quickly after implantation, only the final three follicles would have been exposed to elevated levels of corticosterone during the entire phase of rapid yolk deposition

a majority of rapid yolk deposition (which takes about 5–7 days) for these clutches. However, concentrations of corticosterone in corticosterone-implanted birds were already down to baseline levels by the time those clutches completed (on average 15 days after implantation). Because meiotic segregation completes just hours prior to ovulation (see Chap. 6), corticosterone concentrations in circulation were likely very low at the time that at least the last few eggs in the clutch were completing meiotic segregation prior to their ovulation. This could mean that sex ratio adjustment happens as a result of an effect during rapid yolk deposition, perhaps by altering follicle growth rates (a possibility described in detail in Chap. 6). On the other hand, if enough corticosterone accumulates in yolk to influence meiotic segregation, then there still could be a direct effect at that stage as well.

The quail study (Pike and Petrie 2006) could shed further light on the timing of corticosterone action. In this case, the authors reported the average sex ratio for 10 days following implantation with corticosterone. They further reported that there was no evidence that sex ratio skews became more dramatic towards the end of that 10-day period. Yet, as shown in Fig. 7.3, only the last three eggs collected had been exposed to the treatment during the entire time of follicle growth. This could indicate that, in this case, corticosterone was not acting on sex ratios by influencing the contents of the egg during rapid yolk deposition, because if that were the case, we would have expected the female-biased sex ratio to be driven by a very high proportion of females towards the end of that 10 day period, because eggs laid in the first few days after implantation would have had little to no time to deposit yolk under the influence of the elevated levels of corticosterone. This points to a mechanism by which corticosterone acted directly on the process of meiotic segregation. However, it is also possible that only a short time of corticosterone elevation is required during rapid yolk deposition to exert the effects that would ultimately influence offspring sex.

There was another study that tested the influences of long-term treatment with corticosterone that further supports the idea that corticosterone may act at the time of meiotic segregation. Aslam et al. (2014) provided feed that contained corticosterone to laying hens for 14 days and examined the resulting sex ratio in eggs (Fig. 7.4a). In this case, despite treatment that spanned a similar time period to the previous two studies discussed, corticosterone treatment did not appear to influence sex ratios, though it did alter the relationship between body mass and sex ratios. Corticosterone was provided through the diet and this treatment produced significantly higher concentrations of corticosterone in circulation during the day by day 4 of treatment and extending through at least day 10. However, because birds do not take in food at night, concentrations of corticosterone likely dropped substantially during the night. Given that meiotic segregation often takes place in the very early hours before lights go on, levels of corticosterone would have been much lower at this time than during the time that blood was collected for corticosterone measurement. It is possible that this method of supplying corticosterone did not bias sex ratios because concentrations were not high enough during meiotic segregation to influence the segregation process.

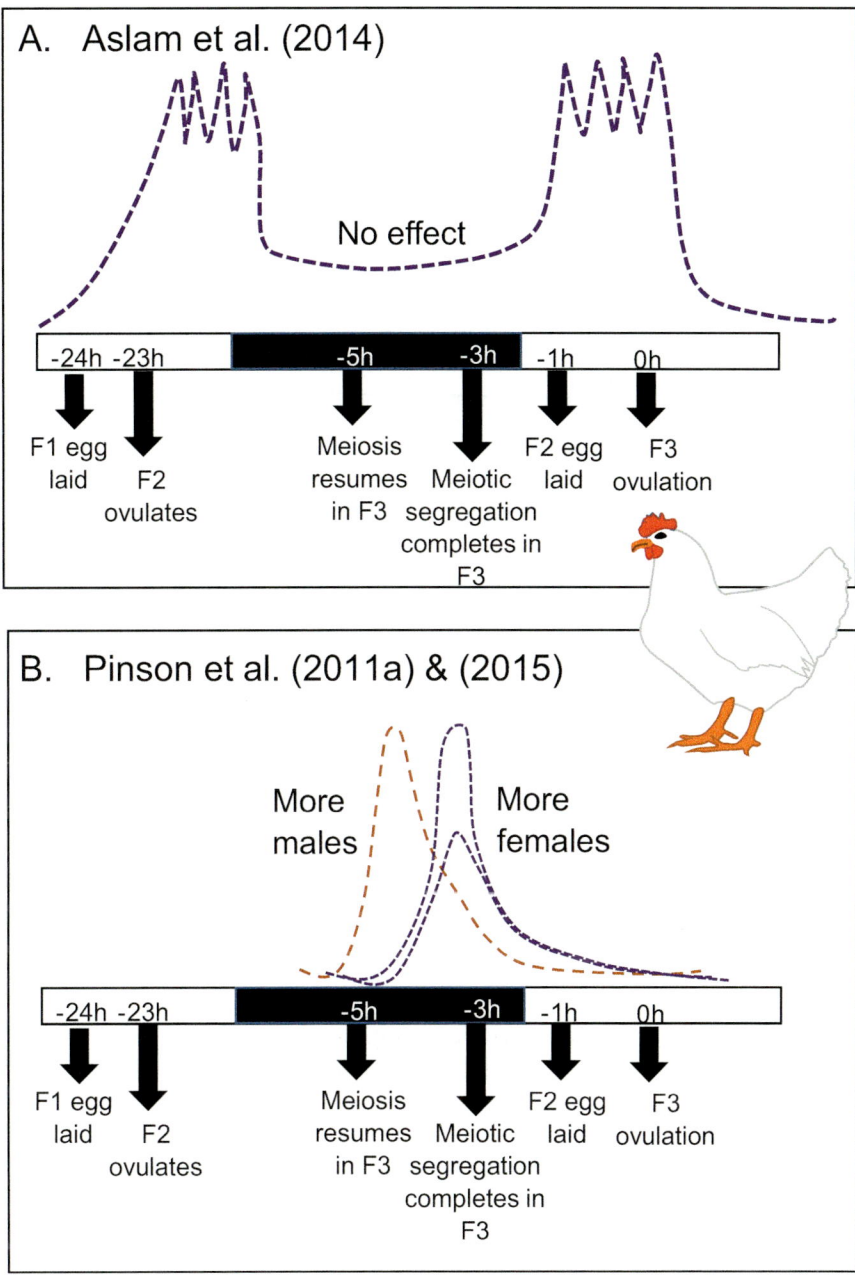

Fig. 7.4 Illustration of the timing of corticosterone treatment relative to the timing of meiotic segregation during three key studies in which corticosterone was administered to white leghorn laying hens (**a**) Corticosterone administered in feed over 14 days—Aslam et al. (2014), (**b**) Corticosterone administered as injections 5 and 4 h prior to ovulation—Pinson et al. (2011a) and Pinson et al. (2015). In both cases, dashed lines indicate the estimated profiles of corticosterone concentrations in circulation, and purple dashed lines indicate profiles that ultimately

In an attempt to further elucidate the target time at which corticosterone acted to skew sex ratios, we experimentally elevated corticosterone concentrations on a shorter timescale, using injections that elevated concentrations for a period of hours rather than implants or feed treatments that elevated concentrations for days (Fig. 7.4b). We used laying hens as our model because they have been selected for their precise timing of ovulation and egg laying; these hens lay eggs at a rate of one egg per day, and ovulation of the next follicle in the sequence takes place approximately 30 min prior to oviposition of the egg that came before it. Because the egg spends 24 h in the reproductive tract after ovulation, we can then use the timing of an egg to predict when the ovulation of future eggs will occur. This allowed us to pinpoint corticosterone treatments to the precise time when sex chromosomes were segregating. For our first attempt at this, we used a pharmacological dose of corticosterone that elevated concentrations well outside of the physiological range, and the result was that hens injected with this high dose of corticosterone just prior to meiotic segregation produced significantly more *male* offspring, rather than the predicted bias towards females (Pinson et al. 2011a). A simultaneous study in zebra finches was conducted in a similar manner and also using a pharmacological dose, and the results were the same—corticosterone-treated birds produced more *male* offspring (Gam et al. 2011). Despite the surprising direction of these effects, these studies tell us that corticosterone does appear to have the ability to influence which way the sex chromosomes segregate during the completion of meiosis I, albeit in the opposite direction as the one we would have predicted.

The next step was then to examine the influences of an elevation of corticosterone within the physiological range. For zebra finches, we triggered a physiological stress response, resulting in an endogenous elevation within the physiological range, just prior to meiotic segregation using a bag stress. This treatment, during which we placed female zebra finches in cloth bags for 5 min, stimulated elevation in corticosterone concentrations, but did not stimulate a bias in sex ratios (Gam and Navara 2016). This treatment may have failed to trigger an effect on sex ratios either because an elevation of corticosterone within the physiological range is not sufficient to do so or because the elevation at this lower level did not last long enough to overlap with the critical time necessary to influence meiotic segregation. To conduct a similar test in laying hens, we continued to rely on injections to

Fig. 7.4 (continued) produced female-biased sex ratios, while the orange dashed line indicates the profile that ultimately produced a male-biased sex ratio. In the Aslam et al. (2014) study (**a**), corticosterone was provided in daily feed, and blood samples taken during the day indicated that concentrations of corticosterone were elevated in plasma at days 4 and 10 after treatment began. However, since hens do not eat during the night, we would expect that concentrations of corticosterone dropped during the night, when meiotic segregation was occurring for each follicle. In the Pinson et al. studies (**b**), an injection containing a pharmacological dose of corticosterone at 5 h prior to ovulation resulted in male-biased sex ratios, while a physiological and a pharmacological dose given at 4 h prior to ovulation resulted in female biases. Note that the individual treatments at 4 h did not produce significant effects on their own, but when combined, did stimulate a significant female bias in offspring sex ratios

maintain better control over the level of corticosterone elevation. We provided high and low physiological doses at 5 h prior to ovulation, the same timing as was used in previous experiments, and there was no influence on offspring sex ratios, similar to what we saw in zebra finches. However, when we treated birds with corticosterone one hour later (4 h prior to ovulation), we did see a sex ratio skew but this time towards *females* (Pinson et al. 2015). Thus, it appears that the timing and the dosage influence not only whether a sex ratio skew occurs but the direction of the sex ratio skew. How this might happen is still unknown, but based on these studies, it appears that there is a short critical window during which corticosterone concentrations must be elevated for sex ratio adjustments to occur.

7.2.4 Influences of Testosterone

It is possible that the observed effects of corticosterone on offspring sex ratios are actually occurring via another downstream hormonal mediator. In the peafowl study mentioned above, sex ratios were not only negatively correlated with corticosterone concentrations, they were also positively correlated with testosterone concentrations. Testosterone concentrations have also been shown to differ between eggs that produce males versus females in some species (e.g., peafowl—Petrie et al. 2001), and when mated with more attractive males, female zebra finches both deposit higher amounts of testosterone into their egg yolks (Gil et al. 1999) and produce a higher proportion of male offspring (Burley 1981). Further, corticosterone is a well-known suppressor of reproduction and the corresponding reproductive hormones (including testosterone) in both male and female birds (Shini et al. 2009). As a result, it is possible that the effects on sex ratio after corticosterone administration actually occur due to downstream changes in circulating levels of reproductive hormones, testosterone in particular, and there are now studies to support this idea (Fig. 7.5).

Veiga et al. (2004) were the first to test this idea experimentally. They implanted spotless starlings with silastic implants containing testosterone and monitored sex ratios for up to 3 years afterward. Females implanted with testosterone produced a significantly higher proportion of male offspring during the year of manipulation and also for 3 years afterward. It is unlikely that testosterone directly manipulated sex ratios by influencing meiotic segregation in this instance, because first, females treated with testosterone delayed laying, perhaps until testosterone concentrations had lowered enough to restore normal reproductive physiology. In addition, when the implants were collected from the birds, there was no hormone remaining in them, and it was unlikely that the hormone in the implants lasted beyond the manipulation year, even though sex ratios produced by implanted birds remained biased at this time. The authors suggest two possibilities for how the treatment exerted such long-term effects on sex ratios. First, the treatment may have permanently influenced testosterone production by the ovaries such that higher concentrations of testosterone were produced by testosterone-implanted birds compared to control birds during all 3 years. Second, the testosterone treatment appeared to affect dominance status in female starlings; testosterone-implanted

Species	Treatment	Outcome
 Peafowl	High endogenous levels of testosterone	More males
Spotless starling 	Silastic implant containing testosterone propionate	More males for 3 years
 Homing pigeon	Silastic implant containing testosterone	More males in 1st eggs for 2 years
Zebra finch 	Injection of testosterone enanthanum after 1st egg	More males in eggs 3, 4, & 5
 White leghorn	Testosterone injection (1.5mg) 5h prior to meiotic segregation	More males in eggs ovulated after treatment
Japanese quail 	Silastic implant containing 2.4mg testosterone	No effect

Fig. 7.5 A summary of studies in birds testing the influences of testosterone on offspring sex ratios

females tended to be more dominant than controls. If this dominance status then remained the same going forward, a resulting increase in aggressive interactions, and, in effect, a higher number of testosterone spikes, could underlie the long-term effects on sex ratios. Neither of these possibilities has yet been tested.

There are additional studies that support the role of testosterone in adjustment of offspring sex ratios in birds. In homing pigeons, an implant containing testosterone stimulated the production of more male offspring in first eggs of the clutch

(Goerlich et al. 2009). As in the starling study, these effects lasted beyond the manipulation year, into 1 year after manipulation. Unlike the starling study, however, the pigeons were held in cages and were not able to participate in conspecific interactions that were hypothesized to influence sex ratios in the starling study. Hence, in this case, it appears more likely that the testosterone treatment permanently influenced the ovarian physiology and/or receptor dynamics in the ovary in such a way that sex ratio skews continue beyond the point when the hormone in the implant ran out.

Finally, Rutkowska and Cichoń (2006) provided even further evidence that testosterone treatment not only has the ability to influence sex ratios but exerts effects beyond when the concentrations of testosterone have dropped to baseline levels. They administered a single injection of testosterone to females just after the appearance of the first egg in the clutch and then monitored the sexes of eggs across the clutch and compared those to the sexes of eggs produced by control females. They showed a striking effect, testosterone-treated females producing significantly more male offspring towards the end of the clutch compared to the beginning. This effect is not likely due to direct influences on meiotic segregation because testosterone concentrations from an injection generally drop within a day.

The three studies above implicate testosterone as a driver of sex ratio adjustment in birds; however, they don't provide much insight into how or when testosterone is acting. Despite the fact that testosterone was not elevated from the treatments at the time of meiotic segregation, it is still possible that passage and storage of that testosterone in the yolk influenced the process of meiotic segregation. Rutkowska and Cichoń (2006) did find an elevation in concentrations of yolk testosterone in the eggs 3, 4, and 5 laid by the testosterone-treated females. Those were the same eggs in which elevations in the proportion of male offspring were observed. Goerlich et al. (2009) found no difference in yolk testosterone concentrations in pigeon eggs from testosterone- and control-treated females but suggest that the outer rings of the yolk could have had elevated levels that were not detected when homogenized with the other layers for the assay. Therefore, the next step was to test whether testosterone exposure at the time of meiotic segregation could influence offspring sex.

We took on this question using a similar design to that described above for corticosterone treatment of laying hens. We treated laying hens with a high dose of testosterone 5 h prior to ovulation, which was likely 1–3 h prior to meiotic segregation of the sex chromosomes. The result was a significant increase in the proportion of male offspring in the eggs ovulated hours after the testosterone injection, significantly higher proportions than controls and than sex ratios in the eggs ovulated a day prior to treatment (Pinson et al. 2011b). Since rapid yolk deposition ceases approximately 24 h prior to ovulation (well before our treatment) (Johnson 2015), the skew did not occur due to differences in yolk testosterone concentrations but could result from a similar mechanism if exposure of the germinal disk from either side is enough to trigger the same process that influences meiotic segregation. The details of how this might occur, however, remain unknown.

7.2.5 Influences of Progesterone

Given the effects of testosterone on sex ratios as well as the fact that two other sex steroids, progesterone and estradiol, are much more prevalent and closely associated with the process of ovulation, it is reasonable to think that progesterone and/or estradiol may also be mediators in the process of sex ratio adjustment in birds. In fact, Correa et al. (2005) were the first group to attempt to target a hormone to the time of meiotic segregation, and they did this by providing injections of progesterone to laying hens. In this case, progesterone was injected 4 h prior to the end of the light phase, well before hens would ovulate, and the injection acted to speed up the process of ovulation, stimulating ovulation of those eggs 6 h after the progesterone injection. They tested the effects of a low (0.25 mg) and a high (2 mg) dose on sexes of the eggs ovulated after injection and found that the high dose significantly reduced the proportion of male offspring in those eggs (25% males were produced in the high progesterone group compared to 61 and 63% in the other two groups). Progesterone concentrations in these hens were significantly elevated at 2 h prior to ovulation when sex chromosomes were likely segregating. This suggests that changes in progesterone concentrations have the ability to influence sex ratios; however, because progesterone is a precursor to other sex steroids, including testosterone, it is unclear whether progesterone was acting directly or via downstream effects of testosterone. Given that elevations of progesterone concentrations potently disrupt and interfere with the timing of the ovulation process in this and other studies (Etches and Cunningham 1976; K. Navara pers obs.), it is unlikely that birds would use elevations of progesterone to influence offspring sex ratios in a natural context. However, this study spurred the idea for testing for hormonal influences at the time of meiotic segregation and also provides support for the idea that downstream mediators, such as other sex steroids, may instead be the mediators used in a natural context.

7.2.6 Influences of Estrogen

We've now seen that two sex steroids (progesterone and testosterone) have the capacity to influence sex ratios in birds, but what about the other major player in the ovulatory process, estrogen? Given that the majority of endocrine-disrupting chemicals that animals are exposed to in the environment are estrogen mimics, it is particularly important for physiological ecologists and toxicologists to know whether these compounds that are already known to influence the reproductive systems of developing offspring and eggshell quality can also influence sex ratios (Giesy et al. 2003). In fact, female-biased sex ratios have been observed at the population level in gull colonies breeding in DDT-contaminated environments (reviewed in Giesy et al. 2003), and it is also now well documented that exposure to contaminants that mimic estrogens can disrupt spindle fiber formation and function in oocytes (e.g., Can and Semiz 2000). Yet, it is unknown whether these skews occurred at the primary or secondary sex ratio level.

During their study examining the influences of progesterone on sex ratios in laying hens, Correa et al. (2005) also measured estradiol concentrations in blood. When included in the full model with all other parameters, estradiol was unrelated to offspring sex ratios, but when included alone in the model, estrogen was negatively correlated with the proportion of male offspring produced, and estradiol concentrations were higher in the group that received the high-dose progesterone injection and produced more female offspring. This is the only evidence to date that estrogen plays a role in the process of sex ratio adjustment in birds. In the Pike and Petrie (2006) study described above, neither estrogen nor fadrozole exerted significant influences on offspring sex ratios, and in an additional study, von Engelhardt et al. (2004) administered 30 mg estradiol injections to female zebra finches on four consecutive days. This treatment significantly elevated estradiol concentrations in female circulation during egg laying but did not influence the sex ratios within those eggs. The results of these studies and the fact that female biases in primary sex ratios are not observed in more avian species that have been exposed to compounds that are estrogen mimics suggest that estrogen does not likely play a role in the process of sex ratio manipulation.

7.3 Evidence that Hormones Influence Sex Ratios in Mammals

While the progress towards understanding the potential hormonal influences on the process of sex ratio adjustment in birds has been steady, evidence for the same in mammalian systems is mostly indirect, and experimental approaches to answer such questions are generally lacking. In three reviews, William James has done an excellent job compiling the evidence that most of the factors that have been observed to influence sex ratios in mammalian systems do so by altering parental concentrations of gonadotropins and sex steroids near the time of conception (James 1996a, b, 2008). As shown in Fig. 7.2 above, many of the cues that skew offspring sex ratios do indeed trigger endocrine responses, particularly in the production of adrenal glucocorticoids and sex steroids within the reproductive system. Below, I will examine the evidence that these hormones play key roles in the adjustment of offspring sex ratios in humans and nonhuman mammals.

7.3.1 Influences of Glucocorticoids

There is now a large body of work showing that exposure to stressful events influences birth sex ratios in humans (reviewed in Chap. 2 and Navara 2010), and there is also evidence of a similar pattern in nonhuman mammals as well (Lane and Hyde 1973; Ideta et al. 2009). In general, stress appears to decrease the proportion of male offspring. Stressful stimuli are transduced into physiological signals via the production of glucocorticoids, primarily in the form of cortisol in most mammals. Yet, there are only a handful of studies that address the potential relationship between glucocorticoid concentrations and sex ratios in mammals (Fig. 7.6). In the first study to test the role of the stress hormone axis in sex ratio adjustment,

Influences of Glucocorticoids on Sex Ratios in Mammals

	Species	Treatment	Outcome
Before Conception	Albino Rat	Injection with long-acting ACTH prior to estrus	Fewer males
	Human	High levels of salivary cortisol	Fewer males
	Human	High cortisol in circulation	No relationship
	Field vole	Corticol in female circulation	No relationship
During Gestation	Golden hamster	Dexamethasone treatment in drinking water	Abolished stress-induced female bias
	Richardson's ground squirrel	High fecal glucocorticoid metabolites	More males
		High bound cortisol in circulation	More males

Fig. 7.6 A summary of studies in mammals testing the influences of glucocorticoids on offspring sex ratios

Geiringer (1961) found that injecting albino rats with a long-acting form of adrenocorticotropic hormone (ACTH), a hormone that stimulates glucocorticoid production from the adrenal glands, reduced the proportion of male offspring in the litters of treated mothers. Much later, in a study of humans, Chason et al. (2012) showed that women who had higher concentrations of cortisol in saliva prior to conception were less likely to produce a male baby. These patterns comply with the multitude of studies that show female biases after exposure to stressful stimuli (see Chaps. 2 and 3).

7.3 Evidence that Hormones Influence Sex Ratios in Mammals

The remaining studies that address the role of glucocorticoids on sex ratio adjustment in mammals, however, show that the story is likely much more complicated than a simple effect of cortisol on the sex determination process. A study in field voles showed no relationship between corticosterone concentrations in blood collected prior to conception and the sex ratios of litters produced afterward (Helle et al. 2008). In another study of humans, Bae et al. (2017) and colleagues tested anxiety levels and salivary cortisol concentrations in women prior to conception, and in this case, there was no association between cortisol and the sexes of babies produced. This could perhaps have been due to the fact that the saliva was collected on day 6 of the menstrual cycle in the Chason et al. (2012) study but on day 1 of the cycle in the Bae et al. (2017) study. Perhaps cortisol concentrations closer to conception are more indicative of the linkage between concentrations of the hormone and the resulting sex of the baby. In the Bae et al. (2017) study, it was instead men who were diagnosed with an anxiety disorder that produced a biased sex ratio; those men had a 76% increase in the chances of producing a male baby. Perhaps cortisol concentrations interact in a different way with sex ratios in men compared to women. Cortisol concentrations were not measured in men, but the relationship itself is the opposite of what we would predict based on the opposite pattern that emerges in the majority of previous literature. In particular, men with high-stress jobs produced *fewer* male offspring in two previous studies (Lyster 1971; Snyder 1961). Cortisol concentrations were not measured in those cases either, however, and it is possible that both jobs may be occupied by a category of men who are somewhat resistant to stress. More studies directly measuring cortisol concentrations in men over time and relating those concentrations to sex ratios produced by those men would be helpful towards determining whether a link is there.

The final studies examine the relationship between glucocorticoid concentrations during gestation and the resulting sex ratio at birth. In golden hamsters, Pratt and Lisk (1990) showed that inducing a social stress caused females to reduce litter sizes and skew litter sex ratios towards females, likely via male-biased fetal death. They then repeated this effect but provided stressed females with dexamethasone in drinking water. Dexamethasone is a synthetic glucocorticoid that acts to inhibit the release of adrenocorticotropic hormone from the pituitary gland, completely abolishing the downstream release of endogenous glucocorticoids from the adrenal glands. This treatment prevented the female bias in females experiencing social stress, indicating that glucocorticoids may be a mediator that controls sex ratios by triggering fetal death in a sex-specific manner.

Two additional studies examine the relationship between gestational glucocorticoids and sex ratios, both in Richardson's ground squirrels. In these two studies, fecal glucocorticoid metabolites and free and bound cortisol in circulation were measured during gestation. While the previous studies were testing whether preconception glucocorticoid concentrations could determine whether a male or female baby was *conceived*, the two studies in ground squirrels tested whether glucocorticoid concentrations predicted sex-specific fetal loss during pregnancy. In the first study, mothers with high amounts of glucocorticoid metabolites in feces produced more male-biased litters (Ryan et al. 2011). In the second study, concentrations of bound cortisol in blood were higher for mothers that produced male-biased litters, while concentrations of free cortisol in blood and amounts of glucocorticoid metabolites in feces were not related to sex ratios within litters (Ryan et al. 2014). It is unclear why fecal glucocorticoid metabolites were not related to sex ratios in both studies, although the authors suggest it may result from the collection method; in the first study, an average over multiple collections was used, while data in the second study were collected from only one fecal sample per animal. Still, it is also unclear why only the concentrations of bound cortisol were associated with sex ratios in the second study, given that cortisol bound to binding globulins is thought to be relatively inactive. Additionally, the fact that higher indicators of glucocorticoid concentrations in these two systems were associated with *male*-biased sex ratios is the opposite of what we would predict given the large number of studies showing that stress during gestation appears to result in the production of a *lower* proportion male offspring.

The studies described above give us only a taste of how glucocorticoids may interact in the process of sex determination in mammals. What the results indicate is that we need to test for influences of glucocorticoids at several stages of gamete and fetal development and test for effects in both males and females. We need to conduct experimental studies that test whether elevating glucocorticoid concentrations above baseline concentrations, as is seen in stressful situations, influences offspring sex. This is a wide open field with a great need for additional study before questions about how glucocorticoids interact with offspring sex ratios in mammals can be answered.

7.3.2 Influences of Estrogen

As in birds, most of the cues that appear to influence offspring sex ratios in mammals have the capacity to directly and/or indirectly modulate the production of reproductive hormones. However unlike in birds, estrogen shows more promise as a mediator of sex ratio adjustment in mammalian systems. Some of the first evidence was indirect; mice fed a diet high in fat produced significantly more male pups (Rosenfeld et al. 2003), and it was later shown that mice on high fat diets have significantly higher concentrations of estradiol in circulation compared to those on low fat diets (Whyte et al. 2007). Since then, more direct tests have been done. Zhang et al. (2006) conducted in vitro fertilization using mouse gametes while incubating with and without estradiol. When estradiol was supplied, the resulting offspring was more likely to be male. In addition, when Holstein dairy cows were treated with estradiol just prior to insemination, this treatment increased the probability of birthing a male by 12.1% (E2 = 63.8% male offspring, Control = 51.8% male offspring) (Emadi et al. 2014).

Unfortunately, not all studies examining the influence of estradiol on sex ratios show skews in the same direction. In 2005, Perret tested urinary estradiol levels both during the follicular phase (i.e., as ovarian follicles were maturing) and at ovulation

7.3 Evidence that Hormones Influence Sex Ratios in Mammals

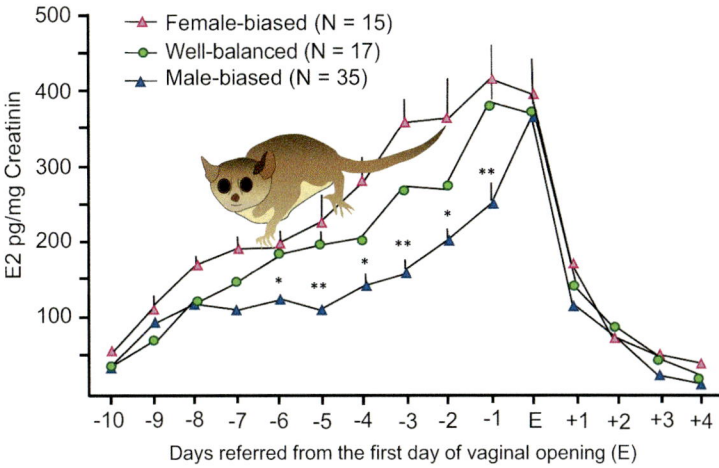

Fig. 7.7 Profiles of urinary estradiol concentrations in gray mouse lemurs during the days before and immediately after vaginal opening and in relation to whether litters produced afterward were female biased, well balanced, or male biased

in gray mouse lemurs (Perret 2005). Resulting sex ratios were positively related to estradiol levels during the follicular phase but not at ovulation (Fig. 7.7). In fact, concentrations of estradiol during this phase were over 150 pg/mg lower when resulting litters were male biased (338 pg mg) compared to when the litters were female biased (181 pg/mg). When litter sex ratios were balanced, concentrations of estradiol in urine were intermediate (281 pg mg). It is unclear why the direction of the skew seen in the lemur study is the opposite of those seen in mice and cows. Perhaps the directional influence of estradiol is species specific. Alternatively, the relationship in lemurs could result from a discrepancy between concentrations in the urine and concentrations in the blood and/or follicular fluid. Interestingly, while the mouse and cow studies showed influences of estradiol at the time of fertilization, the lemur study showed no association between sex ratios and urinary estradiol concentrations at ovulation. However, this study still supports a role for estradiol in the determination of offspring sex ratios in mammals.

How might estradiol be acting on sex ratios in these cases? Emadi et al. (2014) suggest that estradiol treatment could influence the timing between ovulation and fertilization, a factor that has been shown to influence sex ratios in previous studies involving both cows (Martinez et al. 2004) and mice (Krackow 1995). Given that estradiol treatment did not affect oviductal transport of oocytes in cows (Crisman et al. 1980), and the fact that, in the mouse study, fertilization was done directly by the researchers, this possibility is not the likely explanation behind the estradiol-induced sex ratio skews in these two systems. Instead, it is likely that estradiol somehow increased the likelihood that Y-bearing sperm would successfully bind to and fertilize the egg. Perret (2005) pointed out that the development and maintenance of the cumulus oophorus, which functions in sperm sequestering, is influenced by gonadotropin treatment (Bedford and Kim 1993) and, as a result, may also be sensitive to estradiol. Alternatively, it is possible that the estradiol in the medium and/or the follicular fluid influences the sperm, themselves. Human sperm have a biologically active membrane receptor for estradiol and binding this receptor results in a release of intracellular calcium (Luconi et al. 1999). Whether X- and Y-bearing sperm contain different quantities of these receptors remains unknown.

These studies indicate a clear role for estradiol to influence sex ratios at or prior to fertilization in mammals. More work needs to be done to determine whether estradiol is influencing the oocyte or the sperm to affect fertilization in a sex-specific way. In addition, all of these studies measured sex ratios at birth and did not account for potential sex-specific influences on blastocyst survival/loss. Finally, the influences of elevated estradiol concentrations during gestation on sex ratios have never been tested. More work on the potential role of estradiol is badly needed. In addition, given the high number of environmental contaminants that show estrogenic properties, it is important to examine the impacts of estrogen mimics on sex ratios in mammals.

7.3.3 Influences of Testosterone

In Chap. 3, I highlighted the evidence that maternal dominance rank influences the sex ratios produced by those mothers. Given the linkage between dominance and testosterone concentrations in males, it has been suggested that testosterone may also be a candidate in the regulation of sex ratios in mammals. Valerie Grant (2007) provides a nice review addressing this idea. Indeed, women who produce males rank higher on a dominance scale than those who produce females (Grant 1994), and women who are more dominant have higher concentrations of testosterone in circulation (Grant and France 2001). In 2008, Shargal showed in ibexes that dominance rank positively correlated with offspring sex ratios and that the most dominant individuals not only had more sons but also had higher amounts of testosterone in feces (Shargal et al. 2008). Additional indirect evidence for the role of testosterone in the determination of offspring sex in mammals was provided by two studies conducted in house mice and Mongolian gerbils; in both studies, female mice that developed between two male siblings in utero were more likely to produce a higher proportion of male offspring as adults (Vandenbergh and Huggett 1994; Clark and Galef 1995). The female mice that gestated between two male siblings also had shorter anogenital distances at birth compared to other females that did not have males as neighbors in utero, which suggests exposure to higher concentrations of androgens during development. This indicates that androgens can program the physiology of offspring in a way that influences the sex ratios of their offspring. Whether those females produce elevated concentrations of androgens in circulation when conceiving offspring remains unknown; however, these two studies have been cited as supportive evidence of a role for testosterone in the process of mammalian sex determination.

7.3 Evidence that Hormones Influence Sex Ratios in Mammals

Since this time, there have been more direct examinations of how testosterone concentrations relate to mammalian sex ratios. Helle et al. (2008) showed in field voles that females with high testosterone and glucose concentrations produce a higher proportion of male offspring. In this case, it is unclear whether it was the high testosterone or glucose concentrations that were responsible for the influence on sex ratios; however, in a study conducted in mice, injections of flutamide, an androgen receptor blocker, just prior to induced ovulation resulted in female-biased offspring sex ratios (flutamide = 45.3% male, control = 58.4% male) (Gharagozlou et al. 2016). So overall, it appears that high concentrations of testosterone result in male-biased sex ratios, while blocking androgen action reduces the proportion of male offspring.

In an attempt to pinpoint the mechanism by which testosterone might act, Valerie grant and her colleagues performed some elegant work examining the role of testosterone in the fluid contained within the ovarian follicle on the resulting sex of the offspring (Grant and Irwin 2005; Grant et al. 2008) (Fig. 7.8). First, in 2005, they collected primary and subordinate ovarian follicles from cows and aspirated the follicular fluid, granulosa cells, and cumulus–oocyte complexes. Some of the follicular fluid was used to measure testosterone and estradiol concentrations, and the cumulus–oocyte complexes were fertilized in vitro and were then cultured to the 6–8 cell stage, after which the cells were sexed using molecular methods. Follicular concentrations of estradiol were not related to the ultimate sex of the offspring; however, testosterone concentrations in subordinate follicles were; follicles that had higher concentrations of testosterone were more likely to produce a male offspring (Grant and Irwin 2005). The reason that this was only seen in subordinate and not primary follicles is likely because concentrations of testosterone decrease as the follicle nears the point of ovulation. Thus, there may be a sensitive period during which testosterone concentrations in the follicular fluid may influence offspring sex. In an additional study primarily focused on those subordinate, antral follicles, Grant et al. (2008) verified this effect; follicles with higher concentrations of testosterone were significantly more likely to produce male offspring.

This idea has been challenged. García-Herreros et al. (2010) used the same techniques to repeat the experiment and found no significant relationship between testosterone concentrations in follicular fluid and offspring sex; however, Grant and Chamley (2010) argue that if you add the data from the García-Herreros et al. (2010) study to those of the Grant et al. (2008), the result is a significant positive relationship between follicular testosterone concentrations and the proportion of male offspring that result. Still, in two additional studies, oocytes were incubated with exogenously supplied testosterone in attempts to stimulate the production of male blastocysts, and there was no effect of testosterone treatment in either study (Diez et al. 2009; Macaulay et al. 2013). Thus, it is possible that the true effector of sex ratio is an upstream precursor or downstream metabolite of testosterone. This remains to be tested.

Overall, a majority of studies examining influences of testosterone have tested levels prior to conception. To my knowledge, only two studies, one conducted in marmosets and one in ground squirrels, have tested whether androgen levels during

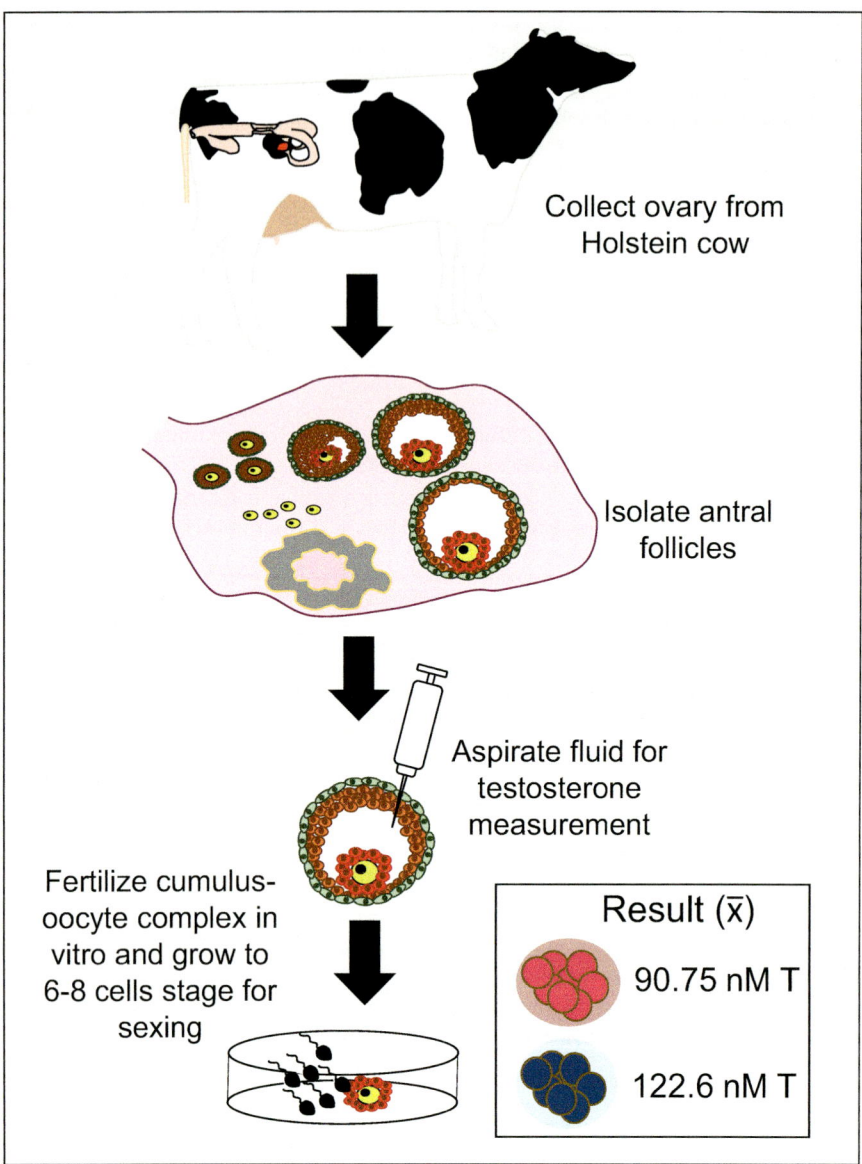

Fig. 7.8 Visual depiction of the experimental design used by Grant et al. (2008). Antral follicles were collected from cows, follicular fluid was aspirated from the follicles, and the cumulus–oocyte complexes were transferred to a dish where they were fertilized using sperm from bulls. At the 6–8 cell stage, blastocysts were collected and sexed molecularly. Results above show the average testosterone concentrations in follicular fluid for follicles that ultimately resulted in male versus female blastocysts

gestation affect embryonic survival in a sex-specific way. No effect was found in either case (French et al. 2010; Ryan et al. 2014). In both of these studies, the phases of gestation were well represented. French et al. (2010) collected urine from the female marmosets 2–5 times per week over the entire course of gestation, and Ryan et al. (2014) collected blood for androgen analyses both early and late in gestation. More studies are needed, however, to test for influences of testosterone during gestation, particularly in species where maternal dominance has been shown to exert an effect on sex ratios.

7.3.4 Evidence that Multiple Reproductive Hormones Act Together

To date, most researchers addressing the role of hormones in sex ratio adjustment have aimed to find a "holy grail" mediator that ties all of the cues known to influence sex ratios in various systems together. Indeed, we have identified some candidate hormones that have the potential to be potent regulators of sex ratios in both avian and mammalian systems. However, some researchers have instead hypothesized a multitier system of regulation, involving multiple mediators that act in concert to adjust sex ratios. An example of this was posited in several reviews by William James (James 1996a, 2004, 2008). James was among the first researchers to suggest that reproductive hormones play a role in the process of sex ratio adjustment, and he further theorizes that these hormones act in both males and females to exert their effects. Specifically, he hypothesized that high concentrations of both testosterone and estradiol in both male and female parents result in the production of more male offspring, while high concentrations of gonadotropins and progesterone have the opposite effect. He postulates that sex ratios correlate positively with R, where R is a function of the equation:

$$(\text{testosterone} + \text{estrogen})/(\text{gonadotrophins} + \text{progesterone})$$

As cited above, there is evidence suggesting that both testosterone and estradiol have the capacity to induce male biases in offspring sex ratios. In addition, occasions during which gonadotropins were provided to men and women for fertility reasons can offer additional insight. For example, men treated with human chorionic gonadotropin (HCG) have elevated concentrations of both testosterone and estradiol and produce male-biased sex ratios (Sas and Szöllösi 1980). While this effect could result from either the testosterone or estradiol elevation independently, James (2008) uses this study as support for the idea that these hormones work in concert. In another study, women treated with clomiphene citrate produce a higher proportion of female offspring. Clomiphene citrate is a chemical that blocks estrogen receptors in the brain and results in elevated levels of the gonadotropins (luteinizing hormone and follicle-stimulating hormone) in the body. In addition, women treated with gonadotropins show a similar female bias in the offspring they produce (James 1985). Still, the fact that high levels of gonadotropins should elevate estradiol concentrations in blood is problematic in terms of supporting the hypothesis.

How, exactly, might these hormones coordinate, in both sexes, to control sex ratios? James provided a potential answer to this as well (James 1997). In males, the epididymis produces a secretory product called glycerylphosphorylcholine (GPC) (Mann and Lutwak-Mann 2012). The role of this molecule is unknown and is unlikely to participate in fertilization or activation of sperm (Jeyendran et al. 1989; Wallace and White 1965). However, the female reproductive tract produces glycerylphosphorylcholine diesterase (GPCD), which splits GPC into free choline (Wallace and White 1965). In men, GPC concentrations are correlated positively with testosterone, and in female rats (Wallace et al. 1964), GPCD concentrations are correlated positively with estrogen and negatively with progesterone concentrations (Cooper et al. 1988). James suggests that this mechanism, controlled by reproductive hormones, may provide a control of offspring sex ratios that is coordinated between males and females. To date, this idea has not been tested empirically. It does, however, point out that we should be considering the possibility that multiple hormones may act together to influence sex ratios in both mammals and birds.

7.4 What About Nonsteroid Hormones?

To date, researchers have focused almost solely on the steroid hormone as mediators of sex determination in both birds and mammals; however, we know that there are multiple other hormonal regulators that interact to control physiological responses to the many cues that have been shown to adjust offspring sex ratios. Below I discuss some other potential regulators of offspring sex ratios in birds and mammals.

7.4.1 Factors that Regulate Blood Glucose

In mammals, there is evidence that circulating glucose concentrations contribute significantly to the determination of offspring sex. For example, Helle et al. (2008) showed that field voles with high concentrations of circulating glucose produce more male offspring, and Mathews et al. (2008) showed in humans that higher intake of energy in general increased the proportion of male offspring born. Cameron et al. (2008) were able to experimentally reduce sex ratios of litters produced by mice while treating them with dexamethasone. While this treatment is known to inhibit glucocorticoid production, as described above, it also reduces circulating glucose concentrations. In the Cameron et al. (2008) study, circulating glucose concentrations were directly related to litter sex ratios, and when glucose concentrations were experimentally reduced by the dexamethasone treatment, fewer male offspring were produced. This suggests that the results in response to dexamethasone treatment described in golden hamsters above Pratt and Lisk (1990) may have resulted from changes in circulating glucose concentrations rather than changes in glucocorticoid production. The link between blood glucose concentrations and sex ratios in birds is less clear.

Given the potential for glucose to be a potent regulator of offspring sex ratios, it makes sense to consider the effects of the hormones that regulate glucose concentrations in circulation. Insulin is the first hormone that comes to mind when thinking about glucose regulation, and mammalian ovarian follicles are sensitive to insulin (Louhio et al. 2000). In fact, there is growing evidence that insulin resistance is involved in a major ovarian dysfunction, polycystic ovary syndrome (Dunaif 1997), and insulin receptors are present and active in mammalian ovaries (Willis and Franks 1995). Thus, insulin could act to coordinate information about food availability with modulation of offspring sex ratios at the level of the ovary.

Perhaps even more tantalizing is the idea that insulin-like growth factors mediate the process of sex ratio adjustment. Insulin-like growth factors (IGF-1 and IGF-II) are molecules that are homologous to pro-insulin in its primary, secondary, and tertiary structures. IGF-1 binds to insulin receptors to help regulate concentrations of glucose in the blood and tissues (Clemmons 2004), and both IGF-1 and IGF-II have well-known functions in controlling the growth and differentiation of granulosa and theca cells in both mammals and birds (Onagbesan et al. 1999) and have been found in the mammalian oocyte as well (Armstrong et al. 2002). In fact, it has been suggested that IGFs are major regulators of follicle selection in mammals (Mazerbourg et al. 2003) and birds (Onagbesan et al. 2009). Thus, there is potential for IGFs to act in the process of sex ratio adjustment during mechanisms that would occur at the level of the ovarian follicle. If we instead consider sex ratio adjustment triggered by the male in mammals, both IGFs and insulin are active in the process of spermatocyte differentiation in the testes (Nakayama et al. 2004). In addition, IGFs are also known to function during early embryonic development (Wang et al. 2006) and are expressed by preimplantation embryos (Doherty et al. 1994) Given that IGFs coordinate the body's energetic reserves with reproductive functions at the level of the ovary, the testis, and the developing embryo, it is a seductive hypothesis that IGF may mediate the process of sex ratio adjustment at more than one developmental level. Studies examining how levels and activity of IGFs vary with offspring sex ratios are needed.

7.4.2 Leptin and Ghrelin

Two of the most striking drivers of sex allocation that emerge in both birds and mammals are the availability of food and the body condition of females. Yet, in studies of sex ratio adjustment, hormones and other factors that control appetite, fat deposition, and other responses to changes in food intake have been virtually ignored. It is now well known that factors produced by the stomach and adipose tissue not only regulate the intake of food and breakdown and storage of energy, they also directly interact with reproductive tissues in both the male and female to regulate reproductive physiology. I will highlight two of these factors below:

Leptin is a hormone that is secreted from adipocytes and acts to increase metabolic rate and decrease appetite in both mammals and birds (Denbow et al. 2000; Barash et al. 1996). Leptin receptors are expressed in the mammalian and avian ovaries (Cassy et al. 2004; Karlsson et al. 1997), and in addition to its actions on metabolism and appetite, leptin regulates reproductive function in both males and females (Barash et al. 1996). In the granulosa and theca cells of the mammalian ovary, leptin acts as an antagonist to IGF-1 (Agarwal et al. 1999; Zachow and Magoffin 1997; Spicer and Francisco 1997). Leptin also directly affects testicular function (Tena-Sempere et al. 2001), is present in seminal fluid, and is secreted by spermatozoa as well (Jope et al. 2003). In fact, there is evidence that leptin can influence sperm function (Lampiao and Du Plessis 2008). It is also known that leptin is secreted by the human endometrium (González et al. 2000) and directs the process of implantation (Yang et al. 2006). As a result, it is easy to see how the pleiotropic effects of this hormone could act to integrate information about food availability and maternal condition into a mechanism to skew offspring sex ratios.

Another hormone known to be intimately involved in both metabolic regulation and reproduction is ghrelin. Ghrelin is produced by the hypothalamus and the stomach and acts as an appetite suppressant. Receptors for ghrelin are present in both the mammalian and avian ovaries (Gaytan et al. 2003; Sirotkin et al. 2006), and ghrelin is now known as a second factor that regulates reproductive function based on energy status within the body (Barreiro and Tena-Sempere 2004). This hormone also influences sperm function and morphology (Kheradmand et al. 2009) and regulates early blastocyst development as well (Steculorum and Bouret 2011). Ghrelin is yet another potential regulator of sex ratios that has, to date, gone unexplored.

7.5 Conclusions

Clearly, our understanding of how hormones are involved in mediating the adjustment of offspring sex ratios is only in its infancy for both mammals and birds. In birds, in which females are heterogametic, hormones are likely acting at the level of the female ovary to influence which sex chromosome is retained in the oocyte. This could occur via influences on the processes of follicle selection and/or growth, or via direct effects on meiotic segregation of sex chromosomes. Many of the hormones identified above show activity during each of those stages. In mammals, where the males are heterogametic, there is a wider range of targets for manipulating sex ratio adjustment. Hormones may act at the level of male sperm production, female receptivity of the egg, blastocyst implantation, and embryonic survival. Many of the hormones described above are active at each level. Given the evidence to date, it appears likely that sex ratio adjustment may not only act at multiple levels in an animal but may also be regulated in different ways by different hormones. As a result, the manipulation of offspring sex ratios is not likely a one-step process. Instead, sex ratio skews likely result from the complex interplay involving a web of hormones that act on several reproductive targets.

References

Agarwal SK, Vogel K, Weitsman SR, Magoffin DA (1999) Leptin antagonizes the insulin-like growth factor-I augmentation of steroidogenesis in granulosa and theca cells of the human ovary. J Clin Endocrinol Metab 84(3):1072–1076

Armstrong D, Baxter G, Hogg C, Woad K (2002) Insulin-like growth factor (IGF) system in the oocyte and somatic cells of bovine preantral follicles. Reproduction 123(6):789–797

Aslam MA, Groothuis TG, Smits MA, Woelders H (2014) Effect of corticosterone and hen body mass on primary sex ratio in laying hen (Gallus gallus), using unincubated eggs. Biol Reprod 90(4):76, 71–79

Bae J, Lynch CD, Kim S, Sundaram R, Sapra KJ, Louis GMB (2017) Preconception stress and the secondary sex ratio in a population-based preconception cohort. Fertil Steril 107(3):714–722

Barash IA, Cheung CC, Weigle DS, Ren H, Kabigting EB, Kuijper JL, Clifton DK, Steiner RA (1996) Leptin is a metabolic signal to the reproductive system. Endocrinology 137(7):3144–3147

Barreiro M, Tena-Sempere M (2004) Ghrelin and reproduction: a novel signal linking energy status and fertility? Mol Cell Endocrinol 226:1):1–1):9

Bedford J, Kim H (1993) Cumulus oophorus as a sperm sequestering device, in vivo. J Exp Zool A Ecol Genet Physiol 265(3):321–328

Bentz AB, Navara KJ, Siefferman L (2013) Phenotypic plasticity in response to breeding density in tree swallows: an adaptive maternal effect? Horm Behav 64(4):729–736

Bentz AB, Becker DJ, Navara KJ (2016a) Evolutionary implications of interspecific variation in a maternal effect: a meta-analysis of yolk testosterone response to competition. R Soc Open Sci 3 (11):160499

Bentz AB, Sirman AE, Wada H, Navara KJ, Hood WR (2016b) Relationship between maternal environment and DNA methylation patterns of estrogen receptor alpha in wild Eastern Bluebird (Sialia sialis) nestlings: a pilot study. Ecol Evol 6(14):4741–4752

Bonier F, Martin PR, Wingfield JC (2007) Maternal corticosteroids influence primary offspring sex ratio in a free-ranging passerine bird. Behav Ecol 18(6):1045–1050

Burley N (1981) Sex ratio manipulation and selection for attractiveness. Science 211(4483):721–722

Cameron EZ, Lemons PR, Bateman PW, Bennett NC (2008) Experimental alteration of litter sex ratios in a mammal. Proc R Soc Lond B Biol Sci 275(1632):323–327

Can A, Semiz O (2000) Diethylstilbestrol (DES)-induced cell cycle delay and meiotic spindle disruption in mouse oocytes during in-vitro maturation. Mol Hum Reprod 6(2):154–162

Cassy S, Metayer S, Crochet S, Rideau N, Collin A, Tesaraud S (2004) Leptin receptor in the chicken ovary: potential involvement in ovarian dysfunction of ad libitum-fed broiler breeder hens. Reprod Biol Endocrinol 2(1):72

Chason RJ, McLain AC, Sundaram R, Chen Z, Segars JH, Pyper C, Louis GMB (2012) Preconception stress and the secondary sex ratio: a prospective cohort study. Fertil Steril 98 (4):937–941

Clark MM, Galef BG (1995) Prenatal influences on reproductive life history strategies. Trends Ecol Evol 10(4):151–153

Clemmons DR (2004) Role of insulin-like growth factor iin maintaining normal glucose homeostasis. Horm Res 62(Suppl 1):77–82

Clutton-Brock TH, Iason GR (1986) Sex ratio variation in mammals. Q Rev Biol 61(3):339–374

Cooper TG, Hing-Heiyeung C, Nashan D, Nieschlag E (1988) Epididymal markers in human infertility. J Androl 9(2):91–101

Correa SM, Adkins-Regan E, Johnson PA (2005) High progesterone during avian meiosis biases sex ratios toward females. Biol Lett 1(2):215–218

Crisman RO, McDonald LE, Thompson FN (1980) Effects of progesterone or estradiol on uterine tubal transport of ova in the cow. Theriogenology 13(2):141–154

Denbow DM, Meade S, Robertson A, McMurtry JP, Richards M, Ashwell C (2000) Leptin-induced decrease in food intake in chickens. Physiol Behav 69(3):359–362

Diez C, Bermejo-Alvarez P, Trigal B, Caamano JN, Munoz M, Molina I, Gutierrez-Adan A, Carrocera S, Martin D, Gomez E (2009) Changes in testosterone or temperature during the in vitro oocyte culture do not alter the sex ratio of bovine embryos. J Exp Zool A Ecol Genet Physiol 311(6):448

Dloniak S, French JA, Holekamp K (2006) Rank-related maternal effects of androgens on behaviour in wild spotted hyaenas. Nature 440(7088):1190–1193

Doherty AS, Temeles GL, Schultz RM (1994) Temporal pattern of IGF-I expression during mouse preimplantation embryogenesis. Mol Reprod Dev 37(1):21–26

Drake AJ, Walker BR, Seckl JR (2005) Intergenerational consequences of fetal programming by in utero exposure to glucocorticoids in rats. Am J Phys Regul Integr Comp Phys 288(1):R34–R38

Dunaif A (1997) Insulin resistance and the polycystic ovary syndrome: mechanism and implications for pathogenesis. Endocr Rev 18(6):774–800

Emadi S, Rezaei A, Bolourchi M, Hovareshti P, Akbarinejad V (2014) Administration of estradiol benzoate before insemination could skew secondary sex ratio toward males in Holstein dairy cows. Domest Anim Endocrinol 48:110–118

Etches R, Cunningham F (1976) The effect of pregnenolone, progesterone, deoxycorticosterone or cortigosterone on the time of ovulation and oviposition in the hen. Br Poult Sci 17(6):637–642

French JA, Smith AS, Birnie AK (2010) Maternal gestational androgen levels in female marmosets (Callithrix geoffroyi) vary across trimesters but do not vary with the sex ratio of litters. Gen Comp Endocrinol 165(2):309–314

Gam A, Navara K (2016) Endogenous corticosterone elevations five hours prior to ovulation do not influence offspring sex ratios in Zebra Finches. Avian Biol Res 9(3):131–138

Gam AE, Mendonça MT, Navara KJ (2011) Acute corticosterone treatment prior to ovulation biases offspring sex ratios towards males in zebra finches Taeniopygia guttata. J Avian Biol 42(3):253–258

García-Herreros M, Bermejo-Álvarez P, Rizos D, Gutiérrez-Adán A, Fahey AG, Lonergan P (2010) Intrafollicular testosterone concentration and sex ratio in individually cultured bovine embryos. Reprod Fertil Dev 22(3):533–538

Gaytan F, Barreiro ML, Chopin LK, Herington AC, Morales C, Pinilla L, Casanueva FF, Aguilar E, Dieguez C, Tena-Sempere M (2003) Immunolocalization of ghrelin and its functional receptor, the type 1a growth hormone secretagogue receptor, in the cyclic human ovary. J Clin Endocrinol Metab 88(2):879–887

Geiringer E (1961) Effect of ACTH on Sex Ratio of the Albino Rat.∗. Proc Soc Exp Biol Med 106(4):752–754

Gharagozlou F, Youssefi R, Vojgani M, Akbarinejad V, Rafiee G (2016) Androgen receptor blockade using flutamide skewed sex ratio of litters in mice. Vet Res Forum 7(2):169

Giesy JP, Feyk LA, Jones PD, Kannan K, Sanderson T (2003) Review of the effects of endocrine-disrupting chemicals in birds. Pure Appl Chem 75(11–12):2287–2303

Gil D, Graves J, Hazon N, Wells A (1999) Male attractiveness and differential testosterone investment in zebra finch eggs. Science 286(5437):126–128

Goerlich VC, Dijkstra C, Schaafsma SM, Groothuis TG (2009) Testosterone has a long-term effect on primary sex ratio of first eggs in pigeons—in search of a mechanism. Gen Comp Endocrinol 163(1):184–192

González RR, Caballero-Campo P, Jasper M, Mercader A, Devoto L, Pellicer A, Simon C (2000) Leptin and leptin receptor are expressed in the human endometrium and endometrial leptin secretion is regulated by the human blastocyst. J Clin Endocrinol Metab 85(12):4883–4888

Grant VJ (1994) Maternal dominance and the conception of sons. Br J Med Psychol 67(4):343–351

Grant VJ (2007) Could maternal testosterone levels govern mammalian sex ratio deviations? J Theor Biol 246(4):708–719

Grant VJ, Chamley LW (2010) Can mammalian mothers influence the sex of their offspring periconceptually? Reproduction 140(3):425–433

Grant VJ, France JT (2001) Dominance and testosterone in women. Biol Psychol 58(1):41–47

Grant VJ, Irwin R (2005) Follicular fluid steroid levels and subsequent sex of bovine embryos. J Exp Zool A Comp Exp Biol 303(12):1120–1125

Grant VJ, Irwin R, Standley N, Shelling A, Chamley L (2008) Sex of bovine embryos may be related to mothers' preovulatory follicular testosterone. Biol Reprod 78(5):812–815

Groothuis TG, Schwabl H (2008) Hormone-mediated maternal effects in birds: mechanisms matter but what do we know of them? Philos Trans R Soc Lond B Biol Sci 363(1497):1647–1661

Hayward LS, Wingfield JC (2004) Maternal corticosterone is transferred to avian yolk and may alter offspring growth and adult phenotype. Gen Comp Endocrinol 135(3):365–371

Helle S, Laaksonen T, Adamsson A, Paranko J, Huitu O (2008) Female field voles with high testosterone and glucose levels produce male-biased litters. Anim Behav 75(3):1031–1039

Ideta A, Hayama K, Kawashima C, Urakawa M, Miyamoto A, Aoyagi Y (2009) Subjecting holstein heifers to stress during the follicular phase following superovulatory treatment may increase the female sex ratio of embryos. J Reprod Dev 55(5):529–533

James WH (1985) The sex ratio of infants born after hormonal induction of ovulation. BJOG Int J Obstet Gynaecol 92(3):299–301

James WH (1996a) Evidence that mammalian sex ratios at birth are partially controlled by parental hormone levels at the time of conception. J Theor Biol 180(4):271–286

James WH (1996b) Further concepts on regulators of the sex ratio in human offspring: Interpregnancy intervals, high maternal age and seasonal effects on the human sex ratio. Hum Reprod 11(1):7–8

James WH (1997) A potential mechanism for sex ratio variation in mammals. J Theor Biol 189 (3):253–255

James WH (2004) Further evidence that mammalian sex ratios at birth are partially controlled by parental hormone levels around the time of conception. Hum Reprod 19(6):1250–1256

James WH (2008) Evidence that mammalian sex ratios at birth are partially controlled by parental hormone levels around the time of conception. J Endocrinol 198(1):3–15

Jeyendran R, Ven H, Rosecrans R, Perez-Pelaez M, Al-Hasani S, Zaneveld L (1989) Chemical constituents of human seminal plasma: relationship to fertility/Chemische Bestandteile des menschlichen Spermaplasmas: Beziehungen zur Fertilität. Andrologia 21(5):423–428

Johnson AJ (2015) Reproduction in the female. In: Scanes CG (ed) Sturkie's avian physiology, 6th edn. Academic Press, Boston, MA, pp 635–665

Jope T, Lammert A, Kratzsch J, Paasch U, Glander HJ (2003) Leptin and leptin receptor in human seminal plasma and in human spermatozoa. Int J Androl 26(6):335–341

Karlsson C, Lindell K, Svensson E, Bergh C, Lind P, Billig H, Carlsson LM, Carlsson B (1997) Expression of functional leptin receptors in the human ovary. J Clin Endocrinol Metab 82 (12):4144–4148

Kheradmand A, Taati M, Babaei H (2009) The effects of chronic administration of ghrelin on rat sperm quality and membrane integrity. Anim Biol 59(2):159–168

Krackow S (1995) Potential mechanisms for sex ratio adjustment in mammals and birds. Biol Rev 70(2):225–241

Lampiao F, Du Plessis SS (2008) Insulin and leptin enhance human sperm motility, acrosome reaction and nitric oxide production. Asian J Androl 10(5):799–807

Lane EA, Hyde TS (1973) Effect of maternal stress on fertility and sex ratio: a pilot study with rats. J Abnorm Psychol 82(1):78

Louhio H, Hovatta O, Sjöberg J, Tuuri T (2000) The effects of insulin, and insulin-like growth factors I and II on human ovarian follicles in long-term culture. Mol Hum Reprod 6 (8):694–698

Luconi M, Muratori M, Forti G, Baldi E (1999) Identification and characterization of a novel functional estrogen receptor on human sperm membrane that interferes with progesterone effects. J Clin Endocrinol Metab 84(5):1670–1678

Lyster W (1971) Three patterns of seasonality in American births. Am J Obstet Gynecol 110 (7):1025–1028

Macaulay AD, Hamilton CK, King WA, Bartlewski PM (2013) Influence of physiological concentrations of androgens on the developmental competence and sex ratio of in vitro produced bovine embryos. Reprod Biol 13(1):41–50

Mann T, Lutwak-Mann C (2012) Male reproductive function and semen: themes and trends in physiology, biochemistry and investigative andrology. Springer Science & Business Media, Berlin

Martinez F, Kaabi M, Martinez-Pastor F, Alvarez M, Anel E, Boixo J, De Paz P, Anel L (2004) Effect of the interval between estrus onset and artificial insemination on sex ratio and fertility in cattle: a field study. Theriogenology 62(7):1264–1270

Mathews F, Johnson PJ, Neil A (2008) You are what your mother eats: evidence for maternal preconception diet influencing foetal sex in humans. Proc R Soc Lond B Biol Sci 275 (1643):1661–1668

Mazerbourg S, Bondy C, Zhou J, Monget P (2003) The insulin-like growth factor system: a key determinant role in the growth and selection of ovarian follicles? a comparative species study. Reprod Domest Anim 38(4):247–258

Nakayama Y, Yamamoto T, Abé SI (2004) IGF-I, IGF-II and insulin promote differentiation of spermatogonia to primary spermatocytes in organ culture of newt testes. Int J Dev Biol 43 (4):343–347

Navara KJ (2010) Programming of offspring sex ratios by maternal stress in humans: assessment of physiological mechanisms using a comparative approach. J Comp Physiol B 180(6):785–796

Navara KJ, Mendonça MT (2008) Yolk androgens as pleiotropic mediators of physiological processes: a mechanistic review. Comp Biochem Physiol A Mol Integr Physiol 150(4):378–386

Navara K, Siefferman L, Hill G, Mendonca M (2006) Yolk androgens vary inversely to maternal androgens in eastern bluebirds: an experimental study. Funct Ecol 20(3):449–456

Onagbesan O, Vleugels B, Buys N, Bruggeman V, Safi M, Decuypere E (1999) Insulin-like growth factors in the regulation of avian ovarian functions. Domest Anim Endocrinol 17(2):299–313

Onagbesan O, Bruggeman V, Decuypere E (2009) Intra-ovarian growth factors regulating ovarian function in avian species: a review. Anim Reprod Sci 111(2):121–140

Perret M (2005) Relationship between urinary estrogen levels before conception and sex ratio at birth in a primate, the gray mouse lemur. Hum Reprod 20(6):1504–1510

Petrie M, Schwabl H, Brande-Lavridsen N, Burke T (2001) Maternal investment: sex differences in avian yolk hormone levels. Nature 412(6846):498–498

Pike TW (2005) Sex ratio manipulation in response to maternal condition in pigeons: evidence for pre-ovulatory follicle selection. Behav Ecol Sociobiol 58(4):407–413

Pike TW, Petrie M (2005) Maternal body condition and plasma hormones affect offspring sex ratio in peafowl. Anim Behav 70(4):745–751

Pike TW, Petrie M (2006) Experimental evidence that corticosterone affects offspring sex ratios in quail. Proc R Soc Lond B Biol Sci 273(1590):1093–1098

Pilz K, Smith H (2004) Egg yolk androgen levels increase with breeding density in the European starling, Sturnus vulgaris. Funct Ecol 18(1):58–66

Pinson SE, Parr CM, Wilson JL, Navara KJ (2011a) Acute corticosterone administration during meiotic segregation stimulates females to produce more male offspring. Physiol Biochem Zool 84(3):292–298

Pinson SE, Wilson JL, Navara KJ (2011b) Elevated testosterone during meiotic segregation stimulates laying hens to produce more sons than daughters. Gen Comp Endocrinol 174 (2):195–201

Pinson SE, Wilson JL, Navara KJ (2015) Timing matters: corticosterone injections 4 h before ovulation bias sex ratios towards females in chickens. J Comp Physiol B 185(5):539–546

Pratt N, Lisk R (1990) Dexamethasone can prevent stress-related litter deficits in the golden hamster. Behav Neural Biol 54(1):1–12

Reed WL, Vleck CM (2001) Functional significance of variation in egg-yolk androgens in the American coot. Oecologia 128(2):164–171

Rosenfeld CS, Grimm KM, Livingston KA, Brokman AM, Lamberson WE, Roberts RM (2003) Striking variation in the sex ratio of pups born to mice according to whether maternal diet is high in fat or carbohydrate. Proc Natl Acad Sci 100(8):4628–4632

Rutkowska J, Cichoń M (2006) Maternal testosterone affects the primary sex ratio and offspring survival in zebra finches. Anim Behav 71(6):1283–1288

Ryan CP, Anderson WG, Gardiner LE, Hare JF (2011) Stress-induced sex ratios in ground squirrels: support for a mechanistic hypothesis. Behav Ecol 23:160–167

Ryan CP, Anderson WG, Berkvens CN, Hare JF (2014) Maternal gestational cortisol and testosterone are associated with trade-offs in offspring sex and number in a free-living rodent (Urocitellus richardsonii). PLoS One 9(10):e111052

Saino N, Ambrosini R, Martinelli R, Calza S, Møller A, Pilastro A (2002) Offspring sexual dimorphism and sex-allocation in relation to parental age and paternal ornamentation in the barn swallow. Mol Ecol 11(8):1533–1544

Sas M, Szöllösi J (1980) Sex ratio of children of fathers with spermatic disorders following hormone therapy. Orv Hetil 121(46):2807

Schwabl H (1993) Yolk is a source of maternal testosterone for developing birds. Proc Natl Acad Sci 90(24):11446–11450

Schwabl H (1997) The contents of maternal testosterone in house sparrow passer domesticus eggs vary with breeding conditions. Naturwissenschaften 84(9):406–408

Shargal D, Shore L, Roteri N, Terkel A, Zorovsky Y, Shemesh M, Steinberger Y (2008) Fecal testosterone is elevated in high ranking female ibexes (Capra nubiana) and associated with increased aggression and a preponderance of male offspring. Theriogenology 69(6):673–680

Shini S, Shini A, Huff G (2009) Effects of chronic and repeated corticosterone administration in rearing chickens on physiology, the onset of lay and egg production of hens. Physiol Behav 98 (1):73–77

Sirotkin A, Grossmann R, María-Peon M, Roa J, Tena-Sempere M, Klein S (2006) Novel expression and functional role of ghrelin in chicken ovary. Mol Cell Endocrinol 257:15–25

Snyder RG (1961) The sex ratio of offspring of pilots of high performance military aircraft. Hum Biol 33(1):1–10

Sockman KW, Schwabl H (1999) Daily estradiol and progesterone levels relative to laying and onset of incubation in canaries. Gen Comp Endocrinol 114(2):257–268

Spicer LJ, Francisco CC (1997) The adipose obese gene product, leptin: evidence of a direct inhibitory role in Ovarian function. Endocrinology 138(8):3374–3379

Steculorum SM, Bouret SG (2011) Developmental effects of ghrelin. Peptides 32(11):2362–2366

Tangalakis K, Lumbers E, Moritz K, Towstoless M, Wintour E (1992) Effect of cortisol on blood pressure and vascular reactivity in the ovine fetus. Exp Physiol 77(5):709–717

Tena-Sempere M, Manna P, Zhang F, Pinilla L, Gonzalez L, Dieguez C, Huhtaniemi I, Aguilar E (2001) Molecular mechanisms of leptin action in adult rat testis: potential targets for leptin-induced inhibition of steroidogenesis and pattern of leptin receptor messenger ribonucleic acid expression. J Endocrinol 170(2):413–423

Vandenbergh JG, Huggett CL (1994) Mother's prior intrauterine position affects the sex ratio of her offspring in house mice. Proc Natl Acad Sci 91(23):11055–11059

Veiga JP, Viñuela J, Cordero PJ, Aparicio JM, Polo V (2004) Experimentally increased testosterone affects social rank and primary sex ratio in the spotless starling. Horm Behav 46(1):47–53

von Engelhardt N, Dijkstra C, Daan S, Groothuis TG (2004) Effects of 17-β-estradiol treatment of female zebra finches on offspring sex ratio and survival. Horm Behav 45(5):306–313

Wallace J, White I (1965) Studies of glycerylphosphorylcholine diesterase in the female reproductive tract. J Reprod Fertil 9(2):163–176

Wallace J, Stone G, White I (1964) The influence of some oestrogens and progestogens on the activity of glycerylphosphorylcholine diesterase in rinsings of the rat uterus. J Endocrinol 29 (2):175–184

Wang T-H, Chang C-L, H-M W, Chiu Y-M, Chen C-K, Wang H-S (2006) Insulin-like growth factor-II (IGF-II), IGF-binding protein-3 (IGFBP-3), and IGFBP-4 in follicular fluid are associated with oocyte maturation and embryo development. Fertil Steril 86(5):1392–1401

Whittingham LA, Schwabl H (2002) Maternal testosterone in tree swallow eggs varies with female aggression. Anim Behav 63(1):63–67

Whyte J, Alexenko A, Davis A, Ellersieck M, Fountain E, Rosenfeld C (2007) Maternal diet composition alters serum steroid and free fatty acid concentrations and vaginal pH in mice. J Endocrinol 192(1):75–81

Wilder RL (1995) Neuroendocrine-immune system interactions and autoimmunity. Annu Rev Immunol 13(1):307–338

Willis D, Franks S (1995) Insulin action in human granulosa cells from normal and polycystic ovaries is mediated by the insulin receptor and not the type-I insulin-like growth factor receptor. J Clin Endocrinol Metab 80(12):3788–3790

Yang Y-J, Cao Y-J, Bo S-M, Peng S, Liu W-M, Duan E-K (2006) Leptin-directed embryo implantation: leptin regulates adhesion and outgrowth of mouse blastocysts and receptivity of endometrial epithelial cells. Anim Reprod Sci 92(1):155–167

Zachow RJ, Magoffin DA (1997) Direct intraovarian effects of leptin: impairment of the synergistic action of insulin-like growth factor-I on follicle-stimulating hormone-dependent estradiol-17β production by rat ovarian granulosa cells. Endocrinology 138(2):847–850

Zhang L, Du W, Chen H, Zhao J, Pei J, Lin X (2006) [Impact of reproductive hormone on mouse embryo sexes] Fen zi xi bao sheng wu xue bao 39(6):573–577

What Went Wrong at Jurassic Park? Modes of Sex Determination and Adaptive Sex Allocation in Reptiles

8

> *There is no unauthorized breeding at Jurassic Park...because all the animals in Jurassic Park are female. We've engineered them that way*
> Dr. Henry Wu, Chief Geneticist at Jurassic Park

When building a live dinosaur exhibit, as in the movie Jurassic Park, producing a population of only females would appear to be a foolproof method of population control. Yet the existence of Jurassic Park II, III, and Jurassic Word illustrate that even Hollywood writers understood that the methods of sex determination employed by reptiles are not simply governed by a pair of sex chromosomes. In fact, when tracing back the evolutionary origins of environmental sex determination (ESD), it becomes clear that dinosaur sex was not strictly subject to chromosomal control; instead dinosaurs likely controlled the sexes of offspring by altering the temperatures at which their eggs were incubated (called temperature-dependent sex determination, or TSD). Changing climates may have even contributed to dinosaur extinction (Miller et al. 2004; Paladino et al. 1989).

This idea highlights the importance of understanding the variety of sex-determination methods found in reptiles. If the process of sex determination is environmentally controlled, then climate change can exert dramatic influences on species that are sensitive to changing temperatures. Reptilian species, in particular, are declining globally, and one of the six identified reasons underlying this decline is climate change (Gibbons et al. 2000). Not only do increasing global temperatures influence habitats and food sources, they may also cause decline by shifting offspring, and thus population, sex ratios in species that exhibit TSD. Global temperatures are expected to increase by 3.5 °C in the next century, and many species that exhibit TSD are already threatened or endangered (Janzen and Paukstis 1991). Furthermore, it is unlikely that many of these species have the capacity to evolve mechanisms to adjust to these higher temperatures quickly enough (Janzen 1994). That being said, it appears that some species may have already evolved mechanisms to prevent sex ratios that are too imbalanced. For example, alligator snapping turtles normally produce more males at cooler temperatures. Yet, when artificially incubating alligator snapping turtle eggs, researchers were never able to induce the production of 100% males, suggesting that, for this species, a method exists to guarantee the production of at least some females even in cooler years (Ewert et al. 1994). It is unclear whether similar mechanisms exist in other species, and whether there is a method to protect against the effects of warm rather than cool temperature extremes. But, to maximize conservation efforts for reptilian species, it is critical that we understand how reptilian sex ratios are determined and how individuals may adaptively alter them in response to surrounding social and environmental conditions.

8.1 The Range of Sex-Determining Systems in Reptiles

Until relatively recently, it was widely believed that the different modes of sex determination observed among vertebrate systems were discrete processes with very different underlying mechanisms. It was thought that TSD evolved from genetic sex determination (GSD) through both the adaptation of thermosensitivity and the loss of one or both sex chromosomes, resulting in a system in which sex is determined entirely by temperature with no genetic influence. We now know that sex-determining systems exist on a continuum, whereby species exhibit different levels of genotype–environment interactions. We have reptilian species that determine sex using only sex chromosomes, like in birds and mammals; we have species that have heteromorphic sex chromosomes whose influence can be overridden by incubation temperature; there are systems that appear to exhibit TSD, but after a closer look, also have genetic influences that drive sex determination; and finally there are systems that appear to use only temperature to determine sex (Fig. 8.1). Sarre et al. (2011) present an excellent review of how this continuum, and

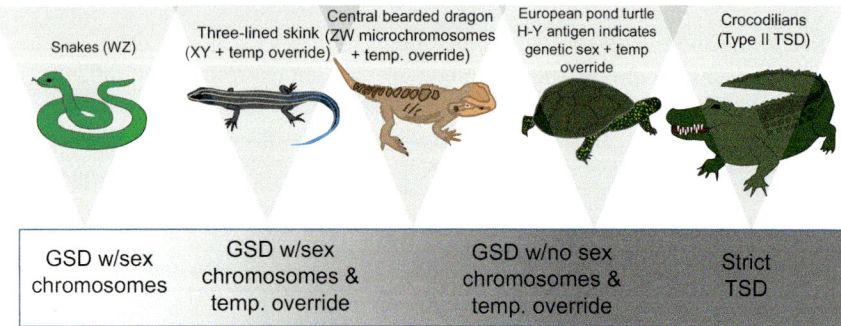

Fig. 8.1 The spectrum of sex-determining mechanisms found in reptilian systems. Reptiles can exhibit strict genetic sex determination (GSD), GSD with a thermal override, GSD without the presences of heteromorphic sex chromosomes with a thermal override, and true temperature-dependent sex determination (TSD)

8.1 The Range of Sex-Determining Systems in Reptiles

transitions between GSD and TSD, may occur, but it has become clear that our understanding of how reptiles determine the sexes of offspring requires more than a concept of which sex is produced at higher or lower temperatures. I will try here to summarize the different modes of sex determination that have been observed among reptilian species, but as documentation of sex determination in additional reptilian species continues, it is likely that even more diversity will be found along the continuum of sex-determination mechanisms.

8.1.1 Reptilian Species with Genetic Sex Determination

Despite the fact that reptiles are publicly known for their ability to adjust offspring sex based on incubation temperatures, many reptilian species still rely on sex chromosomes to determine the sexes of their offspring. For example, all species of snakes studied to date exhibit the female heterogametic WZ form of GSD, some turtles exhibit the male heterogametic XY form, and among lizard species, we see a distribution of both WZ and XY systems (Sarre et al. 2004, 2011). Phylogenetic analysis indicated that GSD is, in fact, the ancestral form of sex determination (Janzen and Krenz 2004). However, we now have examples of species with GSD systems that may also be overridden by temperature cues.

It was initially thought that three-lined skinks operated via a strict GSD system; however it is now clear that the genetic influences on sex can be overridden in this species (Fig. 8.2) Shine et al. (2002). When their eggs were incubated at a cool

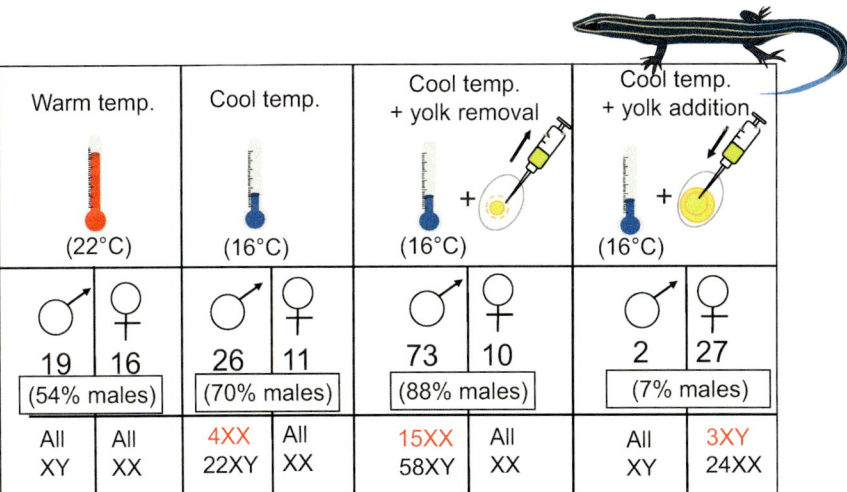

Fig. 8.2 Influences of cool and warm incubation temperatures as well as yolk addition and removal on sex ratios produced by a lizard with GSD (three-lined skinks). Influences of cool and warm incubation temperatures are based on data reported by Shine (2002), and influences of yolk additions and removals (data shown for cold incubation temperatures only) are based on data reported in Radder et al. (2009)

temperature (15 °C), the resulting sex ratio was 70% males. Offspring sex ratios declined with increasing incubation temperatures to reach just below 50% when eggs were incubated at 23 °C. At the time this work was done, the nature of the sex-determination system was unknown, but it was suggested that three-lined skinks were employing a combination of GSD and TSD to determine the sexes of offspring. Further work confirmed this (Radder et al. 2008). The sexes of three-lined skinks are determined by an XY chromosomal system; however, at low temperatures, the phenotypes of genetic XX females are reversed, indicating a temperature-induced override of the sex chromosomes. How this occurs remains unknown.

The story becomes even more interesting when taking egg size into account. Working off of the finding that larger eggs produce females in this species, Radder et al. (2009) conducted an elegant experiment in which they experimentally altered egg contents and thereby influenced offspring sex. When they removed egg yolk from freshly laid eggs, the resulting sex ratio produced was 88% males. Interestingly, this only occurred when eggs were being incubated at cool temperatures. At warm temperatures, the sex ratio of eggs was 69% males and was not significantly different from unmanipulated or sham manipulated controls. In addition, when they added yolk from a larger egg (but not smaller) to eggs incubated at cool temperatures, 93% of the eggs produced female offspring. Genetic analyses confirmed that these ratios occurred via sex-reversal. The fact that only yolk from larger eggs was able to induce this effect suggests that there is a factor in the yolk that is specific to the production of female offspring. What this factor is remains unclear, and the possibility for hormonal involvement will be discussed in Chap. 9.

Central bearded dragons determine the sexes of offspring via a ZW chromosomal system (via micro-sex chromosomes, see below). However, Quinn et al. (2007) found that when eggs were incubated in the laboratory at >32 °C, genetic ZZ males were turned into females (ZZf). The group then sampled wild populations and used molecular markers to determine whether sex reversal also happened in a natural context, and they found that, over 3 years, 6–22% of the females they sampled were genetic males, and these ZZ females produced more eggs per year than concordant ZW females (Holleley et al. 2015). When examined behaviorally, ZZ females were more active and bolder than either concordant ZW females or ZZ males. This is the first evidence of sex-reversal occurring in a wild population of reptiles, and the behavioral differences may act to enhance fitness of individual ZZ females, which would ultimately drive the evolution of new temperature-based sex-determining system. Indeed, pairing ZZ females with ZZ males produced all ZZ offspring whose sex was determined entirely by TSD, a change in the sex-determining mechanism that occurred in only one generation, suggesting a relatively easy transition. In the wild, it would likely take several generations before a complete change in the sex-determining mechanism evolves; however, by observing this system, we may be seeing the evolution of TSD in action.

8.1.2 Genetic Control in the Absence of Heteromorphic Sex Chromosomes

In some species that appear to exhibit pure TSD, that is, they appear to have no identifiable heteromorphic sex chromosomes, there is evidence for an underlying genetic control. For example, two species of sea turtles, green sea turtles and Kemp's Ridley sea turtles, exhibit TSD when eggs are incubated in the laboratory. However, Demas et al. (1990) showed that DNA from both species contains sex-specific sequences. Banded Krait minor (Bkm) is satellite DNA originally sequenced in the snake, and it has also been found throughout the genomes of most eukaryotic species. In many cases, there are concentrations of Bkm on the sex chromosomes. In both Kemp's Ridley and green sea turtles, male-specific Bkm-related fragments were found, suggesting the potential for an underlying genetic mechanism of sex determination. To my knowledge, no further research has been done to test the function of these sequences.

European pond turtles represent another species that typically exhibits TSD, but in which there is evidence that an underlying genetic mechanism of sex determination may exist. H-Y antigen can often be used as a marker of heterogametic sex because it can play an organizing role in the differentiation of the gonad (Fig. 8.3).

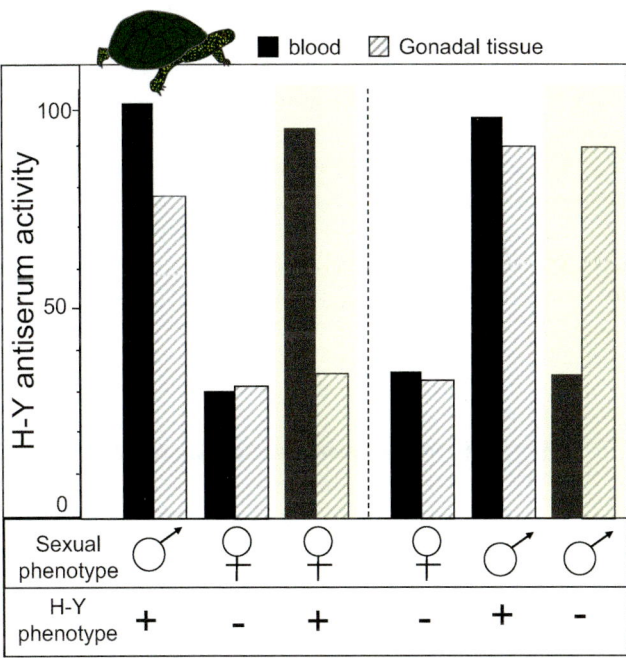

Fig. 8.3 HY-antiserum activity in European pond turtles, a GSD system that shows temperature-induced sex reversal. Sexes were determined based on external phenotype, and an unmatched H-Y phenotype indicates sex-reversal (indicated by highlights around the data from sex-reversed individuals). Sex-reversed individuals had mismatches in HY-antiserum activity between the blood and gonadal tissue. Data reproduced from Zaborski et al. (1982)

In European pond turtles, H-Y antigen is present in the blood and gonads of adult females, but not adult males. However, when eggs are incubated at temperatures that would cause a sex reversal towards males or females, the presence of H-Y antigen in the blood did not match the gonads. These results suggest that H-Y antigen expression in the gonads represents sexual phenotype while H-Y antigen expression in the blood represents sexual genotype (Zaborski et al. 1982, 1988). For example, some females produced at female-producing temperatures had H-Y antigen present in the gonads, but not in the blood, indicating that temperature may have overridden the initial genetic male state. The authors suggest that this indicates the presence of a ZW system of sex determination that underlies the TSD mechanism in European pond turtles.

What do we make of species such as the European pond turtles that appear to have an underlying genetic mechanism of sex determination, but for which no heteromorphic sex chromosomes have yet been identified? While red-eared sliders, for example, show characteristics of a TSD system, Mork et al. (2014) showed that when red-eared slider gonads were cultured by themselves at the pivotal temperature, they had a strong disposition for one sex or the other well before the period when they should have been sensitive to temperature. What then, is controlling the sex? And why do some species appear to have GSD, but appear to lack sex chromosomes? The bearded dragon is a lizard that does not show evidence of TSD, but conventional heteromorphic sex chromosomes also appear to be absent in this species. However, Ezaz et al. (2005) discovered the presence of a heterochromatic microchromosome specific to females, suggesting a ZW system of sex determination. It is possible that reptilian microchromosomes, which are historically considered obsolete, play a role in the sex determination of reptilian systems where conventional sex chromosomes do not appear to be present. In addition, through techniques such as comparative genome hybridization (CGH) and genomic sequencing, we may be able to identify heteromorphic sequences that provide a genetic basis of sex determination even in the absence of clear sex chromosomes. We can then assess how environmental factors like temperature interact with these sex-specific genetic factors to further understand how sex determination is controlled in reptilian species.

8.1.3 True Temperature-Dependent Sex Determination

Of course, there are many species that do appear to determine sex entirely based on temperature. This includes all crocodilians, tuatara, and a majority of turtle species studied to date. A study of 21 species of crocodilians indicated a universal lack of sex chromosomes in this order (Cohen and Gans 1970), and when comparative genome hybridization was used to identify potential sex-specific genetic markers in painted turtles, no such sequences were found (Valenzuela et al. 2014) . These and other species may indeed rely exclusively on temperature to determine offspring sex. There are three known types of TSD (Fig. 8.4): Species that exhibit Type 1A, or MF, produce more males at cool temperatures and more females at warm

8.1 The Range of Sex-Determining Systems in Reptiles

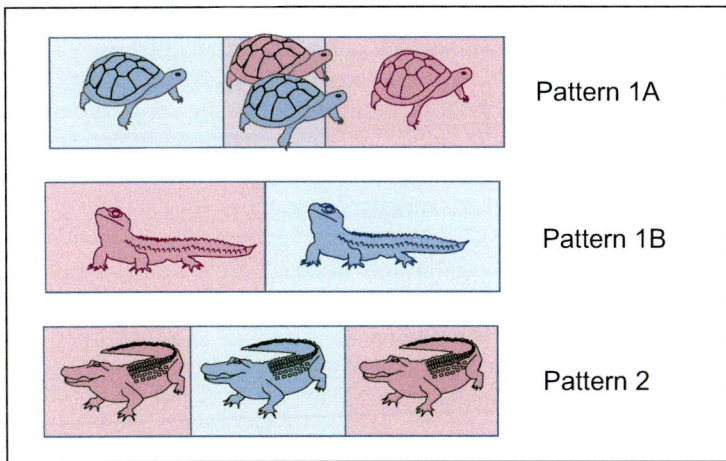

Fig. 8.4 Visual depiction of the three known patterns of temperature-dependent sex determination (TSD)

temperatures. A popular pneumonic used to remember this is "hot chicks and cool dudes," and in most turtles, including all sea turtles, temperatures near 27 °C produce males while females are produced at temperatures near 31 °C (reviewed in Place and Lance 2004). Tuatara, the most ancient reptilian species, are the only ones that show the opposite pattern, Pattern 1A or FM, where more females are produced at low temperatures while more males are produced at high temperatures. The ability to alter sex ratios in this manner in captivity is used in captive breeding programs designed to conserve tuataras in New Zealand. Finally, crocodilians follow a pattern called Type 2 or FMF, in which more females are produced at low and high temperatures, while more males are produced at intermediate temperatures. The mechanism responsible for a pattern which would result in one sex at opposite extremes still remains elusive. In all three patterns of TSD, the influences of temperature on offspring sex occur during a temperature sensitive period (TSP) that varies by species. Additionally, the pivotal temperature, which is defined as the temperature at which sex ratios are 50:50 when incubation temperatures are kept constant, differs among species as well (Mrosovsky and Pieau 1991). Overall, it is clear that, in many species, temperature drives the production of one sex over another. Despite decades of work, however, the mechanisms underlying these effects are still unclear.

Even in cases where there does not appear to be an underlying "default" sex, the key to sex determination is still modification of the genetic code, thus altering of the expression of key genes. Recent work in red-eared sliders showed that levels of DNA methylation at CpG sites near important genes that regulate steroidogenesis were higher at female-inducing than male-inducing temperatures (Matsumoto et al. 2013). Further in some elegant work in American alligators, Parrott et al. (2014) showed that at male-producing temperatures, the gonads showed higher methylation rates of the aromatase promoter and lower methylation rates of *sox9*, a gene likely involved in driving sex determination. In addition, they showed that DNA methyltransferase enzymes were active during critical stages of gonadal development. This work suggests that temperatures may act to differentiate offspring sex by epigenetic modifications, physically modifying DNA to change the rate at which genes can be transcribed, and ultimately driving hormonal pathways of sexual differentiation. Further, work by Warner et al. (2013) in the jacky dragon showed that these epigenetic effects are transgenerational; the incubation temperature that a male experienced significantly affected the sex ratios produced by his progeny, such that offspring of males incubated at 23 °C produced significantly more males than those incubated at 33 °C. The way in which these epigenetic effects may influence hormonal cascades to ultimately drive offspring sex one way or the other will be discussed in detail in Chap. 9.

8.2 Evidence for Adaptive Manipulation of Offspring Sex Ratios in Reptiles that Exhibit TSD

Compared with the large bodies of work exploring the potential adaptive implications of adjusting sex ratios in mammals and birds, the adaptive nature of adjusting sex ratios in reptilian species is relatively understudied. This is perhaps because so many reptilian species are long-lived and do not reach sexual maturity until years after they hatch, which makes observations of long-term reproductive success difficult and also makes linkages between maternal experiences of their environment with what offspring will experience during *their* reproductive attempts nonexistent. Such linkages would be particularly important for forming adaptive predictions regarding how adjusting sex ratios will influence reproductive success further down the line. In addition, to test the influences of TSD on offspring fitness, one would need to produce both sexes at a given temperature, which often does not occur.

8.2.1 Adaptive Models for TSD

Charnov and Bull (1977) were the first to attack reptilian sex ratios in an adaptive sense, and they suggested that TSD may enhance parental fitness by matching the nature of the incubation conditions to offspring sex, such that the environment in which offspring develop affects their fitness in a sex-specific manner (i.e., the Charnov–Bull model) (Charnov and Bull 1977). In 1999, Shine provided an excellent review that used existing literature to outline individual hypotheses for linking TSD to offspring fitness (Fig. 8.5): The first is that females may match sex to egg size (referred to here as the egg size hypothesis). The idea here is that TSD simply acts as a mechanism by which females can match egg size to sex, and this would maximize reproductive success if egg size affects fitness of males and females differently. This idea has been demonstrated in diamondback terrapins, where females hatching from larger eggs reached the minimum size to initiate

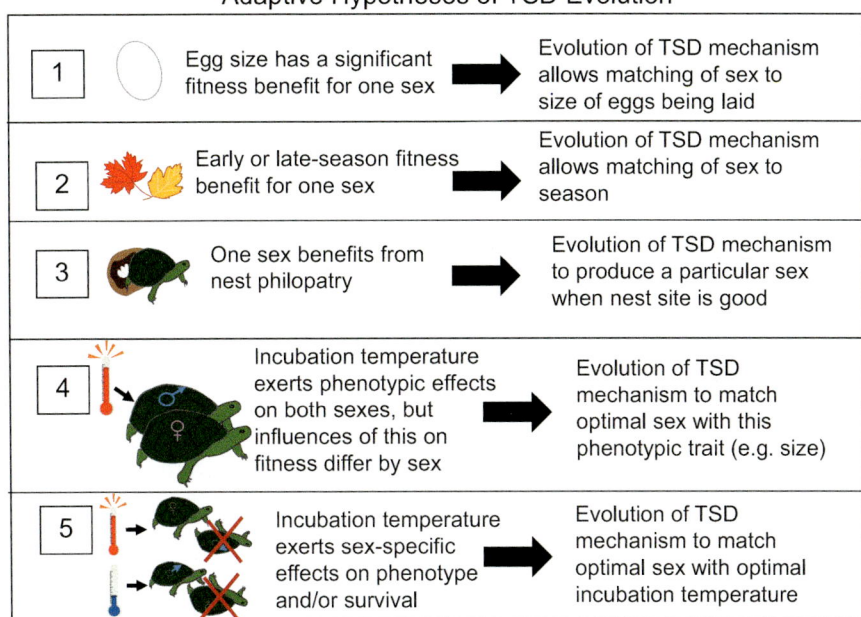

Fig. 8.5 A summary of five adaptive hypotheses based on the Charnov–Bull model of TSD evolution (Charnov and Bull 1977), and summarized by Shine (1999)

reproduction 2–3 years sooner than those hatching from smaller eggs, while males approached the size necessary for reproduction at 3 years old regardless of the size of the egg they hatched from. Thus, egg size affected the fitness of females more than it did males (Roosenburg 1996). However, rather than concluding that the simple link between egg size and fitness drove the evolution of TSD, Roosenburg (1996) uses these data as evidence for a broader hypothesis that encompasses this idea—the maternal-condition-dependent choice hypothesis. This hypothesis states that ESD is a mechanism to bias offspring sex ratio through maternal choice of nest site or incubation position and requires that (1) the maternal effect (in this case, egg size) varies, (2) that this variation in the effect must differentially affect the fitness of males and females (as it clearly appears to in diamondback terrapins), and (3) that the female's reproductive behavior, in this case the place where she places her eggs to incubate, is somehow influenced by information, propagated by her condition and/or physiological state, about the condition of her eggs. In the case of the diamondback terrapin, if a female has a large egg that is ready to be laid, the egg size must somehow be translated into a decision to lay that egg in a particular location, likely via a physiological signal. This idea now has support in other systems. For enymid turtles, egg size explained 59% of the variation in hatching mass for females but none of the variation for males. As in diamond terrapins, this could ultimately translate into substantial variation in the age at first reproduction for females, while this fitness indicator appears to be unrelated to egg size in males

(Roosenburg and Kelley 1996). Work in painted turtles may also support this hypothesis because egg size significantly influences juvenile sizes in this species, and body size is positively related to fecundity in females while this relationship is negative in males (reviewed in Janzen and Morjan 2002). More work needs to be done in additional species to further test this hypothesis.

The second hypothesis highlighted by Shine (1999) is the idea that TSD evolved to allow the female to match the sex of the offspring to the time in the season when the sex produced would be the most successful sex, reproductively speaking. In other words, this idea requires that the time in the season during which the offspring hatches affects the fitness of males and females differently. To date, there is not much support for this idea. In a study of jacky dragons, Harlow and Taylor (2000) found that nests laid earlier in the season tended to produce male-biased sex ratios. They suggested that this pattern was adaptive because males of this relatively short-lived species would be able to grow fast enough to reach the necessary size for reproduction before the time came to hibernate, effectively increasing their reproductive potential. However, these findings were based on only three nests, and further studies refuted these ideas (Warner et al. 2005). Finally, this idea was tested for a species in which there is a documented benefit for females hatching earlier in the season; mallee dragons are short-lived Australian lizards for which early hatching has a significant influence on reproductive success in females, but has no influence in males. For this species, however, there was no seasonal variation in the sexes of offspring produced, suggesting that another adaptive mechanism may be at work. More studies need to be done to examine this hypothesis as a potential adaptive driver of TSD evolution.

The third hypothesis, originally posited by Reinhold (1998), suggests that the influence of incubation temperature, itself, on offspring sex may have sex-specific fitness effects because of sex differences in philopatry, as well as sex differences in the benefit obtained from the quality of the nest site (the nest philopatry hypothesis). For example, if daughters inherit the nest site, then they would benefit more than sons from a high-quality nest site, while the fitness of daughters would also suffer more due to a low-quality nest site than would that of sons. Nest site fidelity has been observed in several reptilian species, including at least six turtles and one crocodile (reviewed in Freedberg and Wade 2001). If this theory holds, then we would expect that mitochondrial diversity between nest sites to be very low. Sea turtles are excellent species on which to test this hypothesis because mitochondrial diversity can be compared between turtles on adjacent beaches, and both green sea turtles and loggerhead turtles exhibit distinct, beach-specific mitochondrial haplotypes (reviewed in Freedberg and Wade 2001). Work in green turtles showed that there was a negative correlation between the genetic similarity among nesting females and the distance between their nest sites (Peare and Parker 1996). This finding indicates that nest sites are, indeed, heritable in this species. If warm nest sites, that produce more females, confer a fitness advantage, then the nest philopatry hypothesis may offer an adaptive explanation for the evolution of TSD.

It is also possible that TSD evolved because incubation temperatures exert direct effects on offspring fitness. This can happen in two ways. In the first, incubation temperature has similar influences on a phenotypic trait in both males and females, but the fitness effect of that trait is sex-specific, and as a result, TSD would be acting to "match" offspring of a particular sex with the optimal trait. While the majority of studies in this area focus on the influences of incubation temperature on size and growth rates, there could be similar influence on the ability to avoid predators, mating behaviors, immune function, etc. Despite many studies examining the effects of incubation temperatures on phenotypic traits, many more studies in many more species are needed to adequately test this hypothesis. In the second possibility, incubation temperature exerts different influences on the same phenotypic trait in males and females. In pine snakes, for example, males suffer higher embryonic mortality at low incubation temperatures while females suffer higher embryonic mortality at high incubation temperatures (Burger and Zappalorti 1988). Work in three-lined skinks, lizards that exhibit a GSD system, showed that females hatching from eggs incubated at warmer temperatures were larger and faster than sister females incubated at cooler temperatures, while the effects of incubation temperatures on males were less consistent (Elphick and Shine 1999). Sex-specific influences on phenotypic traits could be the first step in the evolution of TSD mechanisms.

A final hypothesis that has gained some support in GSD and TSD species suggests that individuals should adjust offspring sex ratios according to the tertiary or operational sex ratios in the population to maximize their changes of breeding and, as a result, their ultimate fitness. Fisher (1930) predicted that selection would favor individuals that bias allocation towards the less abundant sex in a population. However, evidence for this is currently mixed (see work in Jacky dragons and GSD lizards below). Despite these potential adaptive models for the evolution of TSD, the evidence for any one of them is still limited. Ewert and Nelson (1991) propose a possible alternative hypothesis that rather than evolving due to the physiological consequences of incubation temperature on individuals, TSD may have instead evolved because it could produce adaptively biased sex ratios at the population level. This has previously been deemed unlikely because population structures of most TSD reptiles do not appear to be amenable to adaptively skewed sex ratios; however, Ewert and Nelson (1991) generated two hypotheses by which adaptive sex ratio adjustment would drive the evolution of TSD. In the first, termed the Group Structure hypothesis, patterns of TSD may be a by-product of selection for skewed sex ratios in small breeding groups of closely related individuals. Such a situation would mean that TSD species exhibit particularly high levels of inbreeding compared to GSD species, and studies comparing levels of heterozygosity between GSD and TSD species do not support this idea (Burke 1993). Alternatively, Ewert and Nelson (1991) posited that TSD may have evolved to prevent inbreeding, through the production of same-sex clutches. However, this idea comes with several problems, including the fact that simply producing clutches of all one sex would not prevent individuals from breeding with closely related kin from other nests, or even full-sibs from subsequent years (Burke 1993). Thus overall, there is currently little evidence supporting the idea that TSD evolved to adaptively bias sex ratios at the population level.

Tests of Adaptive Models of TSD Evolution in Painted Turtles	
Egg size hypothesis	Egg size correlated to juvenile sizes, but in both sexes
	No relationship b/t egg size and nest temperature or vegetation cover
Sex-specific influences of seasonal timing	Observed seasonal pattern of sex ratios – more males produced early while more females produced late
	No data on the potential fitness effects of this pattern
Natal philopatry hypothesis	Females showed philopatry for vegetation types
	Females did not prefer to nest in open sites, and preferred sites did not produce more female offspring
Direct influences of incubation temps on phenotype	Natural incubation temperatures did not influence hatchling size, plastron color, righting behavior, or delayed-type hypersensitivity response
	Warm incubation temperatures produced turtles with longer carapaces
	Females (produced at warm temps) thrive in warm overwintering temps while males (produced at cold temps) thrive in cool overwintering temps.

Fig. 8.6 A summary of the adaptive models for TSD evolution tested using painted turtles

8.2.2 Case Study: Adaptive TSD in Painted Turtles?

As with studies in birds, it is often helpful to examine how sex ratios within a single species respond to a variety of factors. Painted turtles are among the best studied when it comes to the potential for adaptive sex allocation via TSD (Fig. 8.6). This freshwater turtle species exhibits Type 1A (MF) TSD. Previous studies show that all males are produced at temperatures below 26 °C, both sexes are produced at 28 °C, and all females are produced at temperatures above 29 °C (reviewed in Valenzuela and Lance 2004). Painted turtles are an excellent system in which to examine the potential for an adaptive significance behind TSD because they are easy to study, nesting near ponds and lakes that can easily be reached, and there are painted turtle subspecies that span a large latitudinal gradient, from Canada all the way down to the Southeastern USA. Finally, they inhabit a variety of habitat types, with diversity in vegetation and environmental impacts on the nesting process.

For the incubation site to be considered a phenotypic trait on which selection can act, there must be some genetic variability underlying the choice of nesting sites (Bulmer and Bull 1982). Indeed, individual female painted turtles appear to exhibit repeatable preferences for a particular quantity of vegetation near their nesting sites (reviewed in Janzen and Phillips 2006). Further, these preferences for a particular quantity of vegetation appears to be heritable (González 2004), and the heritability estimates for these preferences were influenced by the temperatures experienced during the previous winters; heritability of nest preferences was significant when the previous winter was warm, while they were not significant when previous winters were cold (McGaugh et al. 2009). These studies suggest that preference for particular nest sites is likely heritable, but there is a high level of complexity in the inheritance of those nest site preferences.

Given the heritable and repeatable nature of nest site choice in this species, we can consider nest site choice a female-specific phenotypic trait on which selection may act. We can then consider studies that examine evidence for each of the adaptive hypotheses listed above. Janzen and Morjan (2002) tested three key conditions of the egg size hypothesis posited originally by Roosenburg (1996): (1) that egg size is correlated with the size of the resulting individual, (2) that incubation temperature must influence growth rates of offspring after hatch, and (3) the covariance between egg size and temperature must be sex-specific. To test this, Janzen and Morjan (2002) examined the effects of initial egg mass and incubation temperature on juvenile survival for 1 year post-hatch and then looked at whether any effects of egg size and temperature were sex-specific. They found that egg size was correlated with juvenile size in *both* sexes and that the incubation temperatures that produced males (26 °C) and females (30 °C) influenced growth rates, with females growing faster than males. They suggest that while there was no sex-specific covariance with either egg size or temperature when it comes to direct growth measures, instead it may be the timing of growth in relation to reproductive maturity that could make this hypothesis plausible; in this species, female fecundity is positively correlated with body size, while males show no such pattern (reviewed in Janzen and Morjan 2002). However, when the same group took their studies to the field and examined natural nests, they found no evidence that egg size varied with either nest temperatures or vegetation cover (Morjan 2003), putting the potential for egg size to be the true adaptive driver of TSD in this species in doubt.

How about the hypothesis that TSD evolved to match offspring sex to the season during which those offspring hatch? Bowden et al. (2000) provide data in support of this idea; at a single incubation temperature of 28 °C, clutches produced early in the season were 72% male while clutches produced later in the season were 78% female. It remains unclear whether there is more of a benefit for one sex to hatch early or late in the season, and given that these turtles take years for both sexes to reach reproductive maturity (Janzen and Morjan 2002), hatching earlier in the breeding season would not provide enough time for either sex to breed before hibernation. The adaptive significance of this seasonal pattern of sex ratios must be examined further in this species.

Valenzuela and Janzen (2001) tested the idea that TSD evolved in response to nest philopatry in this species (Reinhold 1998). They analyzed whether female painted turtles were philopatric to particular microhabitats; whether survival rates, sex ratios, and incubation conditions were repeatable across years in these microhabitats; whether females are less likely to frequent cooler, more vegetated sites that tend to produce more males than warmer, more open sites that tend to produce more females; and finally, whether embryonic mortality rates were higher at male-producing sites and lower at female-producing sites. They found that while females were somewhat philopatric for particular types of vegetation cover, they did not nest more frequently in open sites that tended to produce more females and the nest sites that females preferred did not tend to produce more females. Additionally, nests that had higher embryonic mortality also produce slightly more females, but the magnitude of this effect was extremely small. Overall, these data do not support the nest philopatry hypothesis for the evolution of TSD in this species.

This leaves the two remaining hypotheses which suggest that incubation temperature directly influences phenotypic traits and/or survival of offspring, either with similar influences on both sexes that ultimately differ in their benefit between those sexes or via divergent influences on the two sexes. Paitz et al. (2010) tested the idea that incubation temperature directly influences offspring phenotype in painted turtles; they examined the influences of temperature fluctuations in natural nests on measures of size (hatchling mass and plastron length), plastron color, righting behavior of offspring, and a measure of immune function via a delayed-type-hypersensitivity test using PHA as the antigen. Incubation temperature did not significantly influence any measure of offspring phenotype in this study. However, these authors did not measure long-term growth, and it would also be helpful to measure additional arms of the immune system to get a global view of immunocompetence. To the contrary, Riley et al. (2014) found that hatchlings incubated at warm temperatures had longer carapace lengths. Whether this ultimately influences fitness is unknown.

Spencer and Janzen (2014) suggest that in turtle species that overwinter in the natal environment, the temperature of the natal environment may predict the temperatures experienced while overwintering in the same environment, providing a drive for adjusting sex ratios based on which sex would survive the winter best in that environment. This idea indicates that it may be more appropriate to look for linkages between incubation temperature and traits that influence overwinter survival, as well as growth during and after hibernation. To test this, they collected individuals nesting at male and female-producing temperatures in the wild and transferred two individuals per nest to warm and cold artificial overwintering environments designed to fluctuate as they would in the field. In warm overwintering regimes, males from male-dominated nests (and thus likely incubated at warm temperatures) had higher metabolic rates and consumed more residual yolk than females that hatched from female-dominated nests. The opposite was true in the cool overwintering regime; in this case, females from female-dominated nests (and thus likely incubated at cool temperatures) had higher metabolic rates and consumed more residual yolk than males from male-dominated nests. These results translated into influences on size, because individuals that have higher metabolic rates and consume more residual yolk grow less. So, females in warm overwintering temperatures grew 3% larger from Autumn to Spring than females in cooler overwintering temperatures, while males at cool overwintering temperatures grew 2% larger than males in warm overwintering temperatures. These findings comply with the Charnov–Bull model of sex allocation (Charnov and Bull 1977), because

8.2 Evidence for Adaptive Manipulation of Offspring Sex Ratios in Reptiles...

warmer temperatures produce females, which thrive in warm overwintering conditions, while cooler temperatures produce males, which thrive in cool overwintering temperatures. Whether this ultimately translates into fitness effects remains to be tested.

Where does this leave us in the understanding of how and why TSD evolved in painted turtles? If egg sizes do induce sex-specific effects on the time it takes to reach reproductive maturity, then TSD could have evolved as a mechanism to link egg size with sex. The seasonal patterns in the sexes of offspring produced by this species also warrants more investigation into the adaptive value of using TSD to match offspring sex with a particular time in the season. However, as of now, we do not have overwhelming evidence for a particular model of TSD evolution in painted turtles. Schwanz et al. (2010) showed, using painted turtles as models, that it is important to assess *relative* condition when examining the adaptive significance of a trait, because the assessment of condition may change as the environment changes (Fig. 8.7). An example of relative condition in birds is maternal rank, which changes based on the characteristics of the surrounding individuals. Painted turtles also show evidence of partial relativity. The authors plotted sex ratios versus mean nest temperatures for each year from 1995 to 2006 and drew sigmoidal curves to estimate the inflection points for each year. These were the annual pivotal temperatures, or the temperature at which both males and females were produced. These annual pivotal temperatures correlated positively with mean annual nest temperatures, which means that a nest with an intermediate temperature may

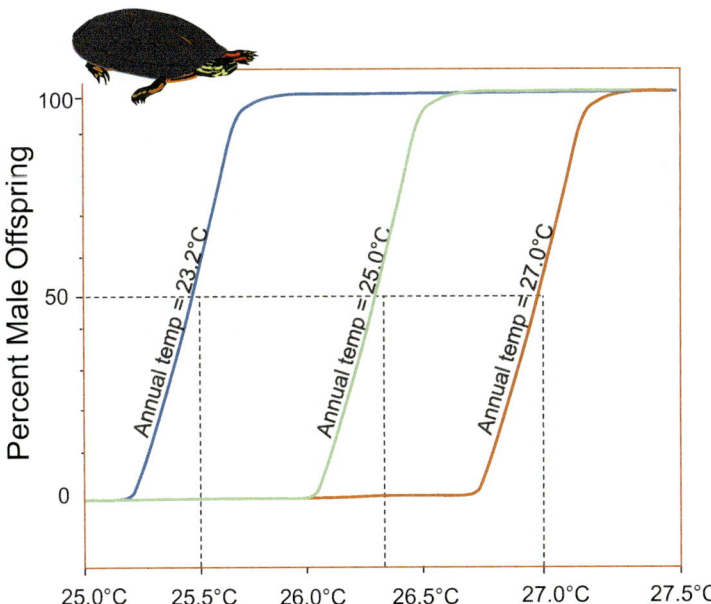

Fig. 8.7 Visual representation of how annual pivotal temperatures change relative to annual ambient temperatures in painted turtles. Data are based on Schwanz et al. (2010)

produce mostly females in a cold year when it is warm relative to the others, but mostly males in a warm year when it is cold relative to the others. This means that adaptive sex allocation may be far more complex that we previously realized, and that we need to take climatic variation into account when examining sex ratio patterns, particularly in painted turtles that occupy large latitudinal ranges.

8.2.3 Evidence for Sex Ratio Adjustment in Other Turtles

Some of the seminal work examining TSD as an adaptive mode of sex ratio adjustment was conducted in turtles. More than 60 turtle species that utilize TSD have been identified; these species exhibit two of the three TSD patterns—Pattern 1A (MF) and Pattern 2 (FMF). In 1991, Ewert and Nelson examined patterns of TSD in 11 species of turtles and suggested that across species, the smaller gender arises at the coolest incubation temperature. For example, TSD pattern 1A appears to be present in species where females are larger than males, while TSD pattern 2 appears to be present in species where males are larger than females (Ewert and Nelson 1991). In both of these cases, females are produced at the warmest temperatures, and thus the difference lies in which sex is produced at the coolest temperature. Perhaps TSD evolved to produce the sex that experiences greater fitness as small individuals in cooler temperatures (reviewed in Ewert et al. 1994). This idea is called the sexual size dimorphism hypothesis and is also supported by work in geckos.

The potential adaptive value of TSD has been discussed in several turtle species beyond painted turtles. Common snapping turtles, for example, show significant family by temperature interactions, indicating potential for sex ratio evolution. This species follows the FMF pattern of TSD, with temperatures below 23 °C producing all females, temperatures from 23 to 27 °C producing all males, and temperatures at or above 29.5 °C producing all females (Rhen and Lang 1995). Shade provided by vegetation cover significantly influences offspring sex ratios (Juliana et al. 2004), and female snapping turtles show preferences for nesting in sandy patches, which have higher success rates and incubation temperatures (Kolbe and Janzen 2002). This behavior satisfies a critical condition of the nest philopatry hypothesis. There is also a good body of evidence that incubation temperatures alter important phenotypic traits in offspring. Rhen and Lang (1995) used an elegant experimental design in which hormone treatments were used to uncouple the effects of sex and incubation temperature on offspring phenotypes to test the influences of incubation temperatures on offspring growth, independently of sex. They showed that growth was enhanced at intermediate incubation temperatures that normally produce males (24 and 26.5 °C). The same research group also showed that energy expenditure for maintenance may be greater for turtles hatching from eggs that were incubated at lower temperatures (Rhen and Lang 1999). In support of this finding, O'Steen and Janzen (1999) found that metabolic rates of hatching snapping turtles were inversely related to temperature. In addition to growth and metabolism, incubation temperatures also affect future nest site choices of offspring in this species; egg temperatures were negatively correlated with the temperatures of the nests chosen by juveniles (O'Steen 1998). These findings indicate that there is potential for adaptation of TSD via direct influences of incubation temperature on offspring phenotype; however, it is not yet known whether the fitness benefits of these effects are sex-specific. It *is* known, however, that males reach sexual maturity earlier and at a larger body size than females, and also that male body size is important in male–male combat during the reproductive season. Further work should explore whether the higher growth rate of males from male-producing incubation temperatures exerts measurable effects on male fitness and how that compares to the influences of high growth rates on female fitness.

Work in sea turtles has instead primarily focused on the impending influences of climate change on operational sex ratios, and as a side effect, has highlighted the potential for these species to utilize TSD in an adaptive manner to keep operational sex ratios balanced. Hays et al. (2010) showed that, globally, a majority of sea turtle species have hatchling sex ratios that are female-biased overall; greater than 50% of these species exceed ratios of 3:1 females:males. However, male sea turtles breed more frequently than females and females have the ability to store sperm, which may bring the balance of the operational sex ratio back to 50:50. TSD may have evolved to maintain this balance in light of sex-specific breeding periodicities. Alternatively, TSD may have evolved as a result of sex ratios at more of a population level. Hawksbill sea turtles show repeatable choices of nest sites and nest in large zones that produce a majority of one sex. Kamel and Mrosovsky (2005) suggest that frequency-dependent selection may be responsible for the evolution of TSD, because females nesting in areas that produce the rarer sex may gain a fitness advantage.

We see from this large body of work in turtles that there is much potential for an adaptive role of TSD, and that this role is likely different among species. While much of this work has focused on sex ratio adjustment at the individual level, there is a dearth of knowledge about how TSD functions at the population level in turtles. Additional studies examining the adaptive significance of TSD on multiple levels are warranted.

8.2.4 Case Study: Adaptive Sex Allocation via TSD in Jacky Dragons

There is a shortage of studies examining the potential for adaptive sex allocation via TSD in oviparous lizards. Jacky dragons are among the lizard species in which the potential adaptive value of TSD has been most extensively studied (Fig. 8.8). Jacky dragons are Australian agamid lizards that lack sex chromosomes and exhibit strict TSD, during which females are produced at low and high incubation temperatures and males are produced at intermediate incubation temperatures (Harlow and Taylor 2000). Warner et al. (2005) used this species to systematically test four of the adaptive hypotheses described above. First, they tested the idea that TSD evolved to link egg size and sex. The sizes of eggs that produced males versus

Tests of Adaptive Models of TSD Evolution in Jacky Dragons	
Egg size hypothesis	Egg size correlated to juvenile sizes, but in both sexes No relationship b/t egg size and nest temperature or vegetation cover
Egg size hypothesis	Sizes of eggs that produced males did not differ from sizes of eggs that produced females
Sex-specific influences of seasonal timing	Warner et al (2005) – no seasonal effect on offspring sex ratios Harlow & Taylor (2000) – more males later in the season, but only based on 3 nests.
Direct influences of incubation temps on phenotype	Warm temperatures led to smaller body sizes and lower body condition in both sexes Minimal influences of temperature on offspring survival Influences at reproductive maturity are unknown

Fig. 8.8 A summary of the adaptive hypotheses for TSD evolution tested in jacky dragons

females did not differ, nor did the optimal egg size, in terms of offspring phenotype, differ based on incubation temperature. This suggests that TSD did not evolve as a way to match sex to the optimal egg size in this species. They then examined morphological and growth data of lizards hatching at each incubation temperature to test two hypotheses: (1) that incubation temperature influences phenotypic traits of both sexes similarly, but that the fitness impacts of that trait differ among the sexes and (2) that incubation temperature exerts different phenotypic influences on the two sexes, leading to sex-specific fitness benefits. While hatchlings from the warm incubation treatment were smaller and had lower body condition than those from eggs incubated at cooler temperatures, these differences were short-lived, as lizards from the warm incubation temperature treatments were ultimately the largest by winter. Additionally, the influences of these differences on survival, at least under artificial conditions, were so small that they wouldn't likely translate into significant fitness effects. However, the researchers did not measure these lizards to the point of reproductive maturity, and as in other studies, it is possible that fitness effects linked to incubation temperature may emerge at that point. Further studies need to be done to test this. Finally, Warner et al. (2005) tested the idea that females match offspring sex to the time of the breeding season that is most optimal for the ultimate fitness of that sex. While they found seasonal variation in egg size, they found no seasonal variation in offspring sex ratios. This finding differs from those of Harlow and Taylor (2000), who found that nests laid earlier in the season tended to produce male-biased sex ratios in this species; however, the latter study was only based on a sample size of three nests.

Warner and Shine (2007) tested whether TSD may have evolved in this species in response to the operational sex ratio in the population. Lizards were maintained in enclosures with either male-biased "populations", simulated by placing a single female with two males, or female-biased "populations", simulated by placing three females together and only introducing a single male for a 5 days period on a monthly basis. Individuals housed in male-biased "populations" produced almost 60% males in first clutches, but then 30–40% males in clutches 2 and 3. In contrast, females housed in female-biased "populations" produced female-biased clutches (30–40% males) in all three consecutive clutches. Warner and Shine (2007) suggest that because the likely success of using current sex ratios to predict operational sex ratios when offspring reach adulthood a year later is low, they may instead be altering sexes according to the optimal conditions for survival during the juvenile period; a male-biased operational sex ratio may indicate that the environment is most suitable for males, and so producing male offspring in that environment may be the better strategy to ensure survival through the juvenile period. This idea still needs to be tested further.

Overall, jacky dragons show potential for adaptive sex allocation via TSD; however, they do not appear to fit any of the current models to explain why TSD may have evolved in this species. This may perhaps be because their system of sex determination is more complex than previously thought. Warner et al. (2007) found that the TSD systems of jacky dragons can be overridden by other factors related to dietary quality. The authors provided females with either a low-quality diet of crickets that were fed only on corn or a high-quality diet of crickets and roaches fed cat food. The high-quality diet was also supplemented with carrots, apples, and leafy greens. When females produced eggs, those eggs were then incubated at 28 °C, a temperature known from previous studies to produce a 50:50 ratio of males:females. Eggs produced by females on the high-quality diet produced female-biased sex ratios, while those on the low-quality diet produced male-biased sex ratios. In addition, this could be an adaptive strategy, as male hatching size is associated with better reproductive success in this species. Thus, more work needs to be done to examine how diet quantity interacts with high and low temperatures in the field and how diet and temperature may interact in an adaptive sense to influence offspring sex in this and potentially other species.

8.2.5 Do Viviparous Lizards Adaptively Allocate Sex via TSD?

It may be surprising to some that must of the evidence for adaptive sex allocation via TSD in lizards come from work in viviparous species. It was previously thought that viviparous lizards could not possibly possess a TSD mechanism because the high temperature extremes that occur during basking would result in highly skewed sex ratios that would ultimately lead to extinction. However, it is becoming increasingly clear that the process of sex determination in viviparous species is sensitive to temperature. One of the first demonstrations of this ability was in southern water skinks, a viviparous species in which no evidence for heteromorphic sex chromosomes has been found. Female water skinks were maintained at 25, 30, or 32 °C for the duration of pregnancy, and the result was a significant effect on the sex ratios of offspring produced; maintaining females at 25 °C resulted in 55% male offspring, 30 °C resulted in 75% males, and 32 °C resulted in 100% male offspring (Robert and Thompson 2001). Since then, additional studies have found similar patterns following artificial incubation at low and high temperatures in other species, including multi-ocellated racerunners (Tang et al. 2012; Zhang et al. 2010), brown forest skinks (Ji et al. 2006), and spotted snow skinks (Wapstra et al. 2004). In each case, more males were produced at higher temperatures. It is possible that while these patterns emerge during artificial incubation at constant temperatures, similar effects may not occur when temperatures are more variable, as under natural circumstances. Indeed, in studies that documented natural sex ratios in wild conditions (Robert and Thompson 2001; Zhang et al. 2010) or where females were allowed to regulate body temperature (Ji et al. 2006), sex ratios did not differ from 50:50. Thus, it is possible that we are seeing the remnants of a TSD mechanism that are not currently being used by viviparous individuals. Work in spotted snow skinks, however, suggests otherwise. For this species, sex ratios varied significantly among years and closely followed field temperatures, suggesting that natural fluctuations in temperature can influence offspring sex ratios in this species.

In addition, the results of a couple of studies show that viviparous lizards may have the ability to adjust sex ratios in an adaptive manner. When adult spotted snow skinks were scarce, females produce male-biased litters, and females that were courted by few males produced mostly males while those courted by many males produced mostly daughters. This work indicates that this viviparous lizard species may adjust sex ratios according to the operational sex ratio, in the direction predicted by (Fisher 1930). Further evidence for this idea was demonstrated in southern water skinks. Females maintained in female-only groups produced only male offspring, while those maintained with many males produced a mix of male and female offspring (Robert et al. 2003). This work indicates that viviparous lizards may be able to use temperature to adjust offspring sex in an adaptive manner, but more work linking TSD to these adaptive patterns is needed, and additional exploration of other adaptive hypotheses is warranted in these species.

8.2.6 What Is Going on in Tuatara?

Thus far, all of the species we have discussed exhibit MF or FMF patterns of TSD; in both of these instances, females are produced at the highest incubation temperatures; however, the reverse, a MF pattern of TSD, is found only in two species of Tuatara. Work on these two species indicates that the pivotal temperature lies between 21 and 23 °C depending on the subspecies (Mitchell et al. 2006),

8.2 Evidence for Adaptive Manipulation of Offspring Sex Ratios in Reptiles...

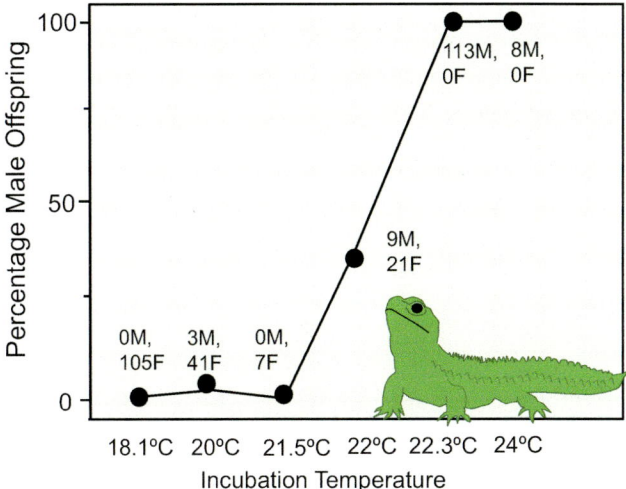

Fig. 8.9 Data in one subspecies of Tuatara collected from a population on Stephens Island showing the range of incubation temperatures that produce different ratios of male:female offspring. Based on data reported in Mitchell et al. (2006)

temperatures above which produce 100% males (Fig. 8.9). Why do these species produce males at high temperatures while all others produce females at high temperatures? One possibility is that the true TSD pattern of this species is FMF, and we simply have not yet examined sex ratios at high enough incubation temperatures for this species. Indeed, a majority of FMF species were initially thought to exhibit the FM pattern of TSD, after which additional studies confirmed a FMF pattern. This initial misconception occurred in some species because females are produced at temperatures that were previously thought to be too high for survival; for example, in geckos, a FM TSD pattern was initially reported and temperatures above 32 °C prevented successful production of young. However, additional studies showed that eggs will successfully hatch at temperatures up to 35 °C, and that eggs incubated above 33 °C were predominantly female (Viets et al. 1993). Additional studies examining a wider range of incubation temperatures in tuatara could ultimately unveil a FMF pattern of TSD in tuatara as well. Indeed, tuatara have been known to hatch from nests that experience repeated exposures to 30°C (reviewed in Mitchell et al. 2006).

Alternatively, perhaps there is an adaptive reason that this species produces males at higher temperatures instead of females. A study employing artificial incubation temperatures of 18°C, 21°C, and 22°C produced 0%, 4%, and 100% males at the respective temperatures. These thermal regimes did not influence hatchling size, but did influence both the length of incubation and the size of juveniles at 10 months old. The time to hatch decreased with increasing incubation temperatures; eggs at the female-producing temperature of 18 °C took 81 additional days to hatch compared to eggs incubated at the male-producing temperature of

22 °C. When juveniles reached 10 months old, those that hatched from eggs incubated at 18 °C were smaller than those incubated at 21 or 22 °C. The fact that incubation temperatures of 21 and 22 °C produced very different sex ratios but juveniles at these incubation temperatures were of similar size at 10 months old indicates that this effect is due to temperature rather than sex. Overall, this results in an earlier hatching time and larger body sizes at 10 months for male tuatara. Given that large male tuatara have better access to mates, this may provide a fitness advantage, but because incubation temperature exerts similar effects on juvenile sizes of male and female tuatara, more work is needed to flush out whether the influences of incubation on size has differential fitness effects on males and females.

8.2.7 Adaptive Sex Allocation in GSD Reptiles

We have now seen evidence that reptiles use temperature as a mechanism to bias sex ratios, but can reptiles that employ genetic mechanisms of sex determination adaptively bias sex as well? There is, in fact, evidence that they can. Side-blotched lizards are polyandrous lizards that exhibit an XY system of GSD. In this system, large males secure better territories, which females then seek out for mating. By manipulating territory quality, Calsbeek and Sinervo (2002) were able to examine the influences of male size on sex allocation in this system, because side-blotched lizards exhibit high rates of multiple paternity and females mated with both small and large males. Offspring that were sired by large males were significantly more likely to be male, while those sired by smaller males were more likely to be female. Additional work showed that large males sired sons with high viability while small males sired daughters with high viability, suggesting that this sex allocation pattern may be adaptive (Calsbeek and Sinervo 2004). In addition, there was no evidence that X and Y sperm production differed based on body size, because males produced sons when they were relatively large, and sons when they were relatively small, compared to other males that co-sired the clutch (Calsbeek and Sinervo 2004). It is likely that females are instead sorting sperm based on a combination of male identity and the sex chromosome carried by the sperm. Perhaps sex-specific sperm storage may be a mechanism by which this occurs. For example, in Australian painted dragons, stored sperm outcompete newly inseminated sperm and are also 23% more likely to produce male offspring. Studies testing for similar phenomena in side-blotched lizards would be helpful.

The brown anole is another species that likely exhibits GSD. The precise mechanism of sex determination is unknown in this species, but other *Anolis* species exhibit forms of GSD, whether XX, XXY, or GSD without heteromorphic sex chromosomes. In 2010, Cox and Calsbeek showed that, like side-blotched lizards, female brown anoles produced a higher proportion of sons via large sires and a higher proportion of daughters via small sires. They then examined fitness of progeny in relation to body size of the sire and found that the maximum fitness was likely achieved by shifting to the production of more sons as sire size increased.

This work provides evidence that female brown anoles adjust offspring sex ratios in an adaptive manner. In additional work, the same research group tested the influence of paternal condition on offspring sex ratios; the number of sons produced increased as paternal condition increased (Cox et al. 2011). This suggests that female brown anoles allocate sex according to multiple assessments of male quality, including both size and condition.

Swedish sand lizards exhibit the WZ form of GSD and also appear to adaptively alter offspring sex ratios. Olsson et al. (2005a) showed that male sand lizards have more DNA fragments at MHC loci, which indicates that they are likely more heterozygous at these loci and also have lower parasite loads when under high levels of physiological stress. When male badge size was experimentally enlarged, females mating with those males produced 10% more daughters than control females (48% versus 38%), and females that were mated to males with more MHC polymorphism produced more daughters as well (Olsson et al. 2005b). The authors suggest that when females are heterogametic, production of daughters is associated with increased risk of inviability due to expression of paternal detrimental recessives on Z chromosome; thus, they should only produce daughters when mated to high-quality partners. More work must be done to determine whether these patterns of sex allocation translate into fitness benefits.

Finally, there is evidence that snakes adaptively adjust the sexes of the offspring they produce, but in these cases, in response to maternal body size. Garter snake females that were large produced male-biased clutches while those that were smaller produced female-biased sex ratios. The adaptive value of this pattern is unclear since large males do not appear to gain an advantage towards acquiring mates (reviewed in Dunlap and Lang 1990). The opposite pattern emerges in two other snake species. For both northern water snakes and beaked sea snakes, large females produce larger offspring that are predominantly female (Weatherhead et al. 1998; Lemen and Voris 1981). The authors suggest that this may be an adaptive mechanism by which producing females of large body size increases reproductive success of those females. However at this point, a fitness advantage of large female body size has not yet been empirically tested.

8.2.8 Conclusions

There is ample evidence that, whether they use a temperature-dependent, genetically based, or mixed mechanism of sex determination, reptiles do have the ability to alter sex ratios in response to environmental and social conditions. Whether these patterns have a significance in an adaptive context, however, remains unknown, and further testing of multiple adaptive hypotheses in the same species at once is needed to fully understand why TSD has evolved and whether reptiles with any sex-determining mechanism adaptively alter sex ratios. Interestingly, there is a dichotomy in terms of how researchers approach the adaptive significance of sex allocation in TSD versus GSD species. A majority of work in TSD species focuses on testing hypotheses regarding how each sex functions in its environment, often forming ideas about how influences of incubation temperatures on early growth measures may correlate with ultimate fitness. The ultimate goal in TSD species appears to be focused on gaining an understanding of why the mechanism of TSD has evolved in those species. In contrast, most work in GSD species focuses on how sex ratios relate to measures of parental quality and size, generating questions about how females may use mate quality to predict which sex would thrive as a result of that quality. So, in GSD systems, the questions center less around how the mechanisms initially evolved (and in fact, the mechanisms of sex allocation in GSD species are currently unknown) and more around the factors that guide sex-allocation decisions in females. Our understanding of the adaptive drivers behind sex allocation in TSD species would benefit from work empirically testing the relationships between parental condition and sex ratios, while work in GSD species would benefit from asking questions about how offspring sex influences fitness in a given set of social or environmental conditions. In addition, given the potential sensitivity of TSD species to changing climatic conditions, it is important to determine vulnerability of each species to long-term temperature elevations in the face of global warming.

References

Bowden R, Ewert M, Nelson C (2000) Environmental sex determination in a reptile varies seasonally and with yolk hormones. Proc R Soc Lond B Biol Sci 267(1454):1745–1749

Bulmer M, Bull J (1982) Models of polygenic sex determination and sex ratio control. Evolution 36:13–26

Burger J, Zappalorti R (1988) Effects of incubation temperature on sex ratios in pine snakes: differential vulnerability of males and females. Am Nat 132(4):492–505

Burke RL (1993) Adaptive value of sex determination mode and hatchling sex ratio bias in reptiles. Copeia 1993(3):854–859

Calsbeek R, Sinervo B (2002) Uncoupling direct and indirect components of female choice in the wild. Proc Natl Acad Sci 99(23):14897–14902

Calsbeek R, Sinervo B (2004) Within-clutch variation in offspring sex determined by differences in sire body size: cryptic mate choice in the wild. J Evol Biol 17(2):464–470

Charnov EL, Bull J (1977) When is sex environmentally determined? Nature 266(5605):828–830

Cohen M, Gans C (1970) The chromosomes of the order Crocodilia. Cytogenetic 9(2):81–105

Cox RM, Calsbeek R (2010) Cryptic sex-ratio bias provides indirect genetic benefits despite sexual conflict. Science 328(5974):92–94

Cox RM, Duryea MC, Najarro M, Calsbeek R (2011) Paternal condtion drives progeny sex ratio bias in a lizard that lacks parental care. Evolution 65(1):220–230

Demas S, Duronslet M, Wachtel S, Caillouet C, Nakamura D (1990) Sex-specific DNA in reptiles with temperature sex determination. J Exp Zool 253(3):319–324

Dunlap KD, Lang JW (1990) Offspring sex ratio varies with maternal size in the common garter snake, Thamnophis sirtalis. Copeia 1990(2):568–570

Elphick MJ, Shine R (1999) Sex differences in optimal incubation temperatures in a scincid lizard species. Oecologia 118(4):431–437

Ewert MA, Nelson CE (1991) Sex determination in turtles: diverse patterns and some possible adaptive values. Copeia 1991:50–69

Ewert MA, Jackson DR, Nelson CE (1994) Patterns of temperature-dependent sex determination in turtles. J Exp Zool 270(1):3–15

Ezaz T, Quinn AE, Miura I, Sarre SD, Georges A, Graves JAM (2005) The dragon lizard Pogona vitticeps has ZZ/ZW micro-sex chromosomes. Chromosom Res 13(8):763–776

Fisher R (1930) Genetical theory of sex allocation. Claredon Press, London

Freedberg S, Wade MJ (2001) Cultural inheritance as a mechanism for population sex-ratio bias in reptiles. Evolution 55(5):1049–1055

Gibbons JW, Scott DE, Ryan TJ, Buhlmann KA, Tuberville TD, Metts BS, Greene JL, Mills T, Leiden Y, Poppy S (2000) The Global Decline of Reptiles, Déjà Vu Amphibians Reptile species are declining on a global scale. Six significant threats to reptile populations are habitat loss and degradation, introduced invasive species, environmental pollution, disease, unsustainable use, and global climate change. Bioscience 50(8):653–666

González JE (2004) Inheritance of nest-site choice in the field in a turtle with temperature-dependent sex determination. Doctoral dissertation. Iowa State University

Harlow PS, Taylor JE (2000) Reproductive ecology of the jacky dragon (Amphibolurus muricatus): an agamid lizard with temperature-dependent sex determination. Austral Ecol 25 (6):640–652

Hays GC, Fossette S, Katselidis KA, Schofield G, Gravenor MB (2010) Breeding periodicity for male sea turtles, operational sex ratios, and implications in the face of climate change. Conserv Biol 24(6):1636–1643

Holleley CE, O'Meally D, Sarre SD, Graves JAM, Ezaz T, Matsubara K, Azad B, Zhang X, Georges A (2015) Sex reversal triggers the rapid transition from genetic to temperature-dependent sex. Nature 523(7558):79–82

Janzen FJ (1994) Climate change and temperature-dependent sex determination in reptiles. Proc Natl Acad Sci 91(16):7487–7490

Janzen FJ, Krenz JG (2004) Which was first, TSD or GSD? In: Valenzuela N, Lance VA (eds) Temperature-dependent sex determination in vertebrates. Smithsonian Books, Washington, DC, pp 121–130

Janzen FJ, Morjan CL (2002) Egg size, incubation temperature, and posthatching growth in painted turtles (Chrysemys picta). J Herpetol 36(2):308–311

Janzen FJ, Paukstis GL (1991) Environmental sex determination in reptiles: ecology, evolution, and experimental design. Q Rev Biol 66(2):149–179

Janzen F, Phillips P (2006) Exploring the evolution of environmental sex determination, especially in reptiles. J Evol Biol 19(6):1775–1784

Ji X, Lin LH, Luo LG, Lu HL, Gao JF, Han J (2006) Gestation temperature affects sexual phenotype, morphology, locomotor performance, and growth of neonatal brown forest skinks, Sphenomorphus indicus. Biol J Linn Soc 88(3):453–463

Juliana JRS, Bowden RM, Janzen FJ (2004) The impact of behavioral and physiological maternal effects on offspring sex ratio in the common snapping turtle, Chelydra serpentina. Behav Ecol Sociobiol 56(3):270–278

Kamel SJ, Mrosovsky N (2005) Repeatability of nesting preferences in the hawksbill sea turtle, Eretmochelys imbricata, and their fitness consequences. Anim Behav 70(4):819–828

Kolbe JJ, Janzen FJ (2002) Impact of nest-site selection on nest success and nest temperature in natural and disturbed habitats. Ecology 83(1):269–281

Lemen CA, Voris HK (1981) A comparison of reproductive strategies among marine snakes. J Anim Ecol 50:89–101

Matsumoto Y, Buemio A, Chu R, Vafaee M, Crews D (2013) Epigenetic control of gonadal aromatase (cyp19a1) in temperature-dependent sex determination of red-eared slider turtles. PLoS One 8(6):e63599

McGaugh SE, Schwanz LE, Bowden RM, Gonzalez JE, Janzen FJ (2009) Inheritance of nesting behaviour across natural environmental variation in a turtle with temperature-dependent sex determination. Proc Biol Sci 277(1685):1219–1226

Miller D, Summers J, Silber S (2004) Environmental versus genetic sex determination: a possible factor in dinosaur extinction? Fertil Steril 81(4):954–964

Mitchell NJ, Nelson NJ, Cree A, Pledger S, Keall SN, Daugherty CH (2006) Support for a rare pattern of temperature-dependent sex determination in archaic reptiles: evidence from two species of tuatara (Sphenodon). Front Zool 3(1):9

Morjan CL (2003) Variation in nesting patterns affecting nest temperatures in two populations of painted turtles (Chrysemys picta) with temperature-dependent sex determination. Behav Ecol Sociobiol 53(4):254–261

Mork L, Czerwinski M, Capel B (2014) Predetermination of sexual fate in a turtle with temperature-dependent sex determination. Dev Biol 386(1):264–271

Mrosovsky N, Pieau C (1991) Transitional range of temperature, pivotal temperatures and thermosensitive stages for sex determination in reptiles. Amphibia-Reptilia 12(2):169–179

O'Steen S (1998) Embryonic temperature influences juvenile temperature choice and growth rate in snapping turtles Chelydra serpentina. J Exp Biol 201(3):439–449

O'Steen S, Janzen FJ (1999) Embryonic temperature affects metabolic compensation and thyroid hormones in hatchling snapping turtles. Physiol Biochem Zool 72(5):520–533

Olsson M, Madsen T, Wapstra E, Silverin B, Ujvari B, Wittzell H (2005a) MHC, health, color, and reproductive success in sand lizards. Behav Ecol Sociobiol 58(3):289–294

Olsson M, Wapstra E, Uller T (2005b) Differential sex allocation in sand lizards: bright males induce daughter production in a species with heteromorphic sex chromosomes. Biol Lett 1(3):378–380

Paitz RT, Clairardin SG, Griffin AM, Holgersson MC, Bowden RM (2010) Temperature fluctuations affect offspring sex but not morphological, behavioral, or immunological traits in the Northern Painted Turtle (Chrysemys picta). Can J Zool 88(5):479–486

Paladino FV, Dodson P, Hammond JK, Spotila JR (1989) Temperature-dependent sex determination in dinosaurs? Implications for population dynamics and extinction. Geol Soc Am Spec Pap 238:63–70

Parrott BB, Kohno S, Cloy-McCoy JA, Guillette LJ (2014) Differential incubation temperatures result in dimorphic DNA methylation patterning of the SOX9 and aromatase promoters in gonads of alligator (Alligator mississippiensis) embryos. Biol Reprod 90(1):2

Peare T, Parker PG (1996) Local genetic structure within two rookeries of Chelonia mydas (the green turtle). Heredity 77(6):619–628

Place AR, Lance VA (2004) The temperature-dependent sex determination drama: same cast, different stars. In: Valenzuela N, Lance VA (eds) Temperature dependent sex determination in vertebrates. Smithsonian Institution, Washington, DC, pp 99–110

Quinn AE, Georges A, Sarre SD, Guarino F, Ezaz T, Graves JAM (2007) Temperature sex reversal implies sex gene dosage in a reptile. Science 316(5823):411–411

Radder RS, Quinn AE, Georges A, Sarre SD, Shine R (2008) Genetic evidence for co-occurrence of chromosomal and thermal sex-determining systems in a lizard. Biol Lett 4(2):176–178

Radder RS, Pike DA, Quinn AE, Shine R (2009) Offspring sex in a lizard depends on egg size. Curr Biol 19(13):1102–1105

Reinhold K (1998) Nest-site philopatry and selection for environmental sex determination. Evol Ecol 12(2):245–250

Rhen T, Lang JW (1995) Phenotypic plasticity for growth in the common snapping turtle: effects of incubation temperature, clutch, and their interaction. Am Nat 146(5):726–747

Rhen T, Lang JW (1999) Incubation temperature and sex affect mass and energy reserves of hatchling snapping turtles, Chelydra serpentina. Oikos 86:311–319

Riley JL, Tattersall GJ, Litzgus JD (2014) Potential sources of intra-population variation in the overwintering strategy of painted turtle (Chrysemys picta) hatchlings. J Exp Biol 217(23):4174–4183

Robert KA, Thompson MB (2001) Sex determination: viviparous lizard selects sex of embryos. Nature 412(6848):698–699

Robert KA, Thompson MB, Seebacher F (2003) Facultative sex allocation in the viviparous lizard Eulamprus tympanum, a species with temperature-dependent sex determination. Aust J Zool 51(4):367–370

Roosenburg WM (1996) Maternal condition and nest site choice: an alternative for the maintenance of environmental sex determination? Am Zool 36(2):157–168

Roosenburg WM, Kelley KC (1996) The effect of egg size and incubation temperature on growth in the turtle, Malaclemys terrapin. J Herpetol 30:198–204

Sarre SD, Georges A, Quinn A (2004) The ends of a continuum: genetic and temperature-dependent sex determination in reptiles. Bioessays 26(6):639–645

Sarre SD, Ezaz T, Georges A (2011) Transitions between sex-determining systems in reptiles and amphibians. Annu Rev Genomics Hum Genet 12:391–406

Schwanz LE, Janzen FJ, Proulx SR (2010) Sex allocation based on relative and absolute condition. Evolution 64(5):1331–1345

Shine R (1999) Why is sex determined by nest temperature in many reptiles? Trends Ecol Evol 14(5):186–189

Shine R (2002) Eggs in autumn: responses to declining incubation temperatures by the eggs of montane lizards. Biol J Linn Soc 76(1):71–77

Shine R, Elphick M, Donnellan S (2002) Co-occurrence of multiple, supposedly incompatible modes of sex determination in a lizard population. Ecol Lett 5(4):486–489

Spencer R-J, Janzen FJ (2014) A novel hypothesis for the adaptive maintenance of environmental sex determination in a turtle. Proc R Soc Lond B Biol Sci 281(1789):20140831

Tang X-L, Yue F, Yan X-F, Zhang D-J, Xin Y, Wang C, Chen Q (2012) Effects of gestation temperature on offspring sex and maternal reproduction in a viviparous lizard (Eremias multiocellata) living at high altitude. J Therm Biol 37(6):438–444

Valenzuela NM, Janzen FJ (2001) Nest-site philopatry and the evolution of temperature-dependent sex determination. Evol Ecol Res 3:779–794

Valenzuela N, Lance V (2004) Temperature-dependent sex determination in vertebrates. Smithsonian Books, Washington, DC

Valenzuela N, Badenhorst D, Montiel EE, Literman R (2014) Molecular cytogenetic search for cryptic sex chromosomes in painted turtles Chrysemys picta. Cytogenet Genome Res 144(1):39–46

Viets BE, Tousignant A, Ewert MA, Nelson CE, Crews D (1993) Temperature-dependent sex determination in the leopard gecko, Eublepharis macularius. J Exp Zool A Ecol Genet Physiol 265(6):679–683

Wapstra E, Olsson M, Shine R, Edwards A, Swain R, Joss JM (2004) Maternal basking behaviour determines offspring sex in a viviparous reptile. Proc R Soc Lond B Biol Sci 271(Suppl 4):S230–S232

Warner DA, Shine R (2007) Reproducing lizards modify sex allocation in response to operational sex ratios. Biol Lett 3(1):47–50

Warner DA, Shine R, Schwenk K (2005) The adaptive significance of temperature-dependent sex determination: experimental tests with a short-lived lizard. Evolution 59(10):2209–2221

Warner DA, Lovern MB, Shine R (2007) Maternal nutrition affects reproductive output and sex allocation in a lizard with environmental sex determination. Proc R Soc Lond B Biol Sci 274(1611):883–890

Warner DA, Uller T, Shine R (2013) Transgenerational sex determination: the embryonic environment experienced by a male affects offspring sex ratio. Sci Rep 3:4

Weatherhead PJ, Brown GP, Prosser MR, Kissner KJ (1998) Variation in offspring sex ratios in the northern water snake (Nerodia sipedon). Can J Zool 76(12):2200–2206

Zaborski P, Dorizzi M, Pieau C (1982) H-Y antigen expression in temperature sex-reversed turtles (Emys orbicularis). Differentiation 22(1–3):73–78

Zaborski P, Dorizzi M, Pieau C (1988) Temperature-dependent gonadal differentiation in the turtle Emys orbicularis: concordance between sexual phenotype and serological H-Y antigen expression at threshold temperature. Differentiation 38(1):17–20

Zhang D-J, Tang X-L, Yue F, Chen Z, Chen Q (2010) Effect of gestation temperature on sexual and morphological phenotypes of offspring in a viviparous lizard, Eremias multiocellata. J Therm Biol 35(3):129–133

The Truth About Nemo's Dad: Sex-Changing Behaviors in Fishes

9

> *Finding Nemo lied to your kids!*
> *How Finding Nemo should have started if it were biologically accurate:*
> *Father and mother clownfish are tending to their clutch of eggs at their sea anemone when the mother is eaten by a barracuda. Nemo hatches as an undifferentiated hermaphrodite (as all clownfish are born) while his father transforms into a female now that his female mate is dead. Since Nemo is the only other clownfish around, he becomes a male and mates with his father (who is now a female). Should his father die, Nemo would change into a female and mate with another male. Although a much different storyline, it still sounds like a crazy adventure!*
>
>   Patrick Cooney, in Fun Fish Fodder

Thus far, I have provided evidence in mammals, birds, and reptiles that parents may adaptively manipulate the sexes of offspring in ways that may maximize offspring survival and/or increase the chances that those offspring will be reproductively successful. In fish systems, however, there are, to my knowledge, no reports of parents that adaptively control the sexes of offspring. This lack of evidence could result from a lack of studies examining the phenomenon, or it could be because sexual differentiation processes in fishes are much more flexible, and in some cases fish can change whether they have ovarian or testicular tissue to best suit environmental conditions. One can see how, in these cases, a mechanism by which parents could predictively skew sex ratios may not be needed, since offspring can adjust their *own* sexes when they need to, and on a timescale much more relevant to when the sex change would effectively influence survival and/or reproductive success. In this chapter, I will highlight the diversity of sex-determining mechanisms in fish systems, introduce the factors that are known to alter sexes of offspring during both embryonic development and adulthood, and discuss the potential that these factors are used for adaptive manipulation of the sexual differentiation process.

© Springer International Publishing AG 2018
K.J. Navara, *Choosing Sexes*, Fascinating Life Sciences,
https://doi.org/10.1007/978-3-319-71271-0_9

9.1 Patterns of Gonadal Differentiation in Fishes

Fishes show enormous diversity in their physical appearances, feeding mechanisms, modes of breeding and parental care, and even methods of locomotion. There are piscine species that are barely recognizable from sea plants, species that practice pouch brooding much like kangaroos, fishes that lure prey in the deepest depths of the ocean using bioluminescent lures, and species that can even fly over water or walk on land. As a result, it is not surprising that the mechanisms by which sex is determined vary among piscine species just as widely. In fact, there are some species in which sex is never truly differentiated and remains flexible throughout life. At the other end of the spectrum are species in which sex is permanently determined, and differentiation occurs early in life, much like it is in mammals and birds (Fig. 9.1). Given that over 250 new species of fishes are discovered every year, it is likely that we still haven't seen all of the modes of sex determination and differentiation that exist in this vertebrate class, and within discussions of every sex determination mode, one can always find numerous species that exhibit modifications of that mode. Multiple excellent reviews have been provided to explain what is known about modes of sex determination and differentiation in fishes and the mechanisms by which those processes occur (e.g., Devlin and Nagahama 2002; Piferrer 2001; Yamamoto 1969). Here, I will provide a

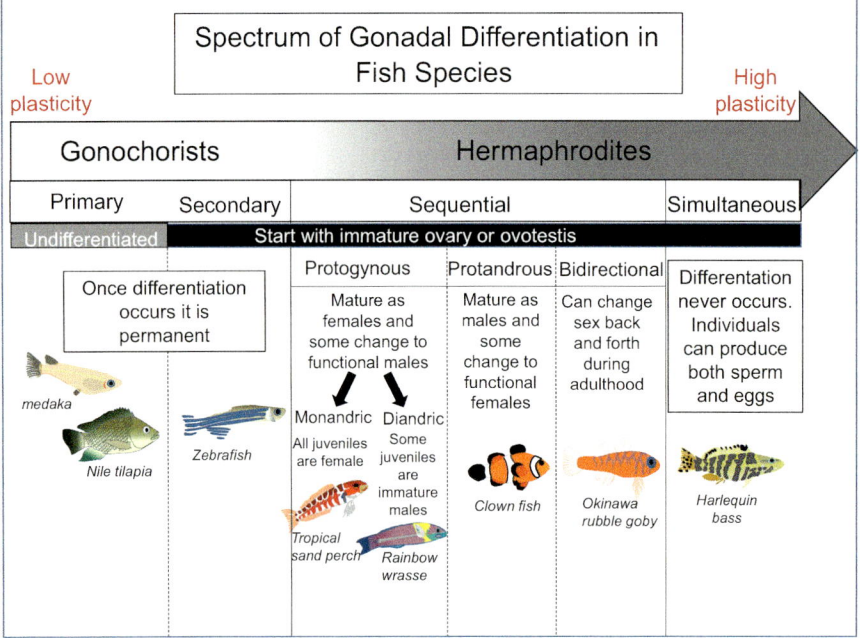

Fig. 9.1 Depicted is the diversity in the system of sex determining modes among piscine species, placed on a plasticity spectrum. Also included are example species for each define system of sex determination

basic overview of identified patterns of sex determination in fishes and encourage readers to refer to those reviews for more detail and additional specific examples. There is often variation among authors in how the terms sex determination and sexual differentiation are defined. These terms are particularly difficult to distinguish in fish systems where sexual characteristics can change throughout life. For the purposes of this chapter, I will use sex determination to define the mechanisms that regulate sexual differentiation, particularly the differentiation of gonadal tissue into functional ovaries or testes.

9.1.1 Gonochoristic Fishes

At one end of the sex determination spectrum lie fish species that are gonochoristic, meaning that sex is determined and gonads are differentiated once in their lives and they remain that sex through the rest of their lives. A majority of fishes, at least 98%, fall into this general category. However, even at this end of the spectrum, there is flexibility both in terms of whether embryonic gonadal tissue starts out truly indifferent, and when during early development sex determination and gonadal differentiation take place. Gonochoristic fishes are divided into two main categories, both containing subcategories within them (Fig. 9.2). Fishes that start out with completely indifferent gonadal tissue that then develops into either testes or ovaries are called primary gonochorists. Medaka and the Nile tilapia are perhaps the most well-known examples of this pattern of sex determination. In medaka, the first sign of sexual differentiation is proliferation of male and female germ cells, which occurs during embryonic development. By just after hatch, female medaka already have meiotically active ovarian tissue (Schartl 2004). In Nile tilapia, the first signs of sexual differentiation can be seen in these fish at 23–26 days *after* hatch with the formation of the ovarian cavity in females or the efferent duct in males, and as in medaka, there appears to be no further sex change after sexual differentiation (Ijiri et al. 2008).

Secondary gonochorists are fish species in which gonads develop into ovaries or, at the very least, contain ovarian tissue, before differentiation into the ultimate functional sex. One of the most well-studied examples of secondary gonochorists is the zebrafish system, in which all individuals develop ovary-like tissue, and starting approximately 21 days after hatch, oocytes undergo apoptosis and spermatocytes begin to develop. In male fish, oocytes completely disappear by day 30 after hatch (von Hofsten and Olsson 2005; Uchida et al. 2002). In cases such as this one, because there is concurrent presence of both ovarian and testicular tissue during the sexual differentiation process, the term "juvenile hermaphrodites" is also used, however while these fish have ovarian and testicular tissue present concurrently at some point in their juvenile lives, at no point do they have *functional* ovarian and testicular tissue present at the same time. In addition, naturally induced adult sex change has never been observed in these species.

For both primary and secondary gonochorists, it should be noted that even though sex determination appears to result in permanent differentiation endpoints that do not change through the remainder of life, there is still plasticity in the process of sex determination for these species, because environmental signals

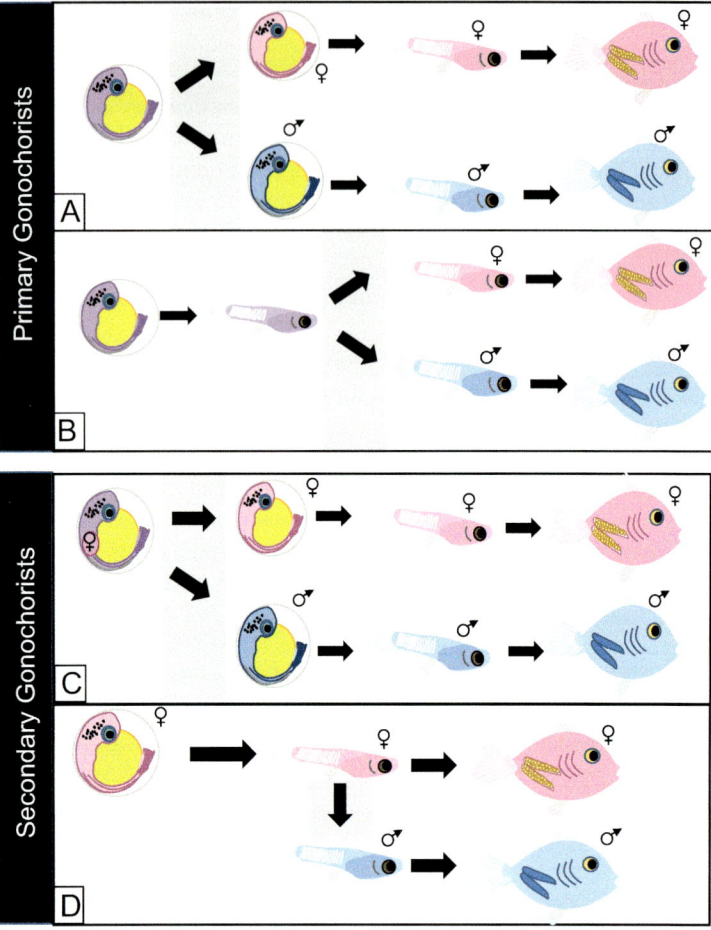

Fig. 9.2 Gonochorists can be divided into two main group: (1) In primary gonochorists (**a** and **b**), gonadal tissues start out undifferentiated (purple) and then differentiates into testes (blue) or ovaries (pink) either during embryonic development (**a**) or after hatch (**b**). (2) In secondary gonochorists (**c** and **d**), undifferentiated gonads either contain some ovarian tissue and then differentiate into functional ovaries or testes (**c**) or gonads initially differentiate into ovaries after which they either continue to produce a fully mature female, or testicular cells begin to take over to produce a mature male (**d**). Just as in primary gonochorists, the final differentiation process can take place either during embryonic development or after hatch. In all four cases, once differentiation has taken place, it is permanent for the remainder of the individual's life

and/or hormones can override the sex determination mechanism to produce a particular sex. This will be explained in more detail below. In addition, a recent study of medaka and tilapia showed that while it has been generally accepted that gonochoristic fishes retain no plasticity after the process of sexual differentiation has completed, treatment with hormones during adulthood can still induce functional sex reversals (Paul and Kuester 1987). Thus, the true level of plasticity within these gonochoristic systems still remains to be determined.

9.1.2 Hermaphroditic Fishes

The next group of fishes is categorized as a separate group, yet one could see where they are simply a step along the spectrum of gonadal differentiation patterns. Fish that are true hermaphrodites have, at some point in their adult lives, both ovarian and testicular tissue present (Fig. 9.3). In one group of hermaphroditic fishes, the sequential hermaphrodites, this time period is often relatively short, ranging from days to months, and marks a sex change from either female to male or male to female. Within this group are subgroups that are defined according to which sex matures first. Protogynous hermaphrodites initially mature into females and then a percentage of those females change to males during adulthood. There is, however, variation even within this group. For some species, called monandric species, gonads differentiate to form only juvenile females with no juvenile males in the population. In this case, the only males in the population exist as a result of sex changes during adulthood. Tropical sand perches are good examples of fish that show monandric protogynous hermaphrotidism. Individuals of this species start with an early undifferentiated gonad that becomes intersexual. All of these gonads differentiate into functional ovaries, and changes in the dominance hierarchy can then trigger select individuals to transition into functional males (Walker et al. 2007; Pandian 2012).

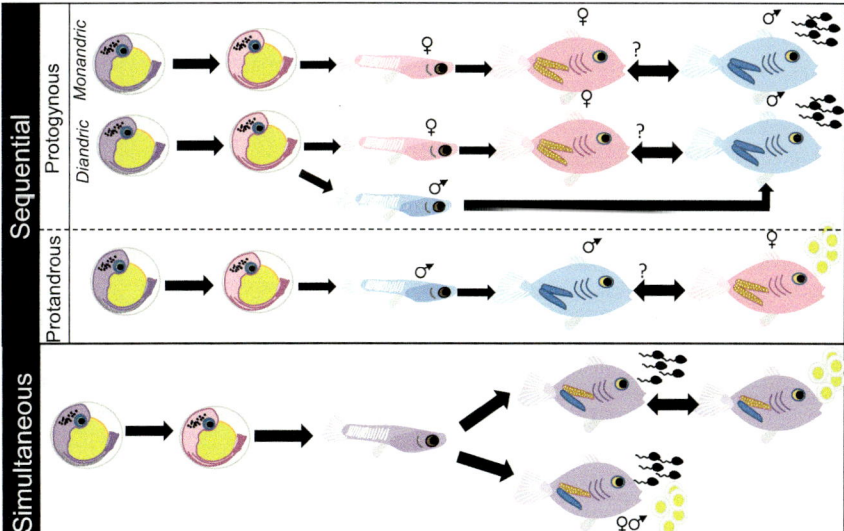

Fig. 9.3 Hermaphroditic species can be divided into two main categories. Sequential hermaphrodites initially mature as one sex and then undergo a sex change into the other. This group contains two subgroups. Protogynous individuals start life as females and then change to male in adulthood. Protandic individuals start life as male and change to female during adulthood. The second main group, simultaneous hermaphrodites, maintain both ovarian and testicular tissue at the same time throughout life and alternate between producing sperm and eggs. This can occur within days or over years depending on the species. All hermaphrodites initially develop ovarian tissue before differentiating into mature males, females, or a mixture of both

For other protogynous hermaphrodites, termed diandric, gonadal differentiation results in both juvenile females *and* males. The females then mature into functional adult females while the males develop into smaller initial phase (IP) males that often resemble immature fish or females. Adult terminal phase (TP) males in the population can either arise from females that transition into functional males or from IP males that grow to take the place of another TP male. Wrasses represent some of the most well-studied examples of diandric protogynous hermaphrodites. For example, in both blue-head and rainbow wrasse, loss of a dominant male triggers females to change sex and/or IP males to grow into TP males (Linde et al. 2011; Godwin et al. 2003). It is unknown whether the mechanisms involved in gonadal differentiation of these two groups of TP males are the same.

While protogynous hermaphrodites initially mature as females, protandric hermaphrodites initially mature into males, and then a percentage of these males change to females later in life. A classic example of this type of system is the one portrayed incorrectly in the movie Finding Nemo: the clown fish (also called the anemone fish). In this system, histological examination of gonads indicated that all individuals initially mature as males and then, loss of the dominant females triggers males in the social group to undergo several stages of gonadal change to achieve complete sex reversal into functional females. Some sequential hermaphrodites are even bidirectional, changing back and forth between male and female during their lifetimes (Manabe et al. 2008; Munday et al. 1998, 2010). The first species identified that is capable of this bi-directional sex change was the Okinawa rubble goby (Fig. 9.4). This system usually consists of a single dominant male. If that male is lost, a female will transition into a functional dominant male; however, if the initial dominant male returns, the individual that changed to male can then transition back into a female.

For sequential hermaphrodites, most sex changes, barring the species that exhibit bi-directional sex changes, are permanent, and even in bi-directional species, sperm and eggs are never produced simultaneously. In contrast, the final group of hermaphrodites, the synchronous hermaphrodites, possess *functional* ovarian and testicular tissue in their gonads simultaneously. In some of these species, both sperm and eggs can be produced in close succession, allowing the individual to take different roles serially. There are even species that can self-fertilize, though reports of those are few (Cole 1997; Harrington 1961; Soto et al. 1992). In others, individuals choose between production of sperm and eggs during life stages. For example, blue-banded gobies show characteristics of both bi-directional sequential and simultaneous hermaphrodites because despite maintaining both testicular and ovarian tissue, they express only one sex depending on their dominance status in the group (St. Mary 1994; Rodgers et al. 2007). This supports the notion that fish exist not in distinct groups of sexual differentiation patterns, but a continuous spectrum that characterizes the evolutionary transition from unmalleability to complete plasticity in reproductive function (Fig. 9.1).

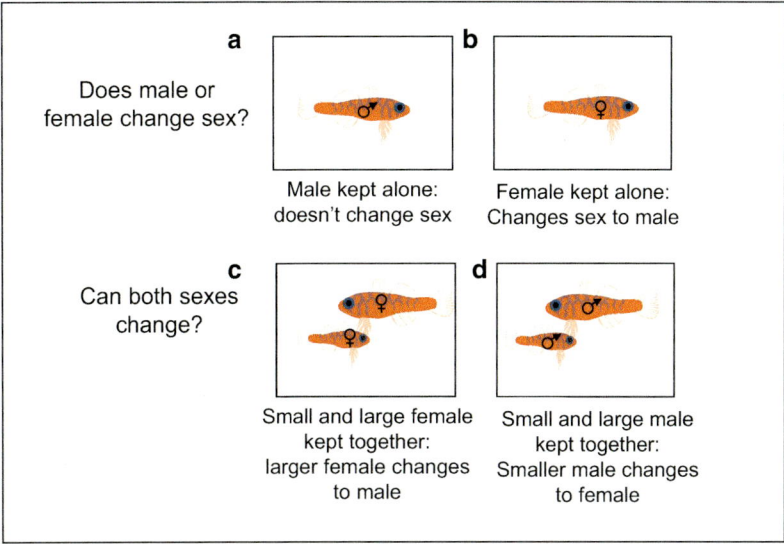

Fig. 9.4 In part of a study conducted by Sunobe and Nakazono (1993), Okinawa rubble gobies were placed in tanks either alone or with larger individuals of the same sex. Males kept alone (**a**) did not change sex, but females kept alone (**b**) changed to males. When small and large females were kept together (**c**), the larger female changed to male and when small and large males were kept together (**d**) the smallest male changed to females. This was the first documented case of bi-directional sex change in a sequential hermaphrodite

9.2 Diversity in Sex-Determining Systems Among Fish Species

Just as the patterns of sexual differentiation lie along a spectrum in fishes, so do the mechanisms of sex determination. Fishes contain representatives of all major sex-determining mechanisms identified in other vertebrate classes. Gonochorists are largely comprised of fish species that use genetic sex determination (or GSD); however, not all of them have distinct sex chromosomes like those seen in mammals and birds. In fact, it is thought that the ancestral system of genetic sex determination started with the interaction of multiple genes in the genome that work together to shape the individual's reproductive phenotype and that as specific genes became more potent regulators of the process, systems in which single genes, and eventually distinct sex chromosomes, arose (Devlin and Nagahama 2002). Thus, it is not surprising to find examples along this theoretical spectrum throughout the fishes. In addition, just as in reptiles that show a gradient of sex-determining systems that range from strictly genetic sex determination (GSD) through strictly environmental sex determination (ESD), a similar gradient exists among the fishes as well. The specific genetic and physiological mechanisms that underlie the processes of sexual differentiation are beyond the scope of this chapter and can be found in excellent reviews such as Devlin and Nagahama (2002), Pandian

(2012), and Guerrero-Estévez and Moreno-Mendoza (2010). Below I will highlight the many types of sex determination systems found in fishes and provide examples of each (Fig. 9.5).

Sex determination in fishes is classified at the most broad level based on whether there is a genetic influence that triggers the production of testes versus ovaries (i.e., GSD), and if there is, how many genes are involved in the process. In a monogenic system, a single gene triggers the process of sex determination. Medaka are fish that display an XY system of sex determination and are the first and only piscine system thus far in which a single sex-determining gene (DMY) has been identified on the Y chromosome (Matsuda et al. 2002). This is one of only two sex determining genes found in all vertebrates; the other is the SRY gene that drives male development in mammals (Pandian 2012). The presence of a single gene that determines sex does not, however, appear to be the case for most fishes that exhibit GSD. Most are instead polygenic, where the effects of multiple, independently segregating sex determining loci act together to determine sex (Moore and Roberts 2013). At the other end of the spectrum, in a very small proportion of fishes, sex is determined entirely via environmental influences with no clear genetic driver guiding the process. This group of fishes will be described in more detail below.

For a majority of piscine species, morphologically distinct sex chromosomes have not been identified (Guerrero-Estévez and Moreno-Mendoza 2010). In fact, Pandian (2012) estimates that evidence for sex chromosomes has only been found for only 6–11% of fish species studied. In the species for which sex chromosomes *have* been identified, there is large diversity in the nature of those sex chromosomes, which sex is heterogametic if any, and how many sex chromosomes are involved. To my knowledge, there are currently nine known patterns of sex determination that involve distinct sex chromosomes in fishes (Fig. 9.5, reviewed in Schultheis et al. 2009; Tave 1986). In four of these systems, the male determines the sex of offspring. The first of these is a basic XY system, much like that found in mammals, where the male can donate an X or a Y chromosome, and a Y results in a male offspring. The second (the XO system) is a variant of this system, but in this case, the Y chromosome is not present. Instead, if a male donates an X chromosome, a female is produced, but if he does not donate a sex chromosome, a male is produced. Interestingly, this indicates that in species that utilize this type of sex determining system, the default sex is likely male, unlike in the XY system displayed in medaka, for example, where the DMY gene must be present for development of a male gonad. The third system of sex determination ($X_1X_1X_2X_2/X_1X_2Y$) is one in which three rather than two sex chromosomes are involved. In this system, females have one more sex chromosome than males do. If the male donates his X_1X_2 chromosome pair, the offspring is female, but if he donates a Y chromosome, the offspring is male. Finally, in the XX/XY_1Y_2 system, the *male* is the one with an additional sex chromosome. In this case, if the male donates his Y_1Y_2 chromosomes, the offspring is male, whereas if he donates an X chromosome, the offspring is female.

Not surprisingly, the systems in which the female is the heterogametic sex show just as much variation as those in which the male is heterogametic. Much like birds, some fish species display a simple ZW system of sex determination, in which the

	System Name	♀	♂	Sex-determining contributions
Male determines sex	XY	XX	XY	Male donates X or Y
	XO	XX	XO	Male donates X or none
	$X_1X_1X_2X_2/X_1X_2Y$	$X_1X_1X_2X_2$	X_1X_2Y	Male donates X or Y
	XX/XY_1Y_2	XX	XY_1Y_2	Male donates X_1X_2 or Y
Female determines sex	WZ	WZ	ZZ	Female donates W or Z
	ZW_1W_2/ZZ	ZW_1W_2	ZZ	Female donates W_1W_2 or Z
	ZO	ZO	ZZ	Female donates Z or none
	YW/YY	YW	YY	Female donates a W or Y
Both can Determine sex	WXY	XX / WY	XY / YY	Female or male can donate W, which counteracts Y
	Autosomal			Combination of genes on autosomes determines sex

Fig. 9.5 The nine known systems of genetic sex determination (GSD) in fishes, including the genetic makeup that makes a male and a female, and what is contributed by an individual to determine the sexes of offspring

female donates a Z chromosome to produce a male offspring and a W chromosome to produce a female offspring. As was seen in the heterogametic male group, there is also a variant in the heterogametic female group where a sex chromosome is absent. In this group (the ZO system), if a female donates a Z chromosome, a male offspring is produced while if she donates no sex chromosome, a female offspring is produced. In addition, there is also a group where the female has an additional sex chromosome. In the ZW_1W_2/ZZ group, if the female donates her W_1W_2 chromosomes, a female offspring is produced while if she donates a Z chromosome, a male offspring results.

There is also one system in which sex chromosomes are present, but both sexes have the capability to determine the sex of the offspring. In the WXY system, inheritance of the Y chromosome from the male generally results in differentiation of testicular tissue. However, if a W chromosome is also inherited, the W acts to block the sex-determining influences of the Y chromosome, resulting instead in female offspring. In this case, only the male has the ability to donate a Y chromosome, but either sex can donate a W chromosome, which is simply a modified version of an X chromosome. Therefore, in this case, either sex can ultimately determine whether offspring are male or female. While these various systems appear to run the gamut of possible combinations of sex chromosomes, there is one more system that appears to be a variant of the WXY system. In the YW/YY system, the X chromosome is not present, and the female determines the sex based on whether she donates a Y or a W chromosome. This system has been observed in platyfish (Schultheis et al. 2009).

More common in piscine systems are cases in which a genetic mechanism drives sex determination, but where distinct sex chromosomes have not yet been identified. In these cases, an autosomal process may guide sex determination, and many genes located on autosomes contribute to the ultimate decision to produce testes or ovaries. However, it is unclear whether sex determination in these cases is completely controlled by genes located on autosomes, or whether these genes act as autosomal *modifiers* of the true sex determining mechanism that perhaps has not yet been identified. In swordtails, for example, it was previously thought that sex was determined based on the interactions of multiple genes located on autosomes; however, later work revealed that these fish may instead utilize a YW/YY female-heterogametic system of sex determination and that genes on autosomes may modify this process (reviewed by Schultheis et al. 2009). In fact, this idea appears to be quite common among fish species, and particularly among poeciliid fishes, which are relatively young in terms of their evolutionary origins. For example, in guppies, males are heterogametic, displaying an XX/XY system of sex determination; however, they also have autosomal sex determining loci as well, and the activities of these genes can result in XX individuals differentiating as males and XY individuals differentiating as females. In addition, in some cases, multiple different mechanisms of sex determination can be present even within a single piscine species. Examples of this include platyfish, which generally exhibit a WXY system of sex determination in lab populations but XY in some natural populations (reviewed in Volff and Schartl 2001). Another example is the zebrafish system (see Fig. 9.6). Thus, even when sex chromosomes are identified, more work must be done to elucidate whether other factors help modulate the process of sex determination and whether multiple different systems may be at work in the same species.

In general, the presence of sex chromosomes is much more common for species that are gonochoristic compared to those that are hermaphroditic. According to Devlin and Nagahama (2002), as of the time that paper was published, over 259 hermaphroditic species of fish had been identified, and distinct sex chromosomes were only identified in seven genera and four species, in many cases only in some populations and/or individuals. This is not surprising given that the ability to change sex requires a level of plasticity that GSD systems would not likely afford. In these systems, as is seen in some reptilian systems, it is likely

9.3 Evidence for Adaptive Sex Ratio Adjustment in Gonochoristic Fishes

> **Case Study: Zebrafish Sex Determination**
>
> Zebrafish are known for the complexity and diversity of sex determining systems that exists even among different strains in this single species. In this system, no distinct sex chromosomes have been found, yet environmental factors appear to have very little influence on the sex determination process. Crosses of different strains of zebrafish resulted in highly variable offspring sex ratios, suggesting a genetic basis for sex determination, however comparison of male and female genomes yielded no evidence of a universal sex-linked factor. This led researchers to conclude that the sex determination system of zebrafish could not be chromosomally based (Liew et al. 2012). Yet, work in wild zebrafish in India indicated that these fish have a WZ system for sex determination (Sharma et al. 1998), and a recent study supported that zebrafish in nature do indeed exhibit a WZ system of sex determination with a sex determinant on chromosome 4, while this sex determinant appears to have been modified during domestication. To make the system even more complex, harsh conditions such as hypoxia, high breeding density, high temperatures, and poor nutrition can modify or override this genetic process of sex determination to result in the production of more male males. Thus the zebrafish system highlights the high level of complexity that characterizes processes of sex determination in fish systems.

Fig. 9.6 The story of sex determination in zebrafish

that an environmental influence is the primary driver of sex determination that provokes the expression of downstream genetic pathways to shape differentiation of gonads and behavior (i.e., ESD). The mechanisms by which ESD likely occurs in reptiles and fishes will be discussed in detail in Chap. 10.

9.3 Evidence for Adaptive Sex Ratio Adjustment in Gonochoristic Fishes

There is virtually no evidence to date that fish exercise maternal or paternal control over offspring sex ratios. However, this may simply be because the idea has not yet been tested. Given that the majority of fish exhibit a system that involves at least a partial genetic component in the sex determination process, there is potential for parental control over offspring sex. Further, given that the areas in which male and female fish breed, and into which offspring will hatch, can vary widely in both environmental parameters such as temperature and PH and social parameters such as density and operational sex ratio, it is unclear why fishes would not exercise a parental control over sex ratios when individuals in other vertebrate classes appear to. Perhaps due to the plasticity that allows sex determining mechanisms to respond to environmental cues in fishes, a parental means of adjusting sex ratio was never needed and, as a result, never selected for. It is unclear in just how many fish species the process of sex determination is responsive to environmental and/or social cues; however, it is becoming clear that in at least some species that appear to exhibit genetic sex determination, this process can be overridden by environmental triggers, much like in reptiles. In this section, I will highlight the factors known to influence the processes of sex determination and sexual differentiation during early development, a majority of which exert their influences in gonochoristic systems.

9.3.1 Influences of Temperature on Sex Ratios in Fish

The environmental cue that appears to exhibit the largest influence on sex determination during early development is water temperature. In a study of Atlantic silverside, Conover and Ross (1982) detected the first evidence that sexual differentiation may be responding to temperature. They found that there was significant seasonal fluctuation in the sex ratios of juvenile fish, with over 70% females being caught in the hot month of July. The proportion of female juveniles then dropped to below 50% by September, when temperatures are cooler. When eggs were collected, fertilized and reared in the lab at constant temperatures and photoperiods, they produced an even number of males and females regardless of what time during the spawning season they were collected. This led Conover and Kynard (1981) to test the influences of temperature on sexual differentiation directly in this model system using three experiments (visualized in Fig. 9.7). In experiment 1, eggs were collected from six females, fertilized and transferred to either warm or cold fluctuating conditions that mimicked natural variation in water temperatures. For the eggs collected from five of six females, maintenance at the cold temperatures produced a significantly higher proportion of female larvae when they were sexed at 40 days after hatch. In experiment 2, eggs were hatched at cool temperatures, and 1 day-old larvae were transferred to either cool or warm temperatures. Once again, those maturing at the cool temperatures were significantly more likely to differentiate into females, and because they were transferred after hatch, this experiment indicates that the period when the process of sexual differentiation is sensitive to temperature is during larval development. Finally, experiment 3 was designed to further pinpoint the time during larval development that is most sensitive to the effects of temperature; eggs were fertilized and hatched in cold temperatures and then upon hatch were randomly assigned to warm or cool temperatures. As development progressed, subgroups of larvae in the cold temperature group were transferred to warm temperatures at 18, 32, 46, 60, and 75 days after hatch, and the proportion of larvae that differentiated into female larvae were significantly higher when larvae were moved at day 46 or after compared to those moved before day 46. This indicates that the sexual differentiation process in Atlantic silverside larvae is sensitive to temperature during late stages of development. This was a

9.3 Evidence for Adaptive Sex Ratio Adjustment in Gonochoristic Fishes

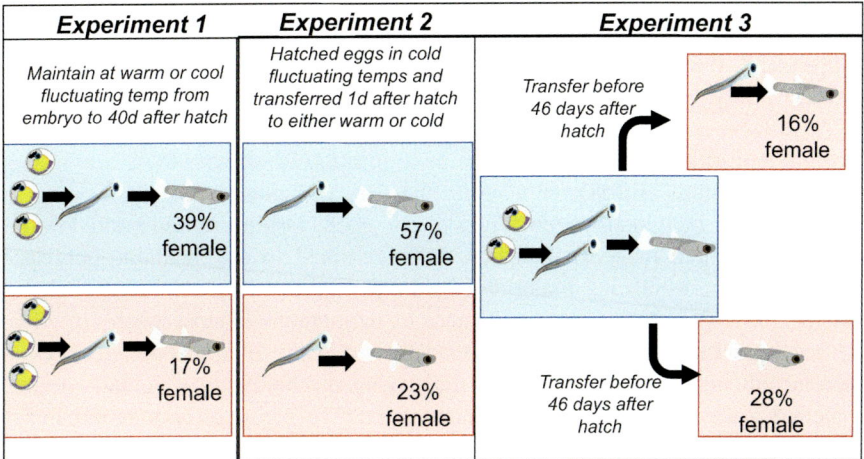

Fig. 9.7 Graphical depiction of a study conducted by Conover and Kynard (1981) testing the influences of temperature on sexual differentiation in Atlantic silverside. In experiment 1, eggs were collected from six females, fertilized using sperm from males, and then raised from embryonic stages to 40 days after hatch in fluctuating temperatures that mimicked either cool (11–19 °C) or warm (17–25 °C) conditions in the wild. Those raised in cold conditions were significantly more likely to differentiate into females. Shown here are average sex ratios for eggs produced by all six females, though there was significant variation in how these temperatures affected eggs from individual females. In experiment 2, to test whether the temperature-sensitive period occurs after hatch, larvae were transferred into the same cold and warm temperature treatments 1 day after hatch. Again, the cold temperature environment resulted in a significantly higher proportion of females. Finally, in experiment 3, to pinpoint when after hatch the temperature-sensitive period occurs, eggs and larvae were initially maintained in cold temperature conditions, and subgroups were transferred to new tanks with either warm or cold temperature conditions at 18, 32, 46, 60, or 75 days after hatch. Those transferred to warm conditions 46 days or later after hatch were significantly less likely to be female compared to those transferred earlier, suggesting that the temperature-sensitive period is late in development

groundbreaking study that countered the earlier belief that fishes do not utilize temperature sex determination.

This seminal study spurred additional studies testing for a potential influence of temperature on the process of sexual differentiation in other species of fish. It is now well known that the process of sexual differentiation is indeed sensitive to temperature in over 60 piscine species. Römer and Beisenherz (1996) studied the influence of temperature in 39 species of cichlid fish by breeding and raising them at 23, 26, or 29 °C. They found that in 33 of those species, temperature exerted a significant influence on the process of sex determination, with a higher proportion of fish differentiating into males at higher temperatures and a higher proportion into females at lower temperatures. In one of these species, the three-striped dwarf cichlid, the authors transferred larvae from a male-determining temperature (29 °C) to a female determining temperature (23 °C) at different developmental stages to determine if there is a sensitive period for the effects of temperature on sexual differentiation. In both cases, the period during which the process of sexual differentiation was sensitive to temperature was between 0 and 800 h after spawning. The longer the larvae were maintained at a male or female-rearing temperature, the larger effect that temperature regime had on the process of sex determination.

As influences of temperature on the sex determination process become obvious in more and more piscine species, the next question is whether these species are utilizing a system of TSD, or whether they are utilizing a GSD system that is modulated by high or low temperatures. In 2008, Ospina-Alvarez and Piferrer conducted a meta-analysis, in which they attempted to use available published data to discern whether a majority of fish species are truly exhibiting TSD, or whether GSD systems are being influenced by temperature in most species (Ospina-Alvarez and Piferrer 2008). They found that in many of the reported cases of thermal influences on sex ratios, the underlying mechanism was actually a GSD system that was manipulated by temperature. True TSD is actually quite rare in fish systems, but when it does occur, increasing temperatures almost always produced male-biased sex ratios. Thus, in comparison with reptile systems, the prevalence of true TSD is much lower, and there is much less variability in the directional patterns of the temperature-induced effects on the sexual differentiation process.

We now know that temperature has the potential to influence sex ratios of developing fish; however, the question remains as to whether the responsiveness of piscine systems of sex determination to temperature is an adaptive mechanism for modulating sex ratios. In a majority of systems testing the influence of temperatures, experiments were conducted in laboratory settings and often utilized temperatures outside of the natural range. To determine whether fish are adjusting sex ratios in response to temperature in an adaptive manner, we must first know (1) whether sex ratio skews occur at temperatures within the natural range of those experienced during the reproductive season, (2) whether there is a genetic basis to this thermal sensitivity, which would provide a means on which selection could act, and (3) if there is a benefit to adjusting sex ratios towards an extreme as temperatures rise or fall.

To answer some of these questions, we can turn back again to work conducted on Atlantic silverside fish. Despite decades of research on how temperature influences sex ratios in fishes, to my knowledge, this system remains the only system in which effects of temperature on sex determination have been demonstrated in the wild; sex ratios are more female biased during warmer months, as mentioned above (Conover and Ross 1982). In addition, the experiment conducted by Conover and Kynard (1981) (described in detail above) utilized temperatures that mimicked natural temperature extremes and fluctuations to demonstrate that warmer temperatures during rearing reduced the proportion of fish that differentiated into females. But what is also known due to work in this system is that there is genetic variation in how potently the sex determination/differentiation process responds to temperature. Through several studies, David Conover and his coauthors have provided extensive evidence that there is a genetic influence underlying TSD in

Atlantic silverside. First, in their seminal study, progeny obtained from different females showed substantial variation in their responses to temperature; for example, while the cold temperature regime produced, on average, 39% female progeny, sex ratios of progeny collected from individual females ranged from 4 to 63% females (Conover and Kynard 1981). Similarly, sex ratios at the cold temperature ranged from 0 to 51.8% among the progeny from the six different females. This indicates that at least some females produce progeny in which the sexual differentiation process is insensitive to temperature. Conover and Heins (1987) then tested for a genetic effect directly by collecting eggs from different families and using artificial insemination techniques. Paternity exerted a significant influence on sex ratios on its own, and there was also a significant paternity by temperature interaction, indicating that the genetic contribution of the father influences the sensitivity of sex ratios to temperature cues. Thus, in this case, a genetic target on which natural selection may act is clearly present.

Is there evidence for selection on this genetic target to manipulate sensitivity of the sex determining system to temperature? Indeed there is. When populations of Atlantic silverside fish were kept for 5–6 years at temperatures that are known to cause highly skewed temperatures, higher and higher proportions of the minority sex were produced over time, establishing a balanced sex ratio. This work provided support for the frequency-dependent selection hypothesis, which predicts that sex ratios will always return to a balanced 50:50 ratio of males to females (Conover and Van Voorhees 1990). Atlantic silverside fish are not the only system in which selection for thermal sensitivity has been shown. Rainbow trout are significantly more likely to differentiate into female offspring when reared at 18 °C compared to 12 °C; however, this relationship was also strongly influenced by genetic parentage with offspring of some parents being more susceptible to temperature than others (Magerhans et al. 2009). After this finding, Magerhans and Hörstgen-Schwark (2010) showed that it is possible to artificially select different temperature sensitivities in offspring; they produced divergent lines of offspring, with 57% of those in the highly sensitive line differentiating into females when reared at 18 °C compared to 44% in the low sensitivity line. This indicates that selection can, in fact, act on the genes responsible for determining how sensitive individuals are to the effects of temperature on sex ratios.

The questions that remain are, does this type of selection occur in nature, and is there a benefit to selecting for TSD? Here the Atlantic silverside steps up again. This species has a range from Florida to Nova Scotia. In the northern range, where the breeding season is short, the influence of temperature on sexual differentiation is not as extreme as in southern environments where the breeding season is long (reviewed in Conover 2004). Similar relationships were seen in a related species, the tidewater silverside (Yamahira and Conover 2003). For these patterns to represent adaptive variation in sensitivity to temperatures, there would need to be a benefit to producing more males at higher temperatures in the environments in which TSD is found. As discussed in Chap. 8, Conover (2004) and Charnov and Bull (1977) postulated that ESD would be selected for if the environmental variable that induced sex ratio skews affected the fitness of males versus females differently.

Conover (2004) suggests that as the length of the breeding season increases, the relative benefit of hatching towards the beginning of that breeding season becomes more substantial, and this could favor selection for TSD in silversides. The specific benefit that males might incur from hatching at the beginning of the season, however, remains unknown. Perhaps, as in reptiles, males hatching at the beginning of the season enjoy adaptive benefits associated with a larger body size at the time of next reproduction. Indeed, body size differs substantially between the sexes in populations where TSD is present (see Conover 2004 for review); however, this idea has not yet been tested empirically.

In a nice review of temperature sex determination in fish systems, Godwin et al. (2003) highlights flounders as species in which TSD may be common. Japanese flounders are particularly well studied in the context of how temperature modulates the process of sexual differentiation during early development. This species utilizes an XX/XY GSD system of sex determination, yet exposing them to lower or higher temperatures than normal during early development results in a high proportion of males. Therefore, this is clearly a species that exhibits a combination GSD/TSD system. Similar effects have also been seen in the southern flounder, indicating that the genus as a whole may contain several species for which temperature is an important controller of the sexual differentiation process (Luckenbach et al. 2003). Further, work in southern flounder indicates a potential adaptive basis for this mechanism of sex determination. Individuals reared at 23 °C (the temperature at which sex ratios are not biased) grew significantly larger than individuals reared at either 18 or 28 °C. Godwin et al. (2003) suggests that, because females must reach larger adult sizes than males of this species, the adjustment of sex ratios towards males when temperatures are suboptimal for growth may serve an ultimate adaptive benefit that is consistent with the Charnov Bull model of sex allocation (Charnov and Bull 1977). In this case, females would suffer more of a detriment at suboptimal temperatures, so it would be beneficial to instead produce males.

These two systems provide potential examples of adaptive sex allocation in (1) a TSD system that has a genetic component (silversides) and (2) a GSD system that involves temperature-induced modification (flounders). The adaptive basis for a role of temperature in any system of sex determination beyond the work in silversides and flounders, however, remains a very large unexplored area in piscine reproductive biology and clearly warrants further work.

9.3.2 Influences of pH on Sex Ratios in Fish

Temperature is not the only environmental variable known to influence the process of sexual differentiation in fishes, though the others are not well studied. Salinity has been studied in only a few cases, and the evidence for an effect of salinity on sex ratios is mixed (reviewed in Chan and Yeung 1983; Abucay et al. 1999). Water acidity, however, is a more likely possibility as a regulator of sex ratios. For example, two studies have shown influence of water acidity on sex ratios in a

9.3 Evidence for Adaptive Sex Ratio Adjustment in Gonochoristic Fishes

total of 10 species (Table 9.1). Rubin (1985) showed that in five species of cichlids and a poeciliid, significantly more males develop in more acidic waters. In fact, 100% of green swordtail broods were male when maintained at a pH of 6.2 compared to 0–3% males in broods maintained at a pH of 7.8. Similarly, in the large study of temperature effects in 39 cichlid species described above, influences of pH were also documented in 20 of those species (Römer and Beisenherz 1996). In all but one of those species, higher pH resulted in a higher proportion of males. In only six of these species was the pattern statistically significant, but this is likely due to sample sizes, because for many of the species in which the pattern was not significant, three or fewer broods per treatment were included. Based on these two studies, it appears that the pH value necessary to trigger a male bias in the sex ratio is species-specific (see Table 9.1). For example, in the Rubin (1985) study, *Pelvicachromis pulcher* produced 96% males when exposed to water at a pH of 5.05, but *Xiphophorus helleri* produced 100% males at a pH only as low as 6.20. In addition, there appears to be variation in the sensitivity of sex ratios to pH among species. For example, *Apistogramma nijsseni* does not appear to be very sensitive at all, varying sex ratios between 45.9 and 58.7% males in response to the same range of pH that caused much larger skews in other closely related species. A study in sea bass showed no influence of pH on sex ratios, but the pH range in this study was a side effect of a density manipulation and ranged in the high density group from 7.1 to 8.0 and in the low density group from 7.4 to 8.0 (Saillant et al. 2003). Perhaps even the lowest pH in this study was too high to trigger a sex ratio skew towards males, because in other studies, higher proportions of female offspring were produced when pH values were at or above 7.0. More studies addressing the potential influence of pH in the processes of sex determination and sexual differentiation are badly needed, particularly because both aquaculture practices and pollution can result in acidification of waters in which fish species live and reproduce.

Is it possible that these sex ratio adjustments in response to pH are adaptive? If the negative consequences of water acidity are more potent for females versus males, this would explain why a higher proportion of males are produced when water is more acidic. It is known that acidity in water affects survival and life spans in fishes (Jonsson and Jonsson 2014); however, to my knowledge, it is unknown if water acidification affects survival and longevity of males and females differently. Natural selection would also favor the production of more males as water becomes more acidic if acidic environments affected reproductive rates of males and females differently. Indeed, it has been shown in some species that females suffer disruption in reproductive function related to low egg production when water pH is low (Fromm 1980; Lee and Gerking 1980), while a meta-analysis conducted on studies of salmonids, cyprinids, and sturgeons indicates that water pH does not influence sperm motility (Alavi and Cosson 2005). To my knowledge, other measures of sperm quality have not been tested in relation to the pH of water. Thus, it is possible that, when pH of water is low, producing a higher proportion of males provides an adaptive benefit in terms of future reproductive success of both parents and offspring. Direct tests would need to be done to confirm this; however, based on

Table 9.1 A list of species for which the influence of pH on the proportion of individuals that differentiated into males has been studied

Species	Percent males @ High pH	Percent males @ Low pH	pH treatments	References
Pelvicachromis pulcher	96	20	5.05 and 6.90	Rubin (1985)
Pelvicachromis subocellatus	89	11	5.40 and 7.00	Rubin (1985)
Pelvicachromis taeniatus	87	11	5.50 and 7.00	Rubin (1985)
Apistogramma borellii	91	9	5.80 and 7.10	Rubin (1985)
Apistogramma borellii	68.4	68.4	4.50 and 6.50	Römer and Beisenherz (1996)
Apistogramma cacatuoides	92	13	5.80 and 7.10	Rubin (1985)
Apistogramma cacatuoides	84.3	43.3	4.50 and 6.50	Römer and Beisenherz (1996)
Xiphophorus helleri	100	1.5	6.20 and 7.80	Rubin (1985)
Apistogramma nijsseni	58.7	45.9	4.50 and 6.50	Römer and Beisenherz (1996)
Apistogramma norberti	71.4	31.1	4.50 and 6.50	Römer and Beisenherz (1996)
Apistogramma paucisquamis	67.2	41.1	4.50 and 6.50	Römer and Beisenherz (1996)
Apistogramma uaupesi	81.7	58.8	4.50 and 5.50	Römer and Beisenherz (1996)
Apistogramma Breitbinden	79.2	68.5	4.50 and 5.50	Römer and Beisenherz (1996)
Apistogramma smaragd	76.2	61.5	4.50 and 6.50	Römer and Beisenherz (1996)
Apistogramma caetei	43.7	14.5	4.50 and 6.50	Römer and Beisenherz (1996)
Apistogramma eunotus	64.3	41.2	4.50 and 6.50	Römer and Beisenherz (1996)
Apistogramma gephyra	65.7	39.2	4.50 and 6.50	Römer and Beisenherz (1996)
Apistogramma gibbiceps	68.9	35.1	4.50 and 6.50	Römer and Beisenherz (1996)
Apistogramma hongsloi	61.6	14.4	4.50 and 6.50	Römer and Beisenherz (1996)
Apistogramma linkei	66.2	49.2	4.50 and 6.50	Römer and Beisenherz (1996)
Apistogramma meinkerii	60.3	45.9	4.50 and 6.50	Römer and Beisenherz (1996)
	59.0	38.0		

(continued)

Table 9.1 (continued)

Species	Percent males @ High pH	Percent males @ Low pH	pH treatments	References
Apistogramma melanogaster			4.50 and 6.50	Römer and Beisenherz (1996)

Data are compiled from two sources: Rubin (1985) and Römer and Beisenherz (1996). While the Römer and Beisenherz (1996) study reported the influences of additional intermdiate pH values, I included only the high and low values here for simplicity

the available data concerning the influences of pH on sex ratios in fishes, it appears that this may potentially be an important regulator of sex ratios that has not yet been fully studied.

9.4 Evidence for Adaptive Sex Ratio Adjustment in Hermaphroditic Fishes

Given that some species of fishes are hermaphroditic and thus have the ability to change their sexes later in life, adaptive adjustment of sex ratios does not have to occur during early development in these cases. While temperature is the major focus of literature on sexual differentiation in gonochoristic species, social factors are coined as the major drivers of sex change in hermaphroditic species. Social factors include the mating system, the individual's place within a dominance hierarchy, the size of the individual in relation to others, and overall, the operational sex ratio or the number of other breeding males and females in the population. Each of these factors has been shown to trigger sex change in hermaphroditic species.

9.4.1 Adaptive Sex Change in Sequential Hermaphrodites

While it was known since the early 1960s that fishes displayed sequential hermaphroditism, the first studies showing that sex change occurred in response to social cues were published in the early 1970s. Fishelson (1970) showed in *Anthias squamipinnis* that housing females with even a single male resulted in advanced degeneration of their ovaries. Then, in 1972, Robertson unveiled an exciting story of socially induced sex change in bluestreak cleaner wrasse (Robertson 1972). In this system, a social unit is made up one dominant male that oversees a harem of 3–6 mature females and several immature individuals. Among those females, there is generally a larger, dominant female, though groups with two co-dominant females have been observed. If the dominant male dies or disappears, and the dominant female can resist intrusions by neighboring males, she changes sex to become the dominant male. This sex change begins within 90 min after the dominant male's death and completes within 2–4 days. We now know that several other systems display similar social structures and similar sex-changing events in response to social cues as well, though there are slight variations to this system. I have already introduced examples of sex change in wrasses above in which a dominant male heads the social group, and loss of that male causes either replacement by an immature male or a sex change by the dominant female (Godwin et al. 2003; Linde et al. 2011). We also know the reverse occurs in species where fish differentiate as males and then change to females if the dominant female dies or disappears. The clown fish system described above is an example of this protandric strategy. In tomato clown fish, the largest nonbreeders maintain gonads that are intermediate between males and females, allowing those nonbreeders to become either male or female depending on the social dynamics of the group (Hattori 1991; reviewd in Devlin and Nagahama 2002). Finally, as mentioned above, some gobiid species can change back and forth between sexes during their lives.

What is the underlying adaptive significance of this striking plasticity seen among sequential hermaphrodites, and what accounts for the variation in which sex differentiates first, and the presence or absence of immature fish, or fish with intermediate sexes? In an attempt to explain the selective basis underlying the evolution of sex change, Ghiselin (1969) generated the size-advantage model, which states that selection would favor an individual that changes sex when it reaches a size at which the new sex could produce more offspring. Warner (1988) points out that there are many factors that affect fecundity, and that all of these factors should be included in a measure of "reproductive value"; this expanded version of the size advantage model thus states that the ability to change sexes at a critical size would be selected for if individuals of the new sex benefits in some way that influences future reproductive success, including through changes in the individual's status, it's place in the local population demography, in relation to sex-specific energy requirements, etc. This version of the model helps to explain the basis for why some species are gonochoristics while others are hermaphroditic and why some hermaphroditic species are protandric while others are protogynandric (Fig. 9.8).

To produce successful offspring and maximize reproductive value, an individual must achieve four goals: (1) It must be able to survive through the juvenile stage, (2) it must be able to get to reproductive maturity as fast as possible, (3) it must be able to secure as many high quality mates as possible, and (4) it must have enough energy to devote all the resources necessary to produce successful offspring. This means that all of the life history characteristics that help define what is necessary to achieve the four goals can help drive the selection for a particular system of sex determination. For example, let's say female of a species have lower mortality rates during the juvenile period. Selection may then result in a system where all juveniles are female and then change to males to reproduce (a protogynous system). At first glance, this scenario could potentially provoke selection for a protogynous system independently of size, but size then comes in to further shape the system in terms of which individuals change sex and when during their lifetimes the change occurs. If, for example, the mating system was polygynous and obtaining a mate was dependent on a large body size, then only the largest males would change sex.

9.4 Evidence for Adaptive Sex Ratio Adjustment in Hermaphroditic Fishes

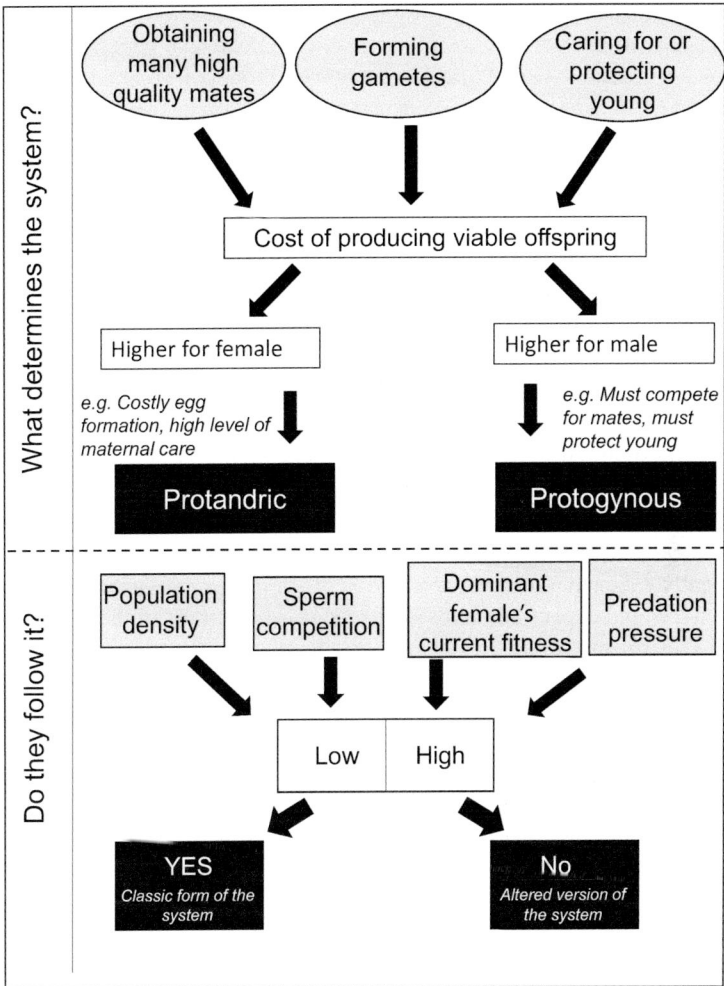

Fig. 9.8 A diagram of the factors that likely contribute to selection for a protogynous or protandric system of hermaphroditism under the size advantage model. If the cost of producing viable offspring is higher for males, then we would expect that individuals would start as female and develop as males only when they have the size/energetic reserves to successfully reproduce, making the system protogynous. If the cost is higher for males, then we would expect a protandric system where individuals start out as males and develop into females for the same reasons. Many factors go into determining how costly the process of producing offspring is for individuals, and we might expect that any one or more of those factors might help drive the formation of a protogynous or protandric system. Even after one of these systems has been selected for, additional factors in the environment surrounding individuals can determine whether those individuals follow the classic protogynous or protandric patterns, or whether they stray from those patterns. Examples are systems where low population densities drive bidirectional sex change, or where high levels of sperm competition or high levels of reproductive success in the dominant female make sex change for that female less profitable than if she remained female

We can also build a scenario in which there is no difference in juvenile mortality between the sexes. A polygynous mating system, alone, can generate selection for males of a particular size if the largest males monopolize matings with many females, and this would select for a protogynous form of hermaphroditism. All individuals would start out as female with potential to reproduce as such, and only the individuals with the size necessary to win a harem of females and reap the reproductive benefit would transition to males. A polyandrous mating system would then generate the opposite effect, a protantric system of hermaphroditism; though just as polyandry is rare compared to polygyny, protandric systems of hermaphroditism are rare compared to protogynous systems. In both of these cases, however, selection for hermaphroditic patterns of sex determination maximizes reproductive potential for the species as a whole because this does not leave individuals that have no potential to obtain a mate and reproduce.

Finally, we can consider the drive for a hermaphrodite system of sex determination from the standpoint of what it takes to successfully produce viable offspring. Factors that may help to shape these needs include the energy expenditure required to form gametes, obtain a mate, and produce offspring, as well as the needs of the offspring themselves. If, for example, the production of large eggs was far more energetically expensive than the production of sperm for individuals of a particular species, we would expect that individuals would spend most of their lives as males producing sperm and would only change into females when they are in good enough condition or had enough energetic reserves to bear the cost associated with producing eggs. We can also shape this in terms of the needs of the offspring. If the offspring require high levels of female investment (maternal care), for example, we would expect that individuals would remain males unless they had enough energetic reserves to bear the cost of that maternal care. In terms of the size advantage hypothesis, in each of these cases, size acts as an indicator of those available energetic reserves.

As is clear from the examples outlined above, each selective pressure acts in concert with others. It is perhaps the factors outlined above that determine the system of sex determination exercised by a majority of individuals within a species. However, whether individuals within a population actually exhibit the expected pattern of sex change appears to depend upon additional factors related to population dynamics. As a result, in some cases, individuals may not change sex when we would expect them to, and in others, an individual that is not the largest of his/her sex in the population may change sex when we would *not* expect it to. Munoz and Warner (2003) provided an addition to the size advantage hypothesis, suggesting that two additional factors may influence whether the largest female changes sex after the loss of a dominant male. The first is the fecundity of the largest female when the dominant male is lost. If the female is much larger than the other females in the population and already has extremely high fecundity in comparison to the other females, then changing her sex to male may not provide a large enough benefit in terms of her reproductive value. In that case, she may not change sex at all. The second factor is the level of sperm competition exhibited within the species. For many hermaphroditic species, levels of sperm competition have been observed to be high, perhaps resulting from pressures associated with alternative mating tactics, such is in the blue-head wrasse system where smaller "sneaker" males are present. High levels of sperm competition can lower the potential fitness benefit associated with a sex change from a female to a male. In this case, we would not expect the female to change sex and would instead expect a smaller female with the highest potential fitness gain over her current fitness potential to change sex instead. Another factor known to inhibit sex change when we would normally expect it to occur is population density. Lutnesky (1994) showed that sex change in angelfish was inhibited when population densities were high, and that in order for sex change to occur, the presence of a smaller female in the population is needed. On the other side, in some cases, females will change sex when the dominant male is still present (reviewed in Munday et al. 2006). Angelfish and spotlight parrotfish represent example species for which females have been observed to change sex to male when still in the presence of the dominant male (Moyer and Zaiser 1984; Van Rooij et al. 1995). There is evidence that parrotfish may do this because nonreproductive males grow faster than dominant males and, thus, may ultimately fare just as well as or better than the dominant male in terms of reproductive success (Van Rooij et al. 1995).

The size advantage model, along with the additions that others have made to it, appears to work well to explain the potential adaptive basis of protogynous and protandric hermaphroditism in piscine species. How, then, do we explain species that change multiple times during their lifetimes? There are two main hypotheses to explain why changing sex multiple times may be adaptive. The first, the risk-of-movement model posits that for some species, the need for a mechanism to change sex more than once may arise due to restricted ability to travel beyond a small range to find a mate of the opposite sex (Nakashima et al. 1995; Munday et al. 1998; reviewed in Avise and Mank 2009). This may occur because predation pressures are extreme, and movement is too risky. Alternatively, the "growth advantage hypothesis" was generated because in at least some species where bidirectional sex change is found, females grow faster and maximal reproductive success is contingent on the sizes of both members of the breeding pair (Kuwamura et al. 1993; reviewed in Munday 2002) (Fig. 9.9). The number of eggs a female can produce is limited by her size, and males can only successfully defend territories if they are large. As a result, when fish first pair, they would be most successful over time if the smallest individual were the faster-growing sex, such that they could reach peak reproductive potential together. Bi-directional sex change would thus be selected for as a method of adjusting individual sex to optimize the pair combination. For example, if two females meet, the larger female should change to male, allowing the combination of a larger male and a smaller female. The opposite would be true if two males met; we would expect the smaller male to change to female to produce the same combination. While this is a plausible hypothesis given the life history traits of the fishes that show bi-directional sex change, the hypothesis that gained support in the only empirical test is the risk of movement model (Munday 2002).

There is one additional adaptive hypothesis that was originally proposed to explain sex change patterns exhibited by simultaneous hermaphrodites, but can

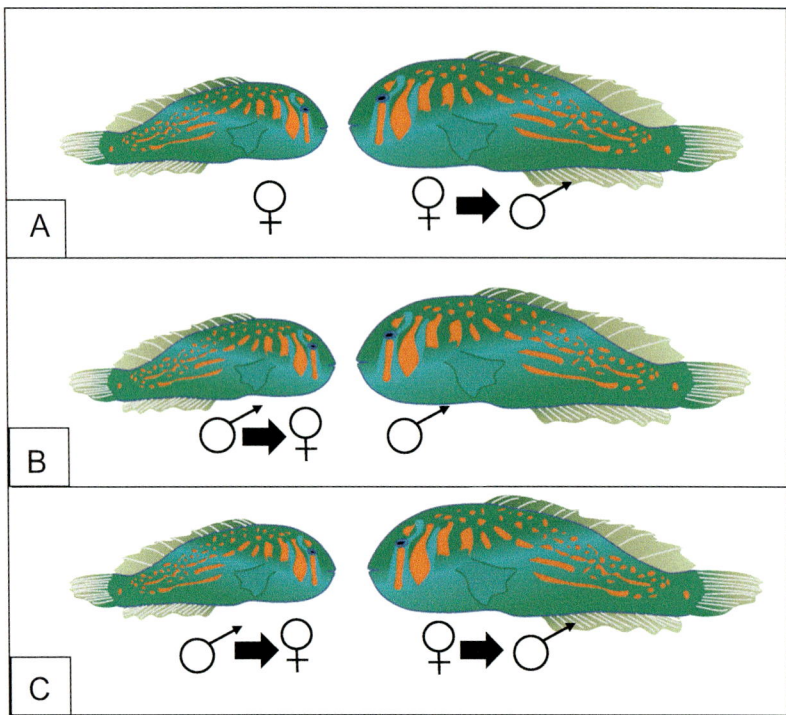

Fig. 9.9 Graphical depiction of the growth advantage hypothesis, using gobies as a model. If, for example, females grow faster and successful reproduction is dependent on an optimal size in both males and females, the best combination for a new pair would be a small, fast-growing female and a large slower growing male. As a result, if two females meet, we would expect that the larger female would change to a male (**a**), if two males meet, the smaller male would change to a female (**b**), and if a small male and a large female meet, both individuals should change sex (**c**)

also apply to bi-directional sex change in sequential hermaphrodites as well; the low density model posits that co-sexuality in these simultaneously hermaphroditic species was potentially selected for due to a scarcity of mates, for example when populations are sparse (Tomlinson 1966; reviewed in Avise and Mank 2009). This hypothesis clearly overlaps with the risk of movement model described above, because a high risk of moving out of a small area due to predation can be one reason that there may be a low density of members of the opposite sex for mating opportunities; however, it is also broad enough to include situations of isolation that are not caused by predation risk. For example, polygynous fish often resort to facultative monogamy when population densities are extremely low (reviewed in Kuwamura et al. 2014). In a study of bluestreak cleaner wrasse, females were removed to create a low density environment. In this new environment, "widowing" of males was observed, and when those males did not readily encounter a new female, they paired with other fish of any sex, and if they were the smaller male in a male–male pair, they changed sex (Kuwamura et al. 2014). Similarly, "widowed" hawkfish living at low population densities also changed sex to obtain a mate (Kadota et al. 2012). In fact, many cases of bi-directional sex change occur in fish species that live in small groups that reside in patches of coral. Thus, this potential adaptive model should be considered more widely to explain bi-directional sex change in fishes.

9.4.2 Adaptive Sex Change in Simultaneous Hermaphrodites

Simultaneous hermaphrodites maintain both ovarian and testicular tissue at the same time during adulthood; however in some cases, the line between bi-directional sex change in sequential hermaphrodites and the sex change patterns in species labeled simultaneous hermaphrodites is blurred. As mentioned, blue-banded gobies, for example, are simultaneous hermaphrodites, by definition containing both ovarian and testicular tissue, but behave like sequential hermaphrodites. Females of this species reallocate their reproductive efforts towards male-typical behaviors and spawning when they reach larger body sizes, which is consistent with the size advantage hypothesis described above for sequential hermaphrodites (St. Mary 1994). Given the similarity between bi-directional sex change in sequential hermaphrodites and sex change patterns in simultaneous hermaphrodites, it is perhaps not surprising that most of the potential adaptive drivers for the former can also be used to explain the latter. As a result, I will not repeat these adaptive explanations for simultaneously hermaphroditic fish. Instead, of major interest in this group, is how individuals allocate energy to male and reproductive functions, given that male and female gonadal tissue coexists within the same individuals.

One of the main models to explain allocation of male versus female reproductive function among simultaneous hermaphrodites is the Local Mate Competition model (Hamilton 1967), modified for simultaneous hermaphrodites (Charnov 1980). The model posits that the number of males in the population contributing competing sperm will determine optimal resource allocation to male reproductive function. Using six species of sea basses, Petersen (1991) showed an expanded model including both the level of sperm competition and the fertilizing efficiency to predict patterns of sex allocation (Charnov 1996).

Unfortunately, this model is a bit too simplistic, because the reproductive decisions of one individual are not necessarily optimal for another individual with which the first might mate. The costs of sperm and egg production are not equal. In fact, it is generally accepted that the cost of producing eggs far exceeds the cost of producing sperm (Williams 1975) and as a result, simultaneous hermaphrodites would benefit most by spending more time in the male rather than the female reproductive role. However, if all individuals did this, then there would be no one with whom those individuals could mate. This could produce direct conflict between two individuals that meet for a mating encounter. As a result, hermaphroditic individuals must often develop cooperative behaviors (Hamilton and Axelrod 1981; Leonard 1993). *Serranid* fishes are the group with the most well-documented cases of simultaneous hermaphroditism, and these fish cooperatively alter the production of sperm and eggs such that both individuals in the mating pair benefit. One way of doing this is called egg trading, which is a behavior observed in black hamlets (Fischer 1980) and chalk bass (Fischer 1984). This behavior begins with pair formation in the late afternoon. The two mates then take part in courtship displays before spawning, and the last fish to display releases eggs while the other member of the pair releases sperm. The fish do not release all of their eggs at once and instead release them in parcels. Each fish alternates releasing parcels that are then fertilized by the other member of the pair, and as a result, each fish gets to fertilize as many eggs as it produces. However, because sperm are so much cheaper, selection should drive individuals to cheat, producing all sperm but never producing eggs for the other member of the pair to fertilize. To prevent this, these simultaneous hermaphrodites appear to play a version of the Prisoner's Dilemma game. In this game, two prisoners are offered a choice: to cooperate with one another and stay silent or to defect and betray the other. There is a higher initial reward for defecting, but if both prisoners defect, they pay a higher price overall. The strategy that is most effective for exploiting this opportunity for cooperation is called TIT FOR TAT (TFT). In this case, both partners cooperate on the first move, and then each partner does whatever the other member chose on the previous move (Axelrod and Dion 1988). A study in six species of *Serranids* showed that these species do appear to use a variant of TFT when deciding whether to cooperate and produce eggs or defect and instead produce only sperm. In this case, the game starts when the first member of the pair releases eggs, and each member releases parcels of eggs if the other does and fails to give up more if the partner does not reciprocate. Indeed, evidence in black hamlets and chalk bass shows that individuals of these species are reluctant to give up egg parcels if its mate fails to reciprocate; they wait longer to release the next parcel of eggs. A problem with this strategy is that in the classic form of TFT, once an individual defects, the pair is locked in a cycle of defection. In simultaneous hermaphrodites, however, the game is more forgiving, and likely has to be to prevent simple mistakes related to errors that occur relatively routinely, such as confusion over who spawned last and variation in rates of egg maturation, from disrupting the process entirely. Instead, these species appear to exercise a more relaxed and forgiving form of TFT, given that they do not stop producing egg parcels completely when a partner fails to release one prior (Fischer 1988). Connor (1992) suggested that while these fishes may appear to be utilizing a TFT strategy, that they may actually be doing something that only resembles TFT. He pointed out that the TFT model does not take into account the fact that individuals have an additional choice: to leave and find another mate. He suggested that by parceling the eggs, the individuals are making it more profitable for the other individual to stay and offer another batch of eggs to their partners than to pay the costs associated with leaving to find another mate. These models are of course not the only ones put forth to explain allocation of reproductive effort between the male and female roles in hermaphrodites (reviewed in Anthes et al. 2006; Schärer et al. 2015), but the others have not yet been observed in or tested for in piscine systems. Much more work is needed to empirically test the driving forces behind how hermaphroditic individuals allocate resource between male and female reproductive function.

9.5 Concluding Remarks

Fishes are clearly the most diverse vertebrate group when it comes to modes of sex determination and patterns of sexual differentiation. While there is an extremely large literature base focused on the mechanisms underlying each mode of sex determination (partially reviewed in Chap. 10, also see Devlin and Nagahama (2002)), discussions of an adaptive nature are sparser. Completely ignored to date are potential parental influences on sex ratios in fishes. Gonochoristic fishes, in particular, could potentially benefit from parental modifications of offspring sexes, given that offspring generally lack the plasticity to adjust sexes themselves. For example, given the clear importance of temperature in systems exhibiting TSD, it would be interesting to see whether, in species that exhibit strict GSD, parents modify the sex ratios of offspring based on the temperatures of the water in which they spawn. Further, it has been shown in some species that exhibit TSD that the influences of temperature on sex are at least partially modulated by stress hormones (Sopinka et al. 2017). It would be interesting to test whether stressful conditions stimulate sex ratio biases in both GSD and TSD species. To my knowledge, there are no studies examining how quality or attractiveness of the sire influences sex ratios in fishes. There is great potential for more work to be done to further understand if and how fishes adjust sex ratios in an adaptive context.

References

Abucay JS, Mair GC, Skibinski DO, Beardmore JA (1999) Environmental sex determination: the effect of temperature and salinity on sex ratio in Oreochromis niloticus L. Aquaculture 173 (1):219–234

Alavi S, Cosson J (2005) Sperm motility in fishes. I. Effects of temperature and pH: a review. Cell Biol Int 29(2):101–110

Anthes N, Putz A, Michiels NK (2006) Sex role preferences, gender conflict and sperm trading in simultaneous hermaphrodites: a new framework. Anim Behav 72(1):1–12

Avise J, Mank J (2009) Evolutionary perspectives on hermaphroditism in fishes. Sex Dev 3 (2–3):152–163

Axelrod R, Dion D (1988) The further evolution of cooperation. Science 242(4884):1385–1390

Chan S, Yeung W (1983) Sex control and sex reversal in fish under natural conditions. Fish Physiol 9:171–222

Charnov EL (1980) Sex allocation and local mate competition in barnacles. Marine Biol Lett 2:53–57

Charnov EL (1996) Sperm competition and sex allocation in simultaneous hermaphrodites. Evol Ecol 10(5):457–462

Charnov EL, Bull J (1977) When is sex environmentally determined? Nature 266:828–830

Cole KS (1997) Gonadal development and sexual allocation in mangrove killifish, Rivulus marmoratus (Pisces: Atherinomorpha). Copeia 1997(3):596–600

Connor RC (1992) Egg-trading in simultaneous hermaphrodites: an alternative to Tit-for-Tat. J Evol Biol 5(3):523–528

Conover DO (2004) Temperature-dependent sex determination in fishes. In: Valenzuela N, Lance V (eds) Temperature-dependent sex determination in vertebrates. Smithsonian Books, Washington, pp 11–20

Conover DO, Heins SW (1987) The environmental and genetic components of sex ratio in Menidia menidia (Pisces: Atherinidae). Copeia 1987(3):732–743

Conover DO, Kynard BE (1981) Environmental sex determination: interaction of temperature and genotype in a fish. Science 213:31

Conover DO, Ross MR (1982) Patterns in seasonal abundance, growth and biomass of the Atlantic silverside, Menidia menidia, in a New England estuary. Estuar Coasts 5(4):275–286

Conover DO, Van Voorhees DA (1990) Evolution of a balanced sex ratio by frequency-dependent selection in a fish. Science 250(4987):1556

Devlin RH, Nagahama Y (2002) Sex determination and sex differentiation in fish: an overview of genetic, physiological, and environmental influences. Aquaculture 208(3):191–364

Fischer EA (1980) The relationship between mating system and simultaneous hermaphroditism in the coral reef fish, Hypoplectrus nigricans (Serranidae). Anim Behav 28(2):620–633

Fischer EA (1984) Egg trading in the chalk bass, Serranus tortugarum, a simultaneous hermaphrodite. Ethology 66(2):143–151

Fischer EA (1988) Simultaneous hermaphroditism, tit-for-tat, and the evolutionary stability of social systems. Ethol Sociobiol 9(2-4):119–136

Fishelson L (1970) Protogynous sex reversal in the fish Anthias squamipinnis (Teleostei, Anthiidae) regulated by the presence or absence of a male fish. Nature 227(5253):90–91

Fromm PO (1980) A review of some physiological and toxicological responses of freshwater fish to acid stress. Environ Biol Fish 5(1):79–93

Ghiselin MT (1969) The evolution of hermaphroditism among animals. Q Rev Biol 44(2):189–208

Godwin J, Luckenbach JA, Borski RJ (2003) Ecology meets endocrinology: environmental sex determination in fishes. Evol Dev 5(1):40–49

Guerrero-Estévez S, Moreno-Mendoza N (2010) Sexual determination and differentiation in teleost fish. Rev Fish Biol Fish 20(1):101–121

Hamilton WD (1967) Extraordinary sex ratios. Science 156(3774):477–488

Hamilton WD, Axelrod R (1981) The evolution of cooperation. Science 211(27):1390–1396

Harrington RW (1961) Oviparous hermaphroditic fish with internal self-fertilization. Science 134(3492):1749–1750

Hattori A (1991) Socially controlled growth and size-dependent sex change in the anemonefish Amphiprion frenatus in Okinawa, Japan. Jpn J Ichthyol 38(2):165–177

Ijiri S, Kaneko H, Kobayashi T, Wang D-S, Sakai F, Paul-Prasanth B, Nakamura M, Nagahama Y (2008) Sexual dimorphic expression of genes in gonads during early differentiation of a teleost fish, the Nile tilapia Oreochromis niloticus. Biol Reprod 78(2):333–341

Jonsson B, Jonsson N (2014) Early environment influences later performance in fishes. J Fish Biol 85(2):151–188

Kadota T, Osato J, Nagata K, Sakai Y (2012) Reversed sex change in the haremic protogynous hawkfish Cirrhitichthys falco in natural conditions. Ethology 118(3):226–234

Kuwamura T, Yogo Y, Nakashima Y (1993) Size-assortative monogamy and paternal egg care in a coral goby Paragobiodon echinocephalus. Ethology 95(1):65–75

Kuwamura T, Kadota T, Suzuki S (2014) Testing the low-density hypothesis for reversed sex change in polygynous fish: experiments in Labroides dimidiatus. Sci Rep 4:srep04369

Lee R, Gerking S (1980) Survival and reproductive performance of the desert pupfish, Cyprinodon n. nevadensis (Eigenmann and Eigenmann), in acid waters. J Fish Biol 17(5):507–515

Leonard JL (1993) Sexual conflict in simultaneous hermaphrodites: evidence from serranid fishes. Environ Biol Fish 36(2):135–148

Liew WC, Bartfai R, Lim Z, Sreenivasan R, Siegfried KR, Orban L (2012) Polygenic sex determination system in zebrafish. PLoS One 7(4):e34397

Linde M, Palmer M, Alós J (2011) Why protogynous hermaphrodite males are relatively larger than females? Testing growth hypotheses in Mediterranean rainbow wrasse Coris julis (Linnaeus, 1758). Environ Biol Fish 92(3):337–349

Luckenbach JA, Godwin J, Daniels HV, Borski RJ (2003) Gonadal differentiation and effects of temperature on sex determination in southern flounder (Paralichthys lethostigma). Aquaculture 216(1):315–327

Lutnesky MM (1994) Density-dependent protogynous sex change in territorial-haremic fishes: models and evidence. Behav Ecol 5(4):375–383

Magerhans A, Hörstgen-Schwark G (2010) Selection experiments to alter the sex ratio in rainbow trout (Oncorhynchus mykiss) by means of temperature treatment. Aquaculture 306(1):63–67

Magerhans A, Müller-Belecke A, Hörstgen-Schwark G (2009) Effect of rearing temperatures post hatching on sex ratios of rainbow trout (Oncorhynchus mykiss) populations. Aquaculture 294 (1):25–29

Manabe H, Matsuoka M, Goto K, Dewa S-I, Shinomiya A, Sakurai M, Sunobe T (2008) Bi-directional sex change in the gobiid fish Trimma sp.: does size-advantage exist? Behaviour 145(1):99–113

Matsuda M, Nagahama Y, Shinomiya A, Sato T (2002) DMY is a Y-specific DM-domain gene required for male development in the medaka fish. Nature 417(6888):559

Moore EC, Roberts RB (2013) Polygenic sex determination. Curr Biol 23(12):R510–R512

Moyer JT, Zaiser MJ (1984) Early sex change: a possible mating strategy of Centropyge angelfishes (Pisces: Pomacanthidae). J Ethol 2(1):63–67

Munday PL (2002) Bi-directional sex change: testing the growth-rate advantage model. Behav Ecol Sociobiol 52(3):247–254

Munday PL, Caley MJ, Jones GP (1998) Bi-directional sex change in a coral-dwelling goby. Behav Ecol Sociobiol 43(6):371–377

Munday PL, Buston PM, Warner RR (2006) Diversity and flexibility of sex-change strategies in animals. Trends Ecol Evol 21(2):89–95

Munday PL, Kuwamura T, Kroon FJ (2010) Bidirectional sex change in marine fishes. In: Cole KS (ed) Reproduction and sexuality in marine fishes: patterns and processes. University of California Press, Berkeley, pp 241–271

Munoz RC, Warner RR (2003) A new version of the size-advantage hypothesis for sex change: incorporating sperm competition and size-fecundity skew. Am Nat 161(5):749–761

Nakashima Y, Kuwamura T, Yogo Y (1995) Why be a both-ways sex changer? Ethology 101 (4):301–307

Ospina-Alvarez N, Piferrer F (2008) Temperature-dependent sex determination in fish revisited: prevalence, a single sex ratio response pattern, and possible effects of climate change. PLoS One 3(7):e2837

Pandian T (2012) Genetic sex differentiation in fish, vol 1. CRC Press, Boca Raton

Paul A, Kuester J (1987) Dominance, kinship and reproductive value in female Barbary macaques (Macaca sylvanus) at Affenberg Salem. Behav Ecol Sociobiol 21(5):323–331

Petersen CW (1991) Sex allocation in hermaphroditic sea basses. Am Nat 138(3):650–667

Piferrer F (2001) Endocrine sex control strategies for the feminization of teleost fish. Aquaculture 197(1):229–281

Robertson D (1972) Social control of sex reversal in a coral-reef fish. Science 177 (4053):1007–1009

Rodgers E, Earley R, Grober M (2007) Social status determines sexual phenotype in the bi-directional sex changing bluebanded goby Lythrypnus dalli. J Fish Biol 70(6):1660–1668

Römer U, Beisenherz W (1996) Environmental determination of sex in Apistogrammai (Cichlidae) and two other freshwater fishes (Teleostei). J Fish Biol 48(4):714–725

Rubin DA (1985) Effect of pH on sex ratio in cichlids and a poeciliid (Teleostei). Copeia 1985 (1):233–235

Saillant E, Fostier A, Haffray P, Menu B, Laureau S, Thimonier J, Chatain B (2003) Effects of rearing density, size grading and parental factors on sex ratios of the sea bass (Dicentrarchus labrax L.) in intensive aquaculture. Aquaculture 221(1):183–206

Schärer L, Janicke T, Ramm SA (2015) Sexual conflict in hermaphrodites. Cold Spring Harb Perspect Biol 7(1):a017673

Schartl M (2004) A comparative view on sex determination in medaka. Mech Dev 121(7):639–645

Schultheis C, Böhne A, Schartl M, Volff J, Galiana-Arnoux D (2009) Sex determination diversity and sex chromosome evolution in poeciliid fish. Sex Dev 3(2–3):68–77

Sharma K, Sharrna O, Tripathi N (1998) Female heterogamety in Danio rerio (Cypriniformes: Cyprinidae). Proc Natl Acad Sci India Sect B 68:123–126

Sopinka N, Capelle P, Semeniuk C, Love O (2017) Glucocorticoids in fish eggs: causes of variation and effects on offspring phenotype. Physiol Biochem Zool 90:15–33

Soto CG, Leatherland JF, Noakes DL (1992) Gonadal histology in the self-fertilizing hermaphroditic fish Rivulus marmoratus (Pisces, Cyprinodontidae). Can J Zool 70(12):2338–2347

St. Mary CMS (1994) Sex allocation in a simultaneous hermaphrodite, the blue-banded goby (Lythrypnus dalli): the effects of body size and behavioral gender and the consequences for reproduction. Behav Ecol 5(3):304–313

Sunobe T, Nakazono A (1993) Sex change in both directions by alteration of social dominance in Trimma okinawae (Pisces: Gobiidae). Ethology 94(4):339–345

Tave D (1986) Genetics for fish hatchery managers. AVI, Westport

Tomlinson J (1966) The advantages of hermaphroditism and parthenogenesis. J Theor Biol 11(1):54–58

Uchida D, Yamashita M, Kitano T, Iguchi T (2002) Oocyte apoptosis during the transition from ovary-like tissue to testes during sex differentiation of juvenile zebrafish. J Exp Biol 205(6):711–718

Van Rooij J, Bruggemann J, Videler J, Breeman A (1995) Plastic growth of the herbivorous reef fish Sparisoma viride: field evidence for a trade-off between growth and reproduction. Mar Ecol Prog Ser 122:93–105

Volff J-N, Schartl M (2001) Variability of genetic sex determination in poeciliid fishes. Genetica 111(1):101–110

von Hofsten J, Olsson P-E (2005) Zebrafish sex determination and differentiation: involvement of FTZ-F1 genes. Reprod Biol Endocrinol 3(1):63

Walker S, Ryen C, McCormick M (2007) Rapid larval growth predisposes sex change and sexual size dimorphism in a protogynous hermaphrodite, Parapercis snyderi Jordan & Starks 1905. J Fish Biol 71(5):1347–1357

Warner RR (1988) Sex change and the size-advantage model. Trends Ecol Evol 3(6):133–136

Williams GC (1975) Sex and evolution, vol 8. Princeton University Press, Princeton

Yamahira K, Conover DO (2003) Interpopulation variability in temperature-dependent sex determination of the tidewater silverside Menidia peninsulae (Pisces: Atherinidae). Copeia 2003(1):155–159

Yamamoto T-O (1969) Sex differentiation. Fish Physiol 3:117–175

Mechanisms of Environmental Sex Determination in Fish, Amphibians, and Reptiles

10

> *Which of these points of view, preformation or epigenesis, we may think more profitable as a working hypothesis is, I believe, the question of the hour. My own preference—or prejudice, perhaps—is for the epigenetic interpretation...*
>
> Thomas Hunt Morgan in Hunt (1907)

We now know that vertebrates determine sex along a plasticity continuum; while birds and mammals appear to be strictly restricted to genetic sex determination (GSD), reptiles, amphibians, and fish are able to adjust sex ratios according to environmental and social cues. Phylogenetic analyses reveal frequent evolutionary transitions between GSD and environmental sex determination (ESD), suggesting that ESD might result from genotypic systems becoming sensitive to external cues such as temperature or social triggers. Studies that have examined the hormonal and gene pathways that may be involved in ESD illustrate that many of the gene targets that are likely involved may be shared across vertebrate systems. What has remained unknown until recently is how these pathways are changed in response to environmental triggers. Recent evidence indicates that environmental cues may cause epigenetic modifications of key genes that are involved in the sex determination pathway. In this chapter, I will compare the hypothesized gene targets that may underlie temperature sex determination (or TSD) in fish, amphibians, and reptiles; discuss the potential role of hormones in the process for all three groups; and highlight the new findings implicating epigenetic mechanisms as the target mechanisms that environmental cues trigger in the process of TSD. In addition, I will also discuss the mechanisms that may underlie sex change in fish in response to social cues.

10.1 Potential Gene Targets in the Process of TSD

In vertebrate species that have distinct sex chromosomes, there is often a single gene or a cluster of genes that trigger the indifferent gonads to begin developing into either testes or ovaries. In mammals, the SRY gene on the Y chromosome directs the indifferent gonad away from the road of development into female ovaries and instead toward development of testes. In birds, it is not the presence of a factor that drives sex determination but the absence of one. The critical factor that directs sex determination in avian systems appears to be the number of copies of the gene DMRT1 on the Z chromosome; when two Z chromosomes and thus two copies of DMRT1 are present, the gonads develop into testes; however, when the individual possesses a W rather than a Z chromosome, the absence of one copy of DMRT1 directs the gonads towards ovarian development instead. Medaka fish seem to exhibit a similar system to mammals; they have an XY chromosomal system in which the Y chromosome contains the gene, DMY, which is necessary to direct development of gonads into testes. In these systems, the sex chromosomes provide a genetic "switch" that either turns off female development and turns on development towards a male, or vice versa.

How, then, do TSD systems, in which sex chromosomes have not been identified, trigger gonadal differentiation towards gonads of one sex or the other? And perhaps even more intriguing, how, in some systems, can temperature override existing genetic triggers? It appears that, in vertebrates, the gene pathways that control differentiation beyond the initial determination of whether gonads will become testes or ovaries are relatively well conserved. What is not clear is how environmental variables, such as temperature, are translated into a physiological signal to influence sex determination. Is there a single gene target that responds to temperature, and whose expression levels determine whether the gonadal fate is testicular or ovarian? Let us consider the ways in which this question could be experimentally approached. Given that the genetic components of the mammalian sex-determining pathway are relatively well described, it would make sense to look for gene homologs in systems that exhibit TSD. Indeed, the genes responsible for directing the differentiation of gonadal tissue appear to be conserved across vertebrate species. Eighteen homologs of sex-determining genes in mammals have been found in reptiles (reviewed in Rhen and Schroeder 2010), though the patterns in which these genes are expressed are not always similar to those seen in mammals, nor are they even consistent among reptiles. In fact, Crews and Bull (2009) make a strong case that there is not a single gene that acts as a master controller in all vertebrate systems, but that instead, there is a "parliamentary" system of genes that act as a network and simultaneously control the downstream process of gonadal differentiation. If this is true, how do we identify which genes play a role? Given that, in many reptilian systems, temperature exerts an all-or-none effect (i.e., an all-male or all-female effect), it seems logical to compare the expression levels of potential gene candidates at male-producing versus female-producing temperatures. Extensive work in red-eared sliders, painted turtles, common snapping turtles, and others has identified key genes that may be involved (see

10.1 Potential Gene Targets in the Process of TSD

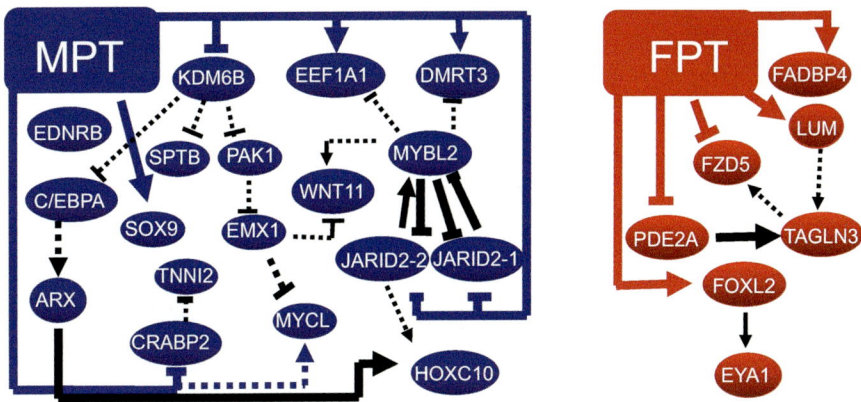

Fig. 10.1 Figure reproduced from Yatsu et al. (2016) showing the predicted gene networks that may change in response to temperature to influence sex determination towards masculinization (blue) or feminization (red) in American alligators

Matsumoto and Crews 2012; Rhen and Schroeder 2010; Shoemaker and Crews 2009 for reviews of these). In a recent study conducted in American alligators, Yatsu et al. (2016) and colleagues conducted transcriptomic analyses of alligator embryos incubated at male-producing and female-producing temperatures. They then predicted the gene–gene and temperature–gene interactions and revealed a gene network that is likely responsive to temperatures (Fig. 10.1). A few of these genes, SOX9, FoxL2, and WNT11, have also emerged as potential gene candidates in turtle studies and have homologs in the mammalian sex-determining pathway as well. Figure 10.2 shows the many genes in addition to those shown in the Yatsu et al. (2016) study that have been shown to be expressed differently in reptiles depending on whether embryos are incubated at a male- or female-producing temperature (also called MPT vs. FPT in much of the literature). The observed functions of those genes, mostly in mammalian systems, are also shown in Fig. 10.2. Based on the fact that several of these genes have known functions in mammalian sex determination and that their expression changes based on thermal regimes in more than one reptilian species, it is possible that two or more of these genes work together in a network to regulate the process of TSD in reptiles.

How do these genes compare to the ones that appear to act during TSD in fish species? It turns out that there is evidence supporting the involvement of similar genes between TSD reptiles and fish that exhibit either TSD or GSD with a TSD override. Shen and Wang (2014) provide an excellent review of this, and Fig. 10.3 shows the gene candidates that were expressed differently between male- and female-producing temperatures in fish species. In particular, DMRT1 and aromatase (also called cyp19a1a) are expressed differently at male- versus female-producing temperatures in multiple reptilian and piscine species. DMRT1 is expressed at higher levels in embryos and larvae maintained at male-producing temperatures, while aromatase is expressed at higher levels when embryos or larvae

Gene	Function	Species	Pattern
SOX9 *(Sry-related HMG-box protein 9)*	Male-determining factor in mammals	Red-eared slider Olive Ridley sea turtle American alligator	Higher at MPT Monomorphic, drops in females after TSP Not detected before sex determination
Sf-1 *(Steroidogenic factor 1)*	Regulates transcription of steroidogenic enzymes	Red-eared slider Painted turtle Snapping turtle American alligator	Higher at MPT Monomorphic Monomorphic or higher at FPT Higher at FPT
DMRT1 *(Doublesex mab3 related transcription factor 1)*	Testicular differentiation in vertebrates	Red-eared slider American alligator	Higher at MPT during late TSP Higher at MPT, low at FPT Increases after shift from FPT to MTP
AMH *(Antimullerian hormone)*	Regression of mullerian ducts	Red-eared slider	Higher at MPT, low at FPT Decreases after shift from MPT to FPT
MALAT1	Non-coding RNA	Red-eared slider	Higher at MPT
C116ORF62	Non-coding RNA	Red-eared slider	Higher at MPT, Decrease after E2 treatment
RSPO1 *(R-spondin 1)*	Regulates Wnt & β-catenin pathways	Red-eared slider	Higher at FPT
Fox-L2 *(Forkhead box protein L2)*	Granulosa cell differentiation in mammals	Red-eared slider Snapping turtle	Higher at FPT during late TSP Increases after shift from MPT to FTP
Aromatase	Conversion of testosterone to estrogen	Red-eared slider American alligator European pond turtle	Higher at FPT Higher at FPT Increases after shift from MPT to FTP
CIRBP *(Cold inducible RNA binding protein)*	mRNA processing, RNA export & translation	Snapping turtle	A-allele induced at FPT

Fig. 10.2 List of the genes whose expression has been shown to differ between male-producing temperatures (MPT) and female-producing temperatures (FPT) in reptilian species. *TSP* temperature-sensitive period. Information in this table was compiled from a review by Shoemaker and Crews (2009) except for the following: DMRT1 and aromatase in American alligators—McCoy et al. (2016), CIRBP in snapping turtles—Schroeder et al. (2016), and MALAT and C116ORF62 in red-eared slider turtles—Chojnowski and Braun (2012)

or maintained at female-producing temperatures. DMRT1 is a critical driver of male sexual development in both invertebrates and vertebrates (Ferguson-Smith 2007), and a recent study suggests that a shift of two amino acids distinguishes the DMRT1 gene of GSD species from that of TSD species. Aromatase, on the other hand, is a key factor involved in ovarian development and thus sexual differentiation into females (Guiguen et al. 2010; Balthazart and Ball 1995). It is responsible for the conversion of testosterone into estrogen, and this could be an important factor regarding the influences of hormones, particularly estrogen, on the process of TSD (see below).

These targeted approaches have successfully identified genes that are expressed differently between male-producing and female-producing temperatures in both reptiles and fishes. However, the simple fact that their expression differs between thermal regimes does not mean that they are drivers of the TSD mechanism. This is because (1) we do not yet know whether these genes are actually involved in the sex

10.1 Potential Gene Targets in the Process of TSD

Gene	Function	Species	Pattern
SOX9 (Sry-related HMG-box protein 9)	Male-determining factor in mammals	Nile tilapia	Strongly expressed in high-temperature treated females (which generally become male) before morphological differentiation
AMH (Antimullerian hormone)	Regression of mullerian ducts	Perjerrey	Higher at MPT, and increases drastically during differentiation.
		Atlantic salmon Japanese flounder Zebrafish	Low levels in undifferentiated gonad, but higher levels in testis compared to ovary
DMRT1 (Doublesex mab3 related transcription factor 1)	Testicular differentiation in vertebrates	Perjerrey	Higher at MPT 2 weeks before testis formation
		Medaka	DMRT1 detected in XX females at sex-reversing temps but not at neutral temps.
		Nile tilapia	Upregulated a critical pd. of sex differentiation in <u>both</u> XX and XY individuals at MPT
		Pufferfish	Involved in degeneration of germ cells in ovary causing sex reversal
Aromatase (Cyp19a1a)	Conversion of testosterone to estrogen	Nile tilapia	Low in both natural and temperature-induced males
		Atlantic halibut	Lower at MTP before differentiation
		Pejerrey	Expression increased at FPT 1 week before differentiation
		European sea bass	No difference b/t MPT and FPT
Fox-L2 (Forkhead box protein L2)	Granulosa cell differentiation in mammals	Japanese flounder	Suppressed at MPT

Fig. 10.3 List of the genes whose expression has been shown to differ between male-producing temperatures (MPT) and female-producing temperatures (FPT) in piscine species. Information in this table was compiled from a review by Shen and Wang (2014)

determination process or whether they are simply thermosensitive and (2) it is often unclear whether the changes in expression of these genes are a driver of sex determination or the result of it. For some genes, for example SOX9 and antimullerian hormone (AMH, also called Mis in some cases), there is evidence in some species that expression of these genes is different between individuals kept at male- and female-producing temperature, but that these expression differences occur after gonads have already been directed towards ovarian or testicular tissue. As a result, these are likely downstream mediators of the sexual differentiation process and are likely not directly triggered by temperature itself. In both fish and reptiles, thermal regimes trigger differences in the expression of genes such as DMRT1 and aromatase during the temperature-sensitive period and before the onset of morphological differentiation of the gonads. In addition, in reptiles, there are studies showing that expression of DMRT1, AMH, FOXL2, and aromatase will change during the temperature-sensitive period if the thermal regime changes. This suggests that the expression of these genes is not only highly sensitive to temperature but also that their expression levels can change during the time that is critical for triggering the direction of the indifferent gonad towards testes or ovaries. This tells us that those genes have the *potential* to be the ones responsible for transduction of temperature into a physiological response that triggers differentiation in one direction or the other.

However, we still don't know if one of these genes responds directly to temperature changes and triggers downstream effects, whether multiple genes respond to temperature and coordinate their effects, or whether *all* of these genes are actually downstream mediators responding to another unidentified gene that responds directly to temperatures. For example, using genetic association and linkage studies, Schroeder et al. (2016) recently demonstrated the potential involvement of a novel gene, CIRBP, short for cold-inducible RNA binding protein, in the process of TSD in snapping turtles. This gene is not one previously known to be involved in the sex determination process and is instead a known moderator of RNA translation and export. However, Schroeder et al. (2016) showed that a particular single CIRBP allele, the A allele, was more highly expressed at female-producing temperatures, while the C allele was not differently expressed based on temperature regime. Additionally, Chojnowski and Braun (2012) conducted a whole embryo screening for genes that were differentially expressed in response to temperature at the temperature-sensitive period for red-eared sliders. They found two noncoding RNAs, MALAT1 and C16ORF62, which were different during the temperature-sensitive period between embryos incubated at male- and female-producing temperatures. What this shows is that it may be helpful to look at genome-wide changes in gene expression to identify candidate genes outside of the known sex determination network to unravel the true trigger(s) that allow the process of sex determination to respond to temperature. What we also need to look out for is the potential for alternative copies of these genes as potential regulators of sex determination. For example, in perjerrey, a Y-linked version of AMH, called AMHY, is expressed much earlier than the autosomal AMH and appears to play a role in the prevention of ovarian development during the sex determination process in this species (Hattori et al. 2012). It is currently unknown whether this gene is expressed differently in response to temperature; however, this finding indicates that it is important to screen for other duplicated copies of genes involved in the sex determination process that may be involved in the responsiveness of this process to temperature cues. Finally, an experimental approach whereby genes are knocked down or out during the temperature-sensitive period would also be helpful in teasing out the roles of the many potential genes that may be involved in TSD. Sifuentes-Romero et al. (2013), for example, developed a technique to silence SOX9 in cultures of gonads collected from sea turtle embryos just prior to the TSP. More approaches like this would be helpful.

10.2 Does TSD Occur via Epigenetic Modifications?

If temperature is, in fact, altering the expression of a gene or a set of genes to push the process of sexual differentiation towards a male or female phenotype, how might this influence be accomplished? The temperature-sensitive periods for both

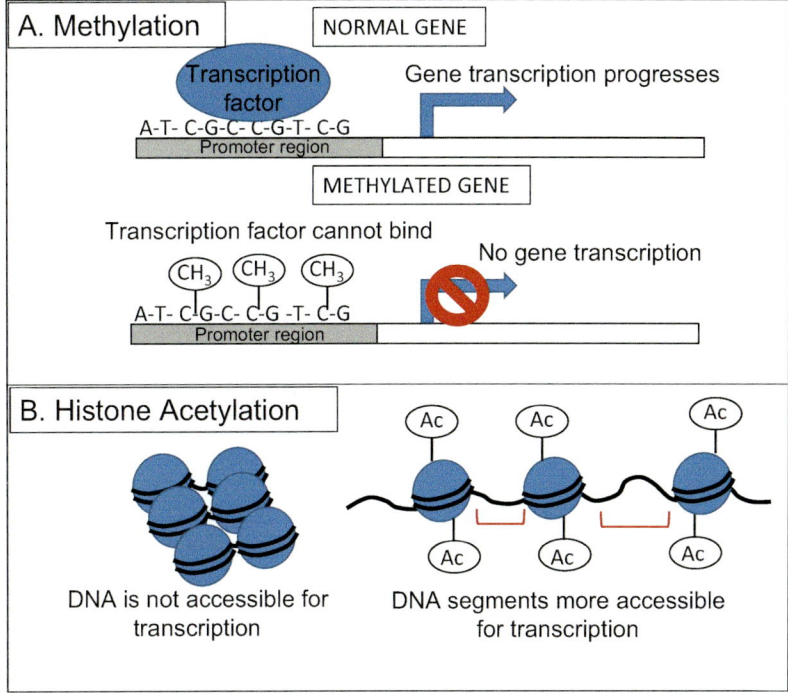

Fig. 10.4 Depiction of two potential modes of epigenetic regulation, including (**a**) methylation of a gene promoter to inhibit gene expression and (**b**) histone acetylation to stimulate gene expression

fish and reptiles occur after the offspring have already inherited their DNA from the mother and the father, so changes in gene expression must be accomplished by modifying the DNA that is already there. Not surprisingly, there are ways to accomplish this by physically modifying portions of the DNA sequence (i.e., epigenetic modifications) (Jaenisch and Bird 2003). Piferrer (2013) provide an excellent review of how epigenetic modifications may function during sex determination in plants, invertebrates, and vertebrates. By altering the conformation of the DNA and/or the accessibility of particular DNA segments for binding of transcription factors or ribosomes, the expression of genes can be up- or downregulated. One key way of accomplishing this is via methylation of CG dinucleotide sites on the promoter region of target genes (Fig. 10.4a). During normal transcription processes, transcription factors bind to the promoter regions to trigger the start of gene transcription. When methyl groups are attached to the CG dinucleotides in the promoter region, the transcription factors can no longer bind, and the expression of the gene is reduced or eliminated. Thus, if temperature were to increase or decrease methylation, this could alter the way that genes are expressed and, as a result, their influences on the process of sex determination.

An alternative epigenetic way of altering the expression of key genes is via histone acetylation or deacetylation. These processes are normal parts of gene regulation whereby acetyl groups are added or removed from the histone core of the nucleosome. When acetyl groups are present, this causes relaxation of the chromatin and increases the accessibility to the DNA segments that would be transcribed (Fig. 10.4b). As a result, histone acetylation tends to increase gene transcription while deacetylation tends to decrease gene transcription. Like methylation, these processes are controlled by enzymes, histone acetytlransferase and histone deacetylase. If the regulation of these enzymes is thermosensitive, this could be a method by which the expression of key genes involved in sex determination is altered.

There is, in fact, evidence in both reptiles and fish that epigenetic regulation may be the underlying mechanism by which temperature influences the process of sex determination. The first evidence of this idea emerged in a fish system. Navarro-Martín et al. (2011) showed in European sea bass that the aromatase (cyp19a) promoter was more highly methylated at higher, male-producing temperatures. Since then, several studies have shown similar findings. In olive flounder, for example, the aromatase promoter exhibited demethylation during early stages of ovarian development, and aromatase expression and estradiol concentrations were reduced by a high (male-producing) temperature treatment during gonadal differentiation (Fan et al. 2017). Similar findings in reptiles mirror these patterns. Matsumoto et al. (2013) found in red-eared sliders that the aromatase promoter in gonads showed higher levels of methylation when eggs were incubated at male-producing temperatures, and Parrott et al. (2014) showed in American alligators that embryos at male-producing temperatures had elevated promoter methylation and decreased expression of the aromatase gene. In addition to aromatase, SOX9 appears to be another potential target of epigenetic modification in TSD species. In American alligators, the SOX9 promoter was more highly methylated and expression was downregulated at *female*-producing temperatures (Parrott et al. 2014). Similarly, in mangrove killifish, the SOX9a promoter was hypermethylated, and expression levels of SOX9a were lower, in fish that developed at lower, female-producing temperatures, though this result was obtained using brain rather than gonadal tissue (Ellison et al. 2015).

While this body of work provides some tantalizing support for the ideas that (1) aromatase and SOX9 may, indeed, play roles in the mechanisms underlying TSD and (2) that epigenetic modifications may be the "signal" by which temperature triggers phenotypic change in gonadal sex, additional questions and challenges remain. First, to date, the approaches at identifying targets of epigenetic regulation related to TSD have been, for the most part, limited to one or two genes. There are two exceptions. The mangrove killifish study involved genome-wide analysis of methylation patterns which indicated multiple genes that may be involved in the process of TSD for this species. However, the authors only tested three of these genes, and while the aromatase and SOX9 genes showed promise as target genes for epigenetic modifications in this process, as mentioned above, this result was obtained using brain rather than gonadal tissue.

Shao et al. (2014) took an elegant approach in an intriguing system, the half-smooth tongue sole; these fish have a ZW sex-determining system, but there are also low percentages (~14%) of ZW "pseudomales" under normal conditions, and increasing the water temperature to 28 °C increases the percentage of pseudomales

10.2 Does TSD Occur via Epigenetic Modifications?

to ~73%. Further, this genotype:phenotype mismatch is heritable. F1 offspring of pseudomales produce very high percentages of pseudomales themselves (~94%). The heritable nature of this genotype:phenotype mismatch provides strong support for an underlying epigenetic influence. To test for the presence of specific genes involved in this mismatch as well as the potential for an epigenetic regulation of those genes, the authors examined DNA methylomes in the gonads of males, females, and temperature-induced pseudomales, as well as F1 pseudomale offspring. They found that a majority of the genes involved in the sex determination pathway in other vertebrates were highly conserved in tongue sole and that the promoter region of DMRT1 was hypermethylated in ovaries of these females, but this hypermethylation was inhibited in ZW pseudomales that had been masculinized by temperature. Expression of this gene is first seen during the period of sex determination, and high levels of expression persist in testes, while expression drops in ovaries. In general, the entire sex determination network contained differentially methylated regions.

These findings in tongue sole raise more questions: given that many genes were methylated in temperature-induced pseudomales, how many of these genes, if any, are truly involved in the mechanism underlying TSD in this species, and in others? In Nile Tilapia, both males and females show higher global levels of methylation after exposure to high temperatures (Sun et al. 2016). Perhaps these genes are simply thermosensitive, yet unrelated to the TSD mechanism. Further, are these methylation patterns a trigger that drives sex determination patterns, or is differential methylation of these genes the *result* of gonadal differentiation? It is known that epigenetic influences are extensively involved in the process of sex determination even in the absence of ESD (Piferrer 2013). In mammals, for example, methylation is present during X chromosome inactivation, a process that does not result in sex reversal and is not selectively triggered by temperature or other environmental conditions (Piferrer 2013). In olive ridley sea turtles, DNA methylation patterns differed between ovaries and testes; however, methylation patterns were similar at the bipotential stage between gonads of animals experiencing male-producing and female-producing temperatures. Therefore, experimental tests are needed to determine what happens to the sex determination process when temperature-induced methylation is inhibited.

Finally, we must also consider *how* temperature would act to stimulate methylation in gonadal genes (Fig. 10.5). Is methylation itself the trigger we have been looking for? As mentioned earlier, enzymes called methyltransferases are responsible for attaching methyl groups to the C-G dinucleotides. It is generally hypothesized that if methyltransferases are involved in the process of TSD, then expression of those methyltransferases would differ between gonads of individuals kept at male-producing and female-producing temperatures. Parrott et al. (2014) found in American alligators that three methyltransferases were expressed in gonadal tissues and that two of these, $DNMT_1$ and $DNMT_{3a}$, changed through the developmental stages. However, for none of the three methyltransferases tested did expression differ based on temperature, and while there was a temperature by stage interaction, the pattern of methyltransferase expression did not appear to coincide with the declines in

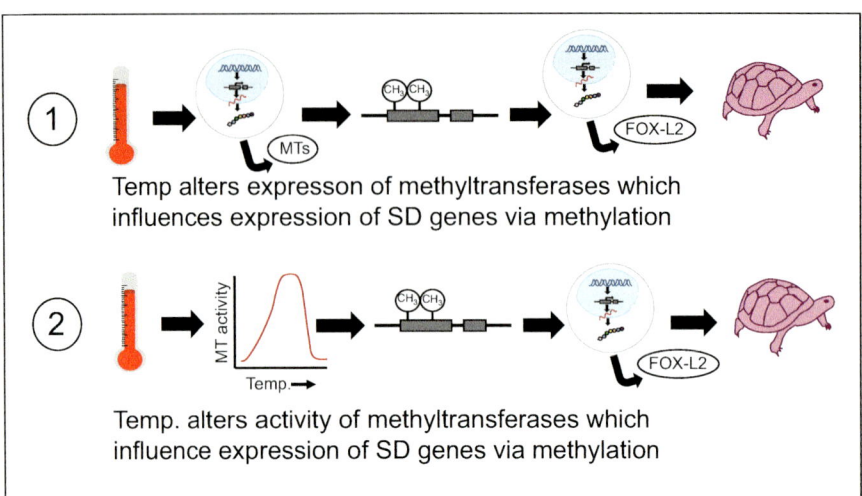

Fig. 10.5 Two potential ways that methylation of key sex-determining genes may be accomplished during TSD, including (1) increasing expression of methyltransferases that then methylate gene promoters or (2) increasing methylation *activity* to accomplish the same outcome

aromatase or SOX9 expression. The authors suggest that methylation in these cases may instead result from locus-specific targeting mechanisms, perhaps involving DNA binding proteins or long noncoding RNAs. Indeed, noncoding RNAs appear to be critical in the sex determination process in birds (Piferrer 2013). Perhaps alternatively, the changes in methylation patterns seen in the studies discussed above result not from changes in *expression* of the methyltransferases but instead from change in enzyme *activity* based on temperature. It is well known that enzymes in general are very sensitive to temperature and that different enzymes have different optimal temperatures at which they are fully active. Future studies should focus on quantifying methyltransferase activity in the bipotential gonads at the male- and female-producing temperature regimes to test whether differential activity could be responsible for the divergent methylation patterns observed on the genes in the sex determination pathway. In addition, the many other possible ways that epigenetic regulation could occur have not yet been addressed. In a study of European sea bass, a genome-wide examination of gene expression patterns in response to temperature showed that expression of a histone methyltransferase (ehmt2) was unaffected by temperature, while four genes involved in other types of transcription regulation were upregulated at high temperatures. Included in this group were two transcriptional repressors (pcgf2 and suz12) and a histone deacetylase (hdac11) (Díaz and Piferrer 2015). Thus, there is much more work to be done to discern how temperature may influence gene expression during TSD. This remains an exciting area of research that seems likely to yield some new understanding into how TSD occurs.

10.3 The Role of Hormones in the Control of TSD

A substantial body of work in fish, frogs, and reptiles indicates that hormones exert significant influences on the process of TSD. Estrogens, for example, seem to exert potent effects in reptiles and fishes. Androgens are now routinely used in fish farming operations to produce fish of the desired sex for human consumption. There is also evidence that glucocorticoids may mediate the process of TSD in some fish species. In many cases, it is difficult to determine precisely what role these hormones may play in the process of TSD, and whether these influences *can* occur before the process of sex determination has already advanced to the point where development towards one sex or the other has been initiated driving the production of those hormones. I will discuss the evidence for the role of estrogens, androgens, and glucocorticoids as drivers of TSD below.

10.3.1 Influences of Estrogens

Given that the aromatase enzyme arises time and again in reptilian species as a potential modulator of TSD, it is perhaps not surprising that estrogen also appears to play a role. In fact, it has been known since the late 1960s that treatment with estradiol causes the embryos of many reptilian species to develop into females, overriding the influences of temperature (reviewed in Wibbels et al. 1994). The first studies to this effect were conducted in tortoises and European pond turtles; treatment of embryos that were incubated at male-producing temperatures stimulated development of ovarian tissue, with some developing normal ovaries (reviewed in Wibbels et al. 1994; Raynaud 1985). Since then, estrogen has been shown to influence sex determination in many reptilian species. For example, in red-eared slider turtles, administration of estradiol to eggs incubated at male-producing temperatures overrode the effects of temperature and resulted in the production of female turtles, while administration of an aromatase inhibitor to eggs incubated at *female*-producing temperatures resulted in the production of *male* turtles. Further, Wibbels et al. (1991) provided evidence that estradiol and temperature may interact to determine sex in turtles; they found that less estrogen is required to reverse the effects of intermediate temperatures than of extreme male-producing temperatures.

These findings are intriguing, but questions remain. Is estrogen production a key responder to or mediator of temperature effects and, if so, where do the precursors and/or the estrogen itself come from, given that gonads are not yet differentiated during most of the TSP? It is possible that estrogens that mediate the influences of temperature are of maternal origin. Janzen et al. (1998) compared yolk hormone levels among five turtles species, two that exhibit TSD (red-eared sliders, snapping turtles, and painted turtles) and two that exhibit strict GSD (northern spiny softshell, smooth softshell); turtles that exhibit TSD had significantly higher concentrations of yolk testosterone, while yolk estradiol was similar across species. This could mean that, in TSD species, testosterone is being made available for conversion into estradiol if aromatase is present in sufficient amounts, such as when eggs are exposed to the female-producing, high temperatures. On the other hand, the work of Bowden et al. (2000) indicates a potential role of yolk estradiol in painted turtles. They showed that yolk estradiol varied seasonally, even within groups produced at either male- or female-producing temperatures and that responses of the sex-determining system differed across the season as well; sex ratios produced at the female-producing temperature of 28 °C were 72% female at the start of the season and 76% female at the end of the season. Further work confirmed that yolk estradiol levels were higher at the end of the season for this species (Bowden et al. 2002). In both alligator snapping turtles and American alligators, the way concentrations of estradiol changed through the temperature-sensitive period differed depending whether eggs were incubated at male- or female-producing temperatures (reviewed in Elf 2003). In the case of the snapping turtle, a species for which incubation at higher temperatures produces more females, eggs incubated at those higher temperatures maintained higher concentrations of estradiol than those incubated at lower temperatures. American alligators produce females at both high and low temperatures, and when yolk estradiol concentrations were sampled, concentrations remained high in eggs incubated at the two female-producing temperatures and lowered in eggs incubated at the intermediate temperatures. Thus, it is striking that eggs retain higher concentrations of estradiol at female-producing temperatures, regardless of whether those female-producing temperatures are high or low. Still, the role, if any, of yolk steroids in the process of TSD remains controversial. As of 2007, of 13 TSD species studied, only 7 showed significant sex differences in yolk steroids at oviposition (reviewed in Radder 2007). Some have suggested that it is the T:E ratio that best correlates with the mode of sex determination; it was suggested that GSD species have T:E ratios >1 while TSD species have T:E ratios <1, but a comparative analysis of TSD and GSD species does not yield a strong pattern to this effect (Radder 2007).

Might *endogenous* estradiol be a player in the process of TSD? This at first seems impossible given that ovarian tissue is considered to be the producer of estrogens, and thus estrogens of gonadal origin could not trigger the process of differentiation of gonads into ovaries. However, it is also known that steroid hormones are produced by the adrenal glands, and in olive ridley sea turtles and painted turtles, the enzyme required to convert steroid molecules into progesterone, androstenedione, and testosterone (3β-hydroxysteroid dehydrogenase or 3β-HSD) was found in the adrenal glands of developing embryos (Merchant-Larios et al. 1989; Thomas et al. 1992). In painted turtles, 17β-HSD, which catalyzed conversion of androstenedione to testosterone, was also found (Thomas et al. 1992). Thus, it is possible that androgens produced by the adrenal gland are the source of the precursors that aromatase enzymes in the gonad ultimately convert to estrogens to further feminize the developing tissue. Still, it is thought that the gonadal–adrenal–kidney complex remains relatively inactive during early stages of development in reptiles (reviewed in Radder 2007), so this idea needs to be further tested.

If estradiol does, indeed, mediate the process of TSD, how might it act on developing gonadal tissue? First, we must consider how the embryos respond to

10.3 The Role of Hormones in the Control of TSD

estrogen. In most vertebrates, estrogen can bind to two receptor subtypes—estrogen receptor α (ERα and also called ESR1) and estrogen receptor β (ERβ and also called ESR2). In American alligators, estrogen treatment overrides the effect of a male-producing temperature, resulting instead in the production of females. Kohno et al. (2015) conducted an elegant experiment in which they treated developing embryos with pharmaceutical agonists that stimulated either ERα *or* ERβ. Only treatment with the ERα agonist overrode the effects of the male-producing temperature in a manner similar to estradiol treatment. Further, in red-eared sliders, both ERα and ERβ are expressed in the developing gonad (reviewed in Ramsey and Crews 2009). Once the gonadal tissue responds to estradiol, what downstream triggers might be activated to guide the process of sex determination? Murdock and Wibbels (2006) showed in red-eared slider turtles that embryos at male-producing temperatures showed a decrease in expression of DMRT1 when they were treated with estradiol during the temperature-sensitive period, indicating that the estradiol might interfere with the normal role of DMRT1 in masculinization of the gonad. Another male-inducing gene, SF-1, is also downregulated in response to estrogen treatment in the same species (Fleming and Crews 2001). Overall, there appears to be a source of estrogen, receptors present in the embryonic gonad to respond to it, and effectors within the sex determination pathway that are sensitive to the influences of this hormone. More work needs to be done to determine specifically where in the TSD pathway estrogen plays its role in reptiles.

Thus far I have discussed the potential role of estrogen primarily in reptiles. However, there is also abundant evidence in fish for a role of estrogen in the TSD mechanism. As in reptiles, aromatase is yet again a gene that is expressed differently in embryos experiencing male- versus female-producing temperatures. Perhaps there is a role for estrogen in TSD that is conserved among vertebrates that exhibit this mode of sex determination. Indeed, there is abundant evidence that aromatase expression changes in response to temperature and that expression of this enzyme mediates the process of sex determination in fishes that exhibit TSD (reviewed in Guiguen et al. 2010). In 1969, Yamamoto (1969) suggested that estrogens were potent feminizers of fish, and it is now well known that exogenous treatment with estradiol causes feminization in a variety of fish species (reviewed in Piferrer 2001). In addition, treatment with aromatase inhibitors caused masculinization in many fish species (reviewed in Guiguen et al. 2010). The sexual differentiation process is clearly sensitive to estrogens in fish, and the enzyme that is necessary to convert testosterone to estradiol is not only present but also varies with temperature during TSD.

While the available evidence points to a role for estrogen during the process of TSD in fish species, studies examining estrogen synthesis in fish are scarce, and those that do exist do not all agree on whether fish can produce estrogens during early embryonic stages. Work in rainbow trout, arctic charr, and medaka shows that whole embryos of these species have the capability to aromatize androgens into estrogens (Khan et al. 1997; Yeoh et al. 1996; Iwamatsu et al. 2006), while Nile tilapia embryos do not exhibit this capability (Rowell et al. 2002). There are few studies in fish examining whether estrogen production by embryos differs with temperature prior to or during sex determination. In one of the only studies to my knowledge, high temperatures during gonadal differentiation reduced both aromatase expression and estradiol concentrations in olive flounder larvae (Fan et al. 2017).

There is now abundant evidence that, as in reptiles, there may be a *maternal* source of estrogens in fish (reviewed in Guiguen et al. 2010) and if temperatures that result in feminization of embryos also rendered them either more responsive to those maternal estrogens or increased the rates at which they take up estrogen from the egg yolks, then this could be a way in which temperature could harness the feminizing effects of estrogens. In this case, then the observed differences in aromatase expression would be a downstream effect. While reptilian studies have tested for the presence of testosterone in egg yolks, to my knowledge, there has been no documentation of testosterone in the eggs of fish. It is unclear whether testosterone has not been tested for, or whether it was not found and such studies went unpublished. However, it is important to consider the possibility that during TSD, increased levels of aromatase at female-producing temperatures may act to convert maternal testosterone into estrogen to influence the sex determination process. There is now increasing evidence that estrogen is a major player in the process of feminization during TSD, particularly in concert with glucocorticoids (see below). However, despite the well-known feminizing role for exogenous estrogen treatment in fish, more work is needed to determine whether and how this hormone acts in the natural process of TSD in the piscine species that exhibit it.

10.3.2 Influences of Androgens

While the potential influences of estrogens described above could also technically be the role of androgens given that estrogens are aromatized androgens, there have been tests of the influences of androgens in the process of TSD in both reptiles and fish. Given that estrogens stimulated feminization at male-producing temperatures, one might expect that androgens would cause masculinization at female-producing temperatures. Indeed, in fish, androgens act as masculinizing agents, and are used as such, in a variety of species (Yamamoto 1969 and reviewed in Devlin and Nagahama 2002). In reptiles, however, treatment with testosterone during the TSP resulted in development of *ovarian* tissue at male-producing temperatures (reviewed in Wibbels et al. 1994). Further work indicated that this effect occurred because embryos were aromatizing the testosterone into estrogen, given that non-aromatizable androgens such as dihydrotestosterone did not induce a similar effect. In fact, treatment with these non-aromatizable androgens stimulated the production of males at female-producing temperatures, as might have been originally expected. However, compared to estradiol's feminizing effects, the masculinizing effects of non-aromatizable androgens like dihydrotestosterone appear weaker, because this hormone did not induce masculinization in all studies (Wibbels and Crews 1992). Despite the well-known masculinizing effects of exogenous testosterone administration in fish species, the potential function of this hormone in the mechanism of TSD has been studied very little in fish. Hattori et al. (2009) found that when perjerrey larvae, which exhibit TSD, were maintained at a male-producing, high temperature, they had higher levels of testosterone and 11-ketotestosterone (11-KT, the main active androgen in fish), and the same group later showed that pejerrey have the ability to produce 11-ketotestosterone at very early embryonic stages (Blasco et al. 2013). However, they hypothesized that the actual upstream driver of the TSD mechanism is cortisol (see below). More work needs to be done to dissect the role of testosterone, if any, in the TSD mechanism of fish.

10.3.3 Influence of Glucocorticoids

As was discussed in previous chapters, there is growing evidence that glucocorticoids are involved in the process of sex ratio adjustment in birds, humans, and nonhuman mammals. Additionally, it appears to make sense in an adaptive context that this hormone, which regulates the responses to environmental and social conditions, might also act as a transducer of those conditions into a physiological signal that may influence the process of sex determination. It is not surprising, then, that studies in fish indicate that glucocorticoids may also serve as a mediator in piscine TSD. In fact, more and more studies indicate that masculinization induced by high temperatures in fishes is actually the result of thermal stress and is mediated by cortisol. Perjerrey are among the most well-studied fish that exhibit TSD. In this system, rearing larvae at high temperatures induces those larvae to differentiate into males. At this masculinizing temperature, those larvae also had higher levels of cortisol, testosterone, and 11-KT (Hattori et al. 2009). When either cortisol or the cortisol agonist, dexamethasone, was administered to larvae during the period of sex determination at an intermediate temperature known to produce equal numbers of males and females, significantly more larvae developed into males. Similar findings were documented in Japanese flounder. This system exhibits and XX/XY system of genetic sex determination, but at high temperatures (27 °C), XX larvae differentiate into males. As in pejerrey, cortisol concentrations were significantly higher in larvae reared at the high, male-producing temperature (Yamaguchi et al. 2010). Cortisol treatment again stimulated sex reversal of XX individuals. Similar effects were also found in medaka (Hayashi et al. 2010). Finally, in support of the idea that high temperatures are acting as a thermal stress, Mankiewicz et al. (2013) exposed southern flounders, which are known to differentiate towards males at high temperatures, to another stressor: a blue background. Larvae exposed to a blue background were masculinized much like those reared at high temperatures and also had higher levels of cortisol.

There are currently three hypotheses for how cortisol may mediate the process of TSD in fish (Fig. 10.6): (1) Cortisol may act to inhibit aromatase expression, thus inhibiting the production of estrogen, which is thought to be essential to the process of feminization. In medaka, when estrogen was supplied to larvae along with cortisol, masculinization did not occur (Kitano et al. 2012). (2) Cortisol may also act by causing apoptosis in primordial germ cells. Yamamoto et al. (2013) showed

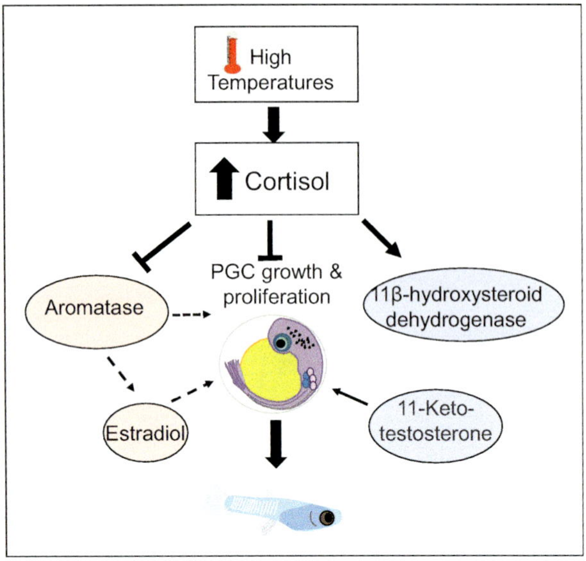

Fig. 10.6 The three hypothesized mechanisms by which temperature may stimulate masculinization of fish larvae through the actions of cortisol. Temperature may act as a thermal stress, increasing cortisol. Cortisol may then (1) inhibit expression of aromatase to reduce estradiol (E2) concentrations and inhibit feminization, (2) cause apoptosis or inhibit proliferation of primordial germ cells (PGCs) to inhibit the development of ovarian tissue, or (3) increase the activity of 11β-hydroxysteroid dehydrogenase (11β-HSD) to trigger conversion of androgens to 11-ketotestosterone (11-KT), the major androgen that appears to stimulate masculinization in fish

that primordial germ cells of pejerrey larvae maintained at high temperatures showed signs of widespread apoptosis, and treatment with estrogen prevented this, while also preventing the masculinization effects induced by high temperatures. In addition, in a recent study, olive flounder larvae reared at higher temperatures showed slower proliferation of primordial germ cells (Wang et al. 2017). Finally, the third hypothesis (3) is that cortisol induces masculinization by stimulating the production of 11-KT. Fernandino et al. (2012) performed an elegant experiment in which they cultured gonadal explants from pejerrey with cortisol and showed that cortisol treatment increased the levels of 11-KT in the culture medium. Fernandino et al. (2013) suggest that the enzyme 11-β-hydroxysteroid dehydrogenase (11βHSD) may be the primary transducer of the temperature influences because treatment of pejerrey larvae with cortisol increased expression of 11βHSD. This enzyme is active in both the glucocorticoid and the androgen synthesis pathways (Fig. 10.7) and is responsive for converting cortisol to its inactive form, cortisone, and for catalyzing the conversion of 11-hydroxytestosterone to 11-KT. At this point, it is unclear whether one of these mechanisms is the predominant one for the influences of temperature and cortisol on sex determination in fish, whether cortisol acts through different pathways in different species, or whether all three of these ideas actually act together as one big pathway in fish.

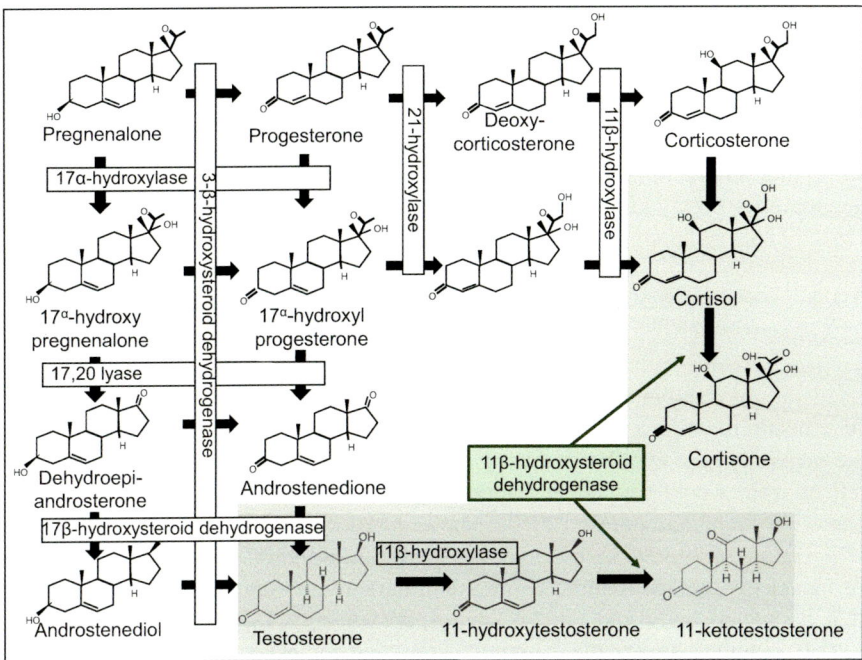

Fig. 10.7 The steroid hormone conversion pathway, with a gray highlight on the portion of the pathway shared by the production of 11-ketotestosterone and the conversion of cortisol to cortisone, both through the actions of 11β-hydroxysteroid dehydrogenase (11β-HSD)

While the influence of cortisol on TSD in fish is becoming more and more convincing, the influences of glucocorticoids in reptiles have been far less well studied, and the few studies that *have* tested for a mediating effect of glucocorticoids in reptilian TSD have produced conflicting results. In 2009, Warner et al. (2009) treated eggs of two lizard species known to exhibit TSD with corticosterone, the main glucocorticoid in reptiles. This treatment resulted in a higher proportion of female hatchlings in jacky dragons but a higher proportion of males in eastern three-lined skinks. Iungman et al. (2015) applied corticosterone to the surfaces of broad-snouted cayman eggs incubated at masculinizing temperatures and showed no effects on sex ratios, and Uller et al. (2009) showed no relationship between concentrations of corticosterone in egg yolks and offspring sex of resulting mallee dragons. However, unlike treatment of rearing water in fish, treatment of reptile eggs may not result in chronic exposure to corticosterone and may instead provide one dose that is more like an acute stress. In addition, it is unclear why the broad-snouted cayman study was conducted at male-producing temperatures, given that in fish, cortisol almost ubiquitously *causes* masculinization. Hence, perhaps it is not surprising that no effect of corticosterone was seen in that study. At this point, glucocorticoids cannot be ruled out as mediators of TSD in reptiles. In further studies, it is important to test the effects of these hormones at both male- and female-producing temperatures. In addition, perhaps stressing mothers in a chronic manner to induce elevated levels of corticosterone throughout the yolk may shed light on how a more chronic exposure to the hormone may influence the process of sex determination at high or low temperatures. Finally, it is important to consider embryonic production of glucocorticoids, and how that changes with incubation temperature. To my knowledge, this has not yet been tested.

10.4 What About the Role of Thermal Fluctuations?

To date, researchers have accepted the paradigm that in most TSD species, higher temperatures result in the production of one sex and lower temperatures in the other. As a result, it seems logical that in studies using artificial incubation, temperatures are generally kept at either stable high or stable low temperatures. Does this truly tell us what would happen in natural conditions? Some studies indicate that the answer is "no." Paitz et al. (2010b) showed that clutches incubated at a constant temperature of 29.1 °C in the laboratory produced a significantly higher proportion of females compared to clutches kept in a natural environment with temperatures that fluctuated daily and ranged from approximately 18–36 °C. Bowden et al. (2014) provided an excellent review discussing the relevance of using constant incubation temperatures to study influences of incubation temperature on offspring sex and other phenotypic traits. Under natural conditions, reptilian eggs experience temperature fluctuations throughout the day, with higher temperatures during the daylight hours and lower temperatures during the nighttime hours. The female's choice of nest site can determine how high and low these extremes go. For example, if a female chooses a spot with a good amount of vegetation cover, the heat of the sun will not as easily reach the area in which the eggs are incubating and the incubation temperatures will not reach highs that are as extreme as those experienced by eggs in an uncovered nest. Similarly, the depth at which a female buries her eggs can also influence how much the incubation temperatures are affected by the heat of the sun. Even the time during the season that a female lays her eggs can affect the magnitude of temperature extremes experienced by the incubating eggs. Studies in the field indicate that mean incubation temperature is not a good predictor of offspring sex ratios (reviewed in Bowden et al. 2014; Bull 1985). To form a better predictor of offspring sex ratios, researchers have developed the constant temperature equivalency (CTE) model, in which effects of daily temperature fluctuations are taken into account during the prediction of resulting sex ratios. For example, by considering the amount of time that incubation temperatures are above or below the mean, Georges et al. (1994) were able to accurately predict the hatchling sex ratios of loggerhead sea turtles in the wild, and a recent model considers the effects when temperatures increase across the incubation period, as they often do as the breeding season progresses or as climate change impacts ambient temperatures (Telemeco et al. 2013).

There have now been several studies testing how the magnitude of temperature fluctuations around the mean influences hatchling sex ratios in reptiles. In those exhibiting Pattern Ia TSD, in which females are produced at cooler temperatures and males at warmer temperatures, greater amplitudes of temperature fluctuations generally result in greater production of females. Paitz et al. (2010a) showed in painted turtles that eggs resulted in 0% females when incubated at a constant temperature of 27 °C but shifted to 100% females when temperatures fluctuated just 8 °C around that same mean. This is not always the case, however. Another study in painted turtles showed that fluctuations of 5 °C around a *male*-producing temperature of 31 °C resulted in more *female* hatchlings (Neuwald and Valenzuela 2011). Studies in jacky dragons, which show pattern II TSD (producing males at intermediate temperatures and females at both extremes), demonstrated a similar divergence in the influences of the magnitude of temperature fluctuations depending on whether the mean incubation temperature is known to be a female-producing temperature. An 8 °C fluctuation in temperature around the female-producing temperature of 25 °C increased the proportion of males, while a similar fluctuation around a temperature known to produce more equal sex ratios (28 °C) increased the proportion of *male* offspring.

These results suggest that the current simplistic models of defining modes of TSD (e.g., Type Ia, Type Ib, Type II) may no longer, by themselves, be sufficient. Bowden et al. (2014) suggest that future work needs to utilize laboratory studies that more realistically mimic field conditions, as well as field studies in which temperature fluctuations are monitored using dataloggers. In addition, while this idea has been examined in reptiles, I was not able to find studies looking at the degree of temperature fluctuations in fishes that exhibit TSD; thus, studies in this area are warranted. New models that incorporate fluctuations of incubation temperatures may help us to further understand both the adaptive significance and mechanisms underlying TSD in species that exhibit this mode of sex determination.

10.5 Mechanisms of TSD in Frogs

There is a conspicuous lack of information provided in this book on patterns of sex ratio adjustment and sex change in amphibians because those patterns have not been well studied. What is known about frogs and other amphibians is that the modes of sex determination found in this group are extremely diverse, at times even varying within the same species! For example, in Japanese wrinkled frogs, some individuals have the XX/XY system of GSD and others that have the ZZ/ZW system depending on the local population (see Hayes 1998; Nakamura 2009 for reviews on amphibian sex determination). What this indicates is that sex chromosomes are still undergoing evolution within amphibians. However, there is some information of mechanisms by which TSD may occur in frogs, and it is known that many of the same sex-determining genes found in mammals, birds, and now reptiles (e.g., aromatase, foxl2, DMRT1) appear also to be present and functional in frogs. Flament (2016) provides an excellent review of sex reversal in amphibians, and I will use that review as a jumping off point to summarize what is known about TSD in these systems below.

Studies conducted more than a century ago indicated that the sex-determining system in common frogs is sensitive to temperature. Tadpoles exposed to high temperatures (25 °C) were significantly more likely to develop gonads that looked morphologically like testes, while those exposed to low temperatures (12 °C) were more likely to develop gonads that looked morphologically like ovaries (Piquet 1930; Witschi 1914). It was unknown whether these individuals actually functioned as males and females, respectively. Since then, further studies have shown that high temperatures induce masculinization in several species, including wood frogs, Japanese brown frogs, American bullfrogs, and common toads. Similarly, in studies of caudate amphibians, high temperature has a masculinizing effect in the ribbed newt and two subspecies of northern crested newts, while cool temperatures exert feminizing effects. Work in the ribbed newt indicates that amphibians with TSD have a TSP much like reptiles; in this species, it is approximately 2 months long.

Studies on the molecular mechanisms underlying TSD in frogs are, to date, sorely lacking. Work in ribbed newts indicates that, as in reptiles and fish, aromatase may be a major player in the process. When ribbed newt larvae were exposed to high masculinizing temperatures, aromatase expression stayed low, even in ZW larvae (Kuntz et al. 2003), and estrogen treatment inhibits the masculinizing effects of high temperatures (Zaborski 1986). Likewise, in salamanders, aromatase expression increased at female-producing temperatures (Sakata et al. 2004). As in reptiles and fish, it is clear that estrogen likely plays a role in the mechanism of TSD, but much more work needs to be done to determine the upstream players and the transducer responsible for responding to temperature in amphibians.

10.6 Mechanisms of Sex change in Hermaphroditic Fish

While to this point, I have focused primarily on the influences of temperature on sex ratios, we can likely gain some insight into the mechanisms by which TSD might work by looking at species that exhibit even more plastic forms of sex determination. While most birds, mammals, and reptiles are gonochoristic, meaning that they develop as one sex and stay that sex for the rest of their lives, many species of fish are able to change sex either once or continuously throughout their lives. In Chap. 9, I highlighted patterns of sex change in hermaphroditic fish. Here, I will briefly summarize what is known about the mechanisms of sex change in fish in an attempt to shed light on whether there is a common mechanism by which different environmental variables influence the process of sex determination in species that display ESD.

While gonochoristic sex determination during embryonic or larval stages involves the triggering of a masculinizing or feminizing cascade, sex change during adulthood instead involves replacement of one tissue type with another (e.g., testicular tissue with ovarian tissue and vice versa). In fish, this can happen just once during an individual's lifetime, or it can happen continuously in simultaneous hermaphrodites that constantly maintain both ovarian and testicular tissue and alter which is functional at a given time.

In sex-changing fish, the endocrine environment appears to drive sex change. The principal androgens and estrogens in fish are 17β-estradiol (E2) and

10.6 Mechanisms of Sex change in Hermaphroditic Fish

11-ketotestosterone (11-KT), and a dramatic shift in the concentrations of these two hormones precedes the restructuring of the gonad towards that of the opposite sex. In protogynous sex change (conversion from a female to a male), there is a drop in E2 levels followed by an increase in concentrations of 11-KT, and the opposite pattern is seen in protandrous sex change (conversion of a male to a female); 11-KT levels drop while E2 levels increase (reviewed in Todd et al. 2016). In general, sex change in fish appears to be driven by social factors, such as changes in the dominance hierarchy. As described in Chap. 9, sex change often happens after loss of the single dominant individual in the population. How this social cue is transduced into a physiological signal that alters the endocrine environment, however, is still a black box. Given that animals respond to social stimuli via neuroendocrine pathways, the process of sex change likely starts in the brain and then transmits to the rest of the body. Perhaps a most obvious possibility would be the regulators of steroid hormone production, gonadotropin-releasing hormone (GnRH) and the gonadotropins LH and FSH. In support of this idea, treatment with either of these induced sex change in some protogynous (e.g., Kobayashi et al. 2006) and protandrous (e.g., Lee et al. 2001) hermaphrodites; however, how patterns of *endogenous* GnRH and gonadotropins may regulate sex change remains unclear and patterns often conflict (reviewed in Todd et al. 2016). It is generally accepted that there are other neurochemicals that react to social cues upstream of GnRH and gonadotropins and may be involved in the mechanism of shifting sex in response to social cues (reviewed in Godwin and Thompson 2012). Such neurochemicals include kisspeptin (Shi et al. 2010), norepinephrine, serotonin, and dopamine (reviewed in Todd et al. 2016). Interestingly, given the discussion above, stress hormones may also be involved in the process of sex change (reviewed in Solomon-Lane et al. 2013). Long-term administration of cortisol stimulated sex change from female to male in three-spot wrasse (Nozu and Nakamura 2015), and spikes in cortisol levels have been found during sex change in both protogynous and protandrous species (Godwin and Thomas 1993; Solomon-Lane et al. 2013).

As with gonochoristic sex determination, sex change also appears to involve many of the same sex-determining genes. Liu (2016) conducted a whole-genome analysis during sex change in bluehead wrasse, a species where sex changes from female to male, and showed that genes related to feminization decline in the gonad while genes involved in masculinization increase. In contrast, in the protandrous black porgy, where sex changes from male to female, expression of male-related genes declines while expression of female-related genes increases. Based on the work in the protogynous bluehead wrasse, it is thought that upregulation of antimullerian hormone (AMH) may be the first step in the process, leading to suppression of aromatase and thus decreases in concentrations of E2. In contrast, there is a good amount of evidence that protandrous sex change is initiated by decreasing expression of DMRT1 (reviewed in Todd et al. 2016).

Finally, just as was discussed in reptiles, there may be a role for epigenetic modifications of the masculinizing and feminizing genes that regulate the process of sex determination and sex change. In protogynous ricefield eels, methylation of the promoter region of the aromatase gene increased as aromatase expression decreased during sex change from female to male, and treating fish with a DNA-methylation inhibitor prevented or reversed the sex change (Zhang et al. 2013). To date, this remains the only species in which epigenetic modifications have been examined in relation to sex change. Much more work is needed in this area.

10.7 What Do We Know and Where Do We Go from Here?

Based on all of the information discussed above, is it possible that there is a conserved mechanism underlying TSD among vertebrates? It is certainly possible to construct a map of potential pathways that may commonly drive the process of TSD in vertebrates (Fig. 10.8). There are clear players, such as aromatase and DMRT1, for example, that emerge as likely players in all three vertebrate classes where TSD is found. In addition, there is growing evidence that epigenetic regulation of key sex-determining genes plays a role in multiple systems as well. However, what we still do not know is how temperature elicits these effects, and whether there is a single cascade of events involving a "holy grail" gene, or whether there are multiple gene targets that are responsive to temperature, as has been suggested by Crews and Bull (2009). There are many elements within the body that may be sensitive to temperatures. Enzymes have optimal temperatures at which they function, and an incubation temperature above or below that optimum could decrease the activity of a critical enzyme in the sex determination process or in the activity of methyltransferases. This, however, would not help to explain why some reptilian species produce females at high incubation temperatures and others males. There are also molecules within the body that are known to respond to temperature and also may act in the process of sex determination. Transient receptor potential cation channels (TRP channels) function as environmental sensors through Ca^{2+} signaling, and one of these channels (TRPV4) is reactive to moderate heat in mammals and also affects expression of genes associated with male development in alligators (Yatsu et al. 2015). Heat shock proteins (HSPs) are proteins that rapidly respond to heat, and Kohno et al. (2010) identified three HSPs (HSP 27, HSP70, and HSP90) that show sexual dimorphism in their expression within American alligator embryos. There are also mechanisms by which noncoding RNAs can change conformation in response to temperature to protect ribosome binding sites (which has earned them the title "RNA thermometers") (Narberhaus et al. 2005). Given that the noncoding RNAs, MALAT1 and C16ORF62, were found at different levels in developing turtles exposed to male- versus female-producing temperatures, this could also be a viable mechanism by which temperature is transduced into a physiological signal to influence sex determination. However, it is difficult to envision a way in which each of these mechanisms would trigger the production of females in response to high temperatures in some species and males in others.

Perhaps part of the reason that no one clear mechanism emerges to explain the process of TSD is because we, as researchers, are focused on how animals are focusing only on how temperature itself is affecting physiological processes associated with sex determination, rather than how a departure from an *optimal*

10.7 What Do We Know and Where Do We Go from Here?

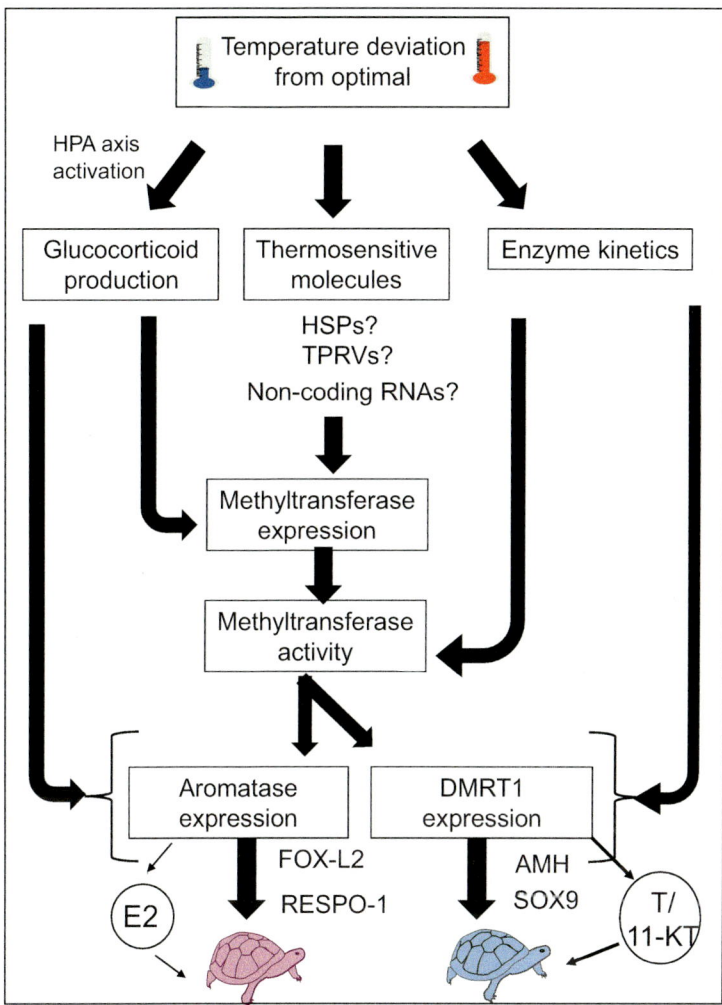

Fig. 10.8 Based on a conglomeration of available information in reptiles and fish, there may be a conserved mechanism responsible for TSD. Outlined here are some potential pathways. Temperature may act as a thermal stress, triggering the production of glucocorticoids that then go on to influence expression of genes involved in sex determination. Alternatively, temperature may trigger reactions of temperature-sensitive molecules such as heat shock proteins (HSPs), transient receptor potential cation channels (TRPs), or noncoding RNAs, which then go on to influence expression of key sex-determining genes perhaps by increasing expression of methyltransferases. Finally, temperature may alter the activity of key enzymes, such as methyltransferases or aromatase, to trigger the downstream effects on sex determination, including changes in expression of the male-determining genes, such SOX9 and AMH, as well as the female-determining genes, such as FOX-L2 and RESPO-1

temperature may exert effects. This could be why the magnitude of temperature fluctuations around a mean influences the sexes of hatchlings in reptiles. As described above, in fish systems, high temperatures are now generally regarded as a thermal stressor, inducing the production of stress hormones and ultimately masculinizing offspring. In fish systems, production of males generally appears to be the default when thermal conditions are stressful. In humans and birds, stress tends to instead induce the production of females. It is possible that, in reptiles, there is also a default direction for sex ratio skews when conditions are stressful, but the perception of which conditions are stressful (hotter or colder than the optimum) differs among species. Additional studies of how glucocorticoids and other molecules that react to heat stress, such as heat shock proteins, need to be examined in the context of TSD to test this idea. Overall, a comparative examination of TSD in different vertebrate classes may point us towards an overarching mechanism that functions similarly among vertebrates.

References

Balthazart J, Ball GF (1995) Sexual differentiation of brain and behavior in birds. Trends Endocrinol Metab 6(1):21–29

Blasco M, Somoza GM, Vizziano-Cantonnet D (2013) Presence of 11-ketotestosterone in pre-differentiated male gonads of Odontesthes bonariensis. Fish Physiol Biochem 39(1):71–74

Bowden R, Ewert M, Nelson C (2000) Environmental sex determination in a reptile varies seasonally and with yolk hormones. Proc R Soc Lond B Biol Sci 267(1454):1745–1749

Bowden RM, Ewert MA, Freedberg S, Nelson CE (2002) Maternally derived yolk hormones vary in follicles of the painted turtle, Chrysemys picta. J Exp Zool A Ecol Genet Physiol 293(1):67–72

Bowden RM, Carter AW, Paitz RT (2014) Constancy in an inconstant world: moving beyond constant temperatures in the study of reptilian incubation. Integr Comp Biol 54(5):830–840

Bull J (1985) Sex ratio and nest temperature in turtles: comparing field and laboratory data. Ecology 66(4):1115–1122

Chojnowski JL, Braun EL (2012) An unbiased approach to identify genes involved in development in a turtle with temperature-dependent sex determination. BMC Genomics 13(1):308

Crews D, Bull JJ (2009) Mode and tempo in environmental sex determination in vertebrates. Semin Cell Dev Biol 20(3):251–255

Devlin RH, Nagahama Y (2002) Sex determination and sex differentiation in fish: an overview of genetic, physiological, and environmental influences. Aquaculture 208(3):191–364

Díaz N, Piferrer F (2015) Lasting effects of early exposure to temperature on the gonadal transcriptome at the time of sex differentiation in the European sea bass, a fish with mixed genetic and environmental sex determination. BMC Genomics 16(1):679

Elf P (2003) Yolk steroid hormones and sex determination in reptiles with TSD. Gen Comp Endocrinol 132(3):349–355

Ellison A, López CMR, Moran P, Breen J, Swain M, Megias M, Hegarty M, Wilkinson M, Pawluk R, Consuegra S (2015) Epigenetic regulation of sex ratios may explain natural variation in self-fertilization rates. Proc R Soc B 282(1819):20151900

Fan Z, Zou Y, Jiao S, Tan X, Wu Z, Liang D, Zhang P, You F (2017) Significant association of cyp19a promoter methylation with environmental factors and gonadal differentiation in olive flounder Paralichthys olivaceus. Comp Biochem Physiol A Mol Integr Physiol 208:70–79

Ferguson-Smith M (2007) The evolution of sex chromosomes and sex determination in vertebrates and the key role of DMRT1. Sex Dev 1(1):2–11

Fernandino JI, Hattori RS, Kishii A, Strüssmann CA, Somoza GM (2012) The cortisol and androgen pathways cross talk in high temperature-induced masculinization: the 11-β-hydroxysteroid dehydrogenase as a key enzyme. Endocrinology 153(12):6003–6011

Fernandino JI, Hattori RS, Acosta ODM, Strüssmann CA, Somoza GM (2013) Environmental stress-induced testis differentiation: androgen as a by-product of cortisol inactivation. Gen Comp Endocrinol 192:36–44

Flament S (2016) Sex reversal in amphibians. Sex Dev 10(5-6):267–278

Fleming A, Crews D (2001) Estradiol and incubation temperature modulate regulation of steroidogenic factor 1 in the developing gonad of the red-eared slider turtle. Endocrinology 142(4):1403–1411

Georges A, Limpus C, Stoutjesdijk R (1994) Hatchling sex in the marine turtle Caretta caretta is determined by proportion of development at a temperature, not daily duration of exposure. J Exp Zool A Ecol Genet Physiol 270(5):432–444

Godwin JR, Thomas P (1993) Sex change and steroid profiles in the protandrous anemonefish Amphiprion melanopus (Pomacentridae, Teleostei). Gen Comp Endocrinol 91(2):144–157

Godwin J, Thompson R (2012) Nonapeptides and social behavior in fishes. Horm Behav 61(3):230–238

Guiguen Y, Fostier A, Piferrer F, Chang C-F (2010) Ovarian aromatase and estrogens: a pivotal role for gonadal sex differentiation and sex change in fish. Gen Comp Endocrinol 165(3):352–366

Hattori RS, Fernandino JI, Kishii A, Kimura H, Kinno T, Oura M, Somoza GM, Yokota M, Strüssmann CA, Watanabe S (2009) Cortisol-induced masculinization: does thermal stress affect gonadal fate in pejerrey, a teleost fish with temperature-dependent sex determination? PLoS One 4(8):e6548

Hattori RS, Murai Y, Oura M, Masuda S, Majhi SK, Sakamoto T, Fernandino JI, Somoza GM, Yokota M, Strüssmann CA (2012) A Y-linked anti-Müllerian hormone duplication takes over a critical role in sex determination. Proc Natl Acad Sci 109(8):2955–2959

Hayashi Y, Kobira H, Yamaguchi T, Shiraishi E, Yazawa T, Hirai T, Kamei Y, Kitano T (2010) High temperature causes masculinization of genetically female medaka by elevation of cortisol. Mol Reprod Dev 77(8):679–686

Hayes TB (1998) Sex determination and primary sex differentiation in amphibians: genetic and developmental mechanisms. J Exp Zool A Ecol Genet Physiol 281(5):373–399

Hunt TM (1907) Sex-determining factors in animals. Science 25(636):382–384

Iungman JL, Somoza GM, Piña CI (2015) Are stress-related hormones involved in the temperature-dependent sex determination of the broad-snouted caiman? South Am J Herpetol 10(1):41–49

Iwamatsu T, Kobayashi H, Sagegami R, Shuo T (2006) Testosterone content of developing eggs and sex reversal in the medaka (Oryzias latipes). Gen Comp Endocrinol 145(1):67–74

Jaenisch R, Bird A (2003) Epigenetic regulation of gene expression: how the genome integrates intrinsic and environmental signals. Nat Genet 33(3s):245

Janzen F, Wilson M, Tucker J, Ford S (1998) Endogenous yolk steroid hormones in turtles with different sex-determining mechanisms. Gen Comp Endocrinol 111(3):306–317

Khan M, Renaud R, Leatherland J (1997) Metabolism of estrogens and androgens by embryonic tissues of Arctic charr, Salvelinus alpinus. Gen Comp Endocrinol 107(1):118–127

Kitano T, Hayashi Y, Shiraishi E, Kamei Y (2012) Estrogen rescues masculinization of genetically female medaka by exposure to cortisol or high temperature. Mol Reprod Dev 79(10):719–726

Kobayashi S, Isotani A, Mise N, Yamamoto M, Fujihara Y, Kaseda K, Nakanishi T, Ikawa M, Hamada H, Abe K (2006) Comparison of gene expression in male and female mouse blastocysts revealed imprinting of the X-linked gene, Rhox5/Pem, at preimplantation stages. Curr Biol 16(2):166–172

Kohno S, Katsu Y, Urushitani H, Ohta Y, Iguchi T, Guillette L Jr (2010) Potential contributions of heat shock proteins to temperature-dependent sex determination in the American alligator. Sex Dev 4(1–2):73–87

Kohno S, Bernhard MC, Katsu Y, Zhu J, Bryan TA, Doheny BM, Iguchi T, Guillette LJ Jr (2015) Estrogen receptor 1 (ESR1; ERα), not ESR2 (ERβ), modulates estrogen-induced sex reversal in the American alligator, a species with temperature-dependent sex determination. Endocrinology 156(5):1887–1899

Kuntz S, Chesnel A, Duterque-Coquillaud M, Grillier-Vuissoz I, Callier M, Dournon C, Flament S, Chardard D (2003) Differential expression of P450 aromatase during gonadal sex differentiation and sex reversal of the newt Pleurodeles waltl. J Steroid Biochem Mol Biol 84(1):89–100

Lee YH, Du JL, Yueh WS, Lin BY, Huang JD, Lee CY, Lee MF, Lau EL, Lee FY, Morrey C (2001) Sex change in the protandrous black porgy, Acanthopagrus schlegeli: a review in gonadal development, estradiol, estrogen receptor, aromatase activity and gonadotropin. J Exp Zool A Ecol Genet Physiol 290(7):715–726

Liu H (2016) Genomic basis of sex change in fish. University of Otago

Mankiewicz JL, Godwin J, Holler BL, Turner PM, Murashige R, Shamey R, Daniels HV, Borski RJ (2013) Masculinizing effect of background color and cortisol in a flatfish with environmental sex-determination. Integr Comp Biol 53(4):755–765

Matsumoto Y, Crews D (2012) Molecular mechanisms of temperature-dependent sex determination in the context of ecological developmental biology. Mol Cell Endocrinol 354(1):103–110

Matsumoto Y, Buemio A, Chu R, Vafaee M, Crews D (2013) Epigenetic control of gonadal aromatase (cyp19a1) in temperature-dependent sex determination of red-eared slider turtles. PLoS One 8(6):e63599

McCoy JA, Hamlin HJ, Thayer L, Guillette LJ, Parrott BB (2016) The influence of thermal signals during embryonic development on intrasexual and sexually dimorphic gene expression and circulating steroid hormones in American alligator hatchlings (Alligator mississippiensis). Gen Comp Endocrinol 238:47–54

Merchant-Larios H, Fierro IV, Urruiza BC (1989) Gonadal morphogenesis under controlled temperature in the sea turtle Lepidochelys olivacea. Herpetol Monogr 3:43–61

Murdock C, Wibbels T (2006) Dmrt1 expression in response to estrogen treatment in a reptile with temperature-dependent sex determination. J Exp Zool B Mol Dev Evol 306(2):134–139

Nakamura M (2009) Sex determination in amphibians. Semin Cell Dev Biol 3:271–282

Narberhaus F, Waldminghaus T, Chowdhury S (2005) RNA thermometers. FEMS Microbiol Rev 30(1):3–16

Navarro-Martín L, Viñas J, Ribas L, Díaz N, Gutiérrez A, Di Croce L, Piferrer F (2011) DNA methylation of the gonadal aromatase (cyp19a) promoter is involved in temperature-dependent sex ratio shifts in the European sea bass. PLoS Genet 7(12):e1002447

Neuwald JL, Valenzuela N (2011) The lesser known challenge of climate change: thermal variance and sex-reversal in vertebrates with temperature-dependent sex determination. PLoS One 6(3):e18117

Nozu R, Nakamura M (2015) Cortisol administration induces sex change from ovary to testis in the protogynous wrasse, Halichoeres trimaculatus. Sex Dev 9(2):118–124

Paitz RT, Clairardin SG, Griffin AM, Holgersson MC, Bowden RM (2010a) Temperature fluctuations affect offspring sex but not morphological, behavioral, or immunological traits in the Northern Painted Turtle (Chrysemys picta). Can J Zool 88(5):479–486

Paitz RT, Gould AC, Holgersson MC, Bowden RM (2010b) Temperature, phenotype, and the evolution of temperature-dependent sex determination: how do natural incubations compare to laboratory incubations? J Exp Zool B Mol Dev Evol 314(1):86–93

Parrott BB, Kohno S, Cloy-McCoy JA, Guillette Jr LJ (2014) Differential incubation temperatures result in dimorphic DNA methylation patterning of the SOX9 and aromatase promoters in gonads of alligator (Alligator mississippiensis) embryos. Biol Reprod 90(1):2, 1–11

Piferrer F (2001) Endocrine sex control strategies for the feminization of teleost fish. Aquaculture 197(1):229–281

Piferrer F (2013) Epigenetics of sex determination and gonadogenesis. Dev Dyn 242(4):360–370

Piquet J (1930) Détermination du sexe chez les Batraciens en fonction de la température. Universite de Geneve

Radder RS (2007) Maternally derived egg yolk steroid hormones and sex determination: review of a paradox in reptiles. J Biosci 32:1213–1220

Ramsey M, Crews D (2009) Steroid signaling and temperature-dependent sex determination—Reviewing the evidence for early action of estrogen during ovarian determination in turtles. Semin Cell Dev Biol 3:283–292

Raynaud A (1985) Embryonic development of the genital system. In: Gans S, Billeu F (eds) Biology of the reptilia. Wiley, New York, pp 149–299

Rhen T, Schroeder A (2010) Molecular mechanisms of sex determination in reptiles. Sex Dev 4(1-2):16–28

Rowell CB, Watts SA, Wibbels T, Hines GA, Mair G (2002) Androgen and estrogen metabolism during sex differentiation in mono-sex populations of the Nile tilapia, Oreochromis niloticus. Gen Comp Endocrinol 125(2):151–162

Sakata N, Tamori Y, Wakahara M (2004) P450 aromatase expression in the temperature-sensitive sexual differentiation of salamander (Hynobius retardatus) gonads. Int J Dev Biol 49(4):417–425

Schroeder AL, Metzger KJ, Miller A, Rhen T (2016) A novel candidate gene for temperature-dependent sex determination in the common snapping turtle. Genetics 203(1):557–571

Shao C, Li Q, Chen S, Zhang P, Lian J, Hu Q, Sun B, Jin L, Liu S, Wang Z (2014) Epigenetic modification and inheritance in sexual reversal of fish. Genome Res 24(4):604–615

Shen Z-G, Wang H-P (2014) Molecular players involved in temperature-dependent sex determination and sex differentiation in Teleost fish. Genet Sel Evol 46(1):26

Shi Y, Zhang Y, Li S, Liu Q, Lu D, Liu M, Meng Z, Cheng CH, Liu X, Lin H (2010) Molecular identification of the Kiss2/Kiss1ra system and its potential function during 17alpha-methyltestosterone-induced sex reversal in the orange-spotted grouper, Epinephelus coioides. Biol Reprod 83(1):63–74

Shoemaker CM, Crews D (2009) Analyzing the coordinated gene network underlying temperature-dependent sex determination in reptiles. Semin Cell Dev Biol 3:293–303

Sifuentes-Romero I, Merchant-Larios H, Milton SL, Moreno-Mendoza N, Díaz-Hernández V, García-Gasca A (2013) RNAi-mediated gene silencing in a gonad organ culture to study sex determination mechanisms in sea turtle. Genes 4(2):293–305

Solomon-Lane TK, Crespi EJ, Grober MS (2013) Stress and serial adult metamorphosis: multiple roles for the stress axis in socially regulated sex change. Front Neurosci 7:1–12

Sun L-X, Wang Y-Y, Zhao Y, Wang H, Li N, Ji XS (2016) Global DNA methylation changes in Nile tilapia gonads during high temperature-induced masculinization. PLoS One 11(8):e0158483

Telemeco RS, Abbott KC, Janzen FJ (2013) Modeling the effects of climate change–induced shifts in reproductive phenology on temperature-dependent traits. Am Nat 181(5):637–648

Thomas EO, Light P, Wibbels T, Crews D (1992) Hydroxysteroid dehydrogenase activity associated with sexual differentiation in embryos of the turtle Trachemys scripta. Biol Reprod 46(1):140–145

Todd EV, Liu H, Muncaster S, Gemmell NJ (2016) Bending genders: the biology of natural sex change in fish. Sex Dev 10(5–6):223–241

Uller T, Hollander J, Astheimer L, Olsson M (2009) Sex-specific developmental plasticity in response to yolk corticosterone in an oviparous lizard. J Exp Biol 212(8):1087–1091

Wang X, Liu Q, Xiao Y, Yang Y, Wang Y, Song Z, You F, An H, Li J (2017) High temperature causes masculinization of genetically female olive flounder (Paralichthys olivaceus) accompanied by primordial germ cell proliferation detention. Aquaculture 479:808–816

Warner DA, Radder RS, Shine R (2009) Corticosterone exposure during embryonic development affects offspring growth and sex ratios in opposing directions in two lizard species with environmental sex determination. Physiol Biochem Zool 82(4):363–371

Wibbels T, Crews D (1992) Specificity of steroid hormone-induced sex determination in a turtle. J Endocrinol 133(1):121–129

Wibbels T, Bull J, Crews D (1991) Synergism between temperature and estradiol: a common pathway in turtle sex determination? J Exp Zool A Ecol Genet Physiol 260(1):130–134

Wibbels T, Bull JJ, Crews D (1994) Temperature-dependent sex determination: a mechanistic approach. J Exp Zool A Ecol Genet Physiol 270(1):71–78

Witschi E (1914) Experimentelle Untersuchungen über die Entwicklungsgeschichte der Keimdrüsen von Rana temporaria. Arch Mikrosk Anat 85(1):A9–A113

Yamaguchi T, Yoshinaga N, Yazawa T, Gen K, Kitano T (2010) Cortisol is involved in temperature-dependent sex determination in the Japanese flounder. Endocrinology 151(8): 3900–3908

Yamamoto T-O (1969) Sex differentiation. Fish Physiol 3:117–175

Yamamoto Y, Hattori R, Kitahara A, Kimura H, Yamashita M, Strüssmann C (2013) Thermal and endocrine regulation of gonadal apoptosis during sex differentiation in pejerrey Odontesthes bonariensis. Sex Dev 7(6):316–324

Yatsu R, Miyagawa S, Kohno S, Saito S, Lowers RH, Ogino Y, Fukuta N, Katsu Y, Ohta Y, Tominaga M (2015) TRPV4 associates environmental temperature and sex determination in the American alligator. Sci Rep 5:18581

Yatsu R, Miyagawa S, Kohno S, Parrott BB, Yamaguchi K, Ogino Y, Miyakawa H, Lowers RH, Shigenobu S, Guillette LJ (2016) RNA-seq analysis of the gonadal transcriptome during Alligator mississippiensis temperature-dependent sex determination and differentiation. BMC Genomics 17(1):77

Yeoh C-G, Schreck CB, Fitzpatrick MS, Feist GW (1996) In vivo steroid metabolism in embryonic and newly hatched steelhead trout (Oncorhynchus mykiss). Gen Comp Endocrinol 102(2):197–209

Zaborski P (1986) Temperature and estrogen dependent changes of sex phenotype and HY antigen expression in gonads of a newt. Prog Clin Biol Res 217:163–169

Zhang Y, Zhang S, Liu Z, Zhang L, Zhang W (2013) Epigenetic modifications during sex change repress gonadotropin stimulation of cyp19a1a in a teleost ricefield eel (Monopterus albus). Endocrinology 154(8):2881–2890

Printed by Printforce, the Netherlands